Studies in Computational Intelligence 479

Editor-in-Chief

Prof. Janusz Kacprzyk
Systems Research Institute
Polish Academy of Sciences
ul. Newelska 6
01-447 Warsaw
Poland
E-mail: kacprzyk@ibspan.waw.pl

For further volumes:
http.//www.springer.com/series/7092

Ngoc Thanh Nguyen, Tien Van Do,
and Hoai An Le Thi (Eds.)

Advanced Computational Methods for Knowledge Engineering

 Springer

Editors
Prof. Ngoc Thanh Nguyen
Division of Knowledge Management
Systems
Institute of Informatics (I-32)
Wroclaw University of Technology
Wroclaw
Poland

Prof. Hoai An Le Thi
LITA - UFR MIM
Université de Lorraine – Metz
Metz
France

Prof. Tien Van Do
Department of Telecommunications
Budapest University of Technology
and Economics
Budapest
Hungary

ISSN 1860-949X   ISSN 1860-9503   (electronic)
ISBN 978-3-319-00292-7   ISBN 978-3-319-00293-4   (eBook)
DOI 10.1007/978-3-319-00293-4
Springer Heidelberg New York Dordrecht London

Library of Congress Control Number: 2013934353

Printed on acid-free paper

Springer is part of Springer Science+Business Media (www.springer.com)

Preface

This volume contains the extended versions of papers presented at the *1st International Conference on Computer Science, Applied Mathematics and Applications* (ICCSAMA 2013) held on 9-10 May, 2013 in Warsaw, Poland. The conference is co-organized by Division of Knowledge Management Systems (Wroclaw University of Technology, Poland) and Laboratory of Theoretical & Applied Computer Science (Lorraine University, France) in cooperation with IEEE SMC Technical Committee on Computational Collective Intelligence and the Analysis, Design and the Development of ICT Systems (AddICT) Laboratory (Budapest University of Technology and Economics, Hungary).

The aim of ICCSAMA 2013 is to bring together leading academic scientists, researchers and scholars to discuss and share their newest results in the fields of Computer Science, Applied Mathematics and their applications. These two fields are very close and related to each other. It is also clear that the potentials of computational methods for knowledge engineering and optimization algorithms are to be exploited, and this is an opportunity and a challenge for researchers.

After the peer review process, 29 papers have been selected for including in this volume. Their topics revolve around Computational Methods, Optimization Techniques, Knowledge Engineering and have been partitioned into 5 groups: *Advanced Optimization Methods and Their Applications, Queueing Theory and Applications, Computational Methods for Knowledge Engineering, Knowledge Engineering with Cloud and Grid Computing,* and *Logic Based Methods for Decision Making and Data Mining.*

ICCSAMA 2013 clearly generated a significant amount of interaction between members of both communities on Computer Science and Applied Mathematics, and we hope that these discussions have seeded future exciting development at the interface between computational methods, optimization and engineering.

The materials included in this book can be useful for researchers, Ph.D. and graduate students in Optimization Theory and Knowledge Engineering fields. It is the hope of the editors that readers can find many inspiring ideas and use them to their research. Many such challenges are suggested by particular approaches and models presented in individual chapters of this book.

We would like to thank all authors, who contributed to the success of the conference and to this book. Special thanks go to the members of the Steering and Program Committees for their contributions to keeping the high quality of the selected papers. Cordial thanks are due to the Organizing Committee members for their efforts and the organizational work.

Finally, we cordially thank Springer for supports and publishing this volume.

We hope that ICCSAMA 2013 significantly contributes to the fulfilment of the academic excellence and leads to greater success of ICCSAMA events in the future.

May 2013   Ngoc Thanh Nguyen
                                                                          Tien Van Do
                                                                        Hoai An Le Thi

ICCSAMA 2013 Organization

General Chair

Nguyen Ngoc Thanh   Wroclaw University of Technology, Poland

Program Chairs

Le Thi Hoai An   Lorraine University, France
Nguyen Van Thoai   Trier University, Germany
Nguyen Hung Son   Warsaw University, Poland
Tien Van Do   Budapest University of Technology and
         Economics, Hungary

Doctoral Track Chair

Nguyen Anh Linh   Warsaw University, Poland

Organizing Chair

Dang Ngoc Han   Le Quy Don Society in Poland

Publicity Chair

Le Hoai Minh   Lorraine University, France

Organizing Committee

Bui Dang Trinh   Warsaw University of Technology, Poland
Tran Trong Hieu   Wroclaw University of Technology, Poland

Steering Committee

Le Thi Hoai An	Lorraine University, France (Co-chair)
Nguyen Ngoc Thanh	Wroclaw University of Technology, Poland (Co-chair)
Pham Dinh Tao	INSA Rouen, France
Nguyen Van Thoai	Trier University, Germany
Pham Duc Truong	University of Birmingham, UK
Nguyen Hung Son	Warsaw University, Poland
Alain Bui	Université de Versailles-St-Quentin-en-Yvelines, France
Nguyen Anh Linh	Warsaw University, Poland
Tien Van Do	Budapest University of Technology and Economics, Hungary
Tran Dinh Viet	Slovak Academy of Sciences, Slovakia

Program Committee

Bui Alain	Université de Versailles-St-Quentin-en-Yvelines, France
Nguyen Thanh Binh	International Institute for Applied Systems Analysis (IIASA), Austria
Dao Thi Thu Ha	University of Versailles Saint-Quentin-en-Yvelines, France
Tien Van Do	Budapest University of Technology and Economics, Hungary
Ha Quang Thuy	Vietnam National University, Vietnam
Le Chi Hieu	University of Greenwich, UK
Le Nguyen-Thinh	Clausthal University of Technology, Germany
Le Thi Hoai An	Lorraine University, France
Luong Marie	Université Paris 13, France
Ngo Van Sang	University of Rouen, France
Nguyen Anh Linh	Warsaw University, Poland
Nguyen Benjamin	University of Versailles Saint-Quentin-en-Yvelines, France
Nguyen Duc Cuong	International University VNU-HCM, Vietnam
Nguyen Hung Son	Warsaw University, Poland
Nguyen Ngoc Thanh	Wroclaw University of Technology, Poland
Nguyen Van Thoai	Trier University, Germany
Nguyen Viet Hung	Laboratory of Computer Sciences Paris 6, France

Nguyen-Verger Mai K	Cergy-Pontoise University, France
Pham Cong Duc	University of Pau and Pays de l'Adour, France
Pham Dinh Tao	INSA Rouen, France
Pham Duc Truong	University of Birmingham, UK
Phan Duong Hieu	Université Paris 8, France
Tran Chi	University of Warmia and Mazury in Olsztyn, Poland
Tran Dinh Viet	Slovak Academy of Sciences, Slovakia
Tran Hoai Linh	Hanoi University of Science and Technology, Vietnam
Trinh Anh Tuan	Budapest University of Technology and Economics, Hungary
Truong Trong Tuong	Cergy-Pontoise University, France

Contents

Part II: Queueing Theory and Applications

Part III: Computational Methods for Knowledge Engineering

Part IV: Knowledge Engineering with Cloud and Grid Computing

Part V: Logic Based Methods for Decision Making and Data Mining

Part I
Advanced Optimization Methods and Their Applications

Solution Methods for General Quadratic Programming Problem with Continuous and Binary Variables: Overview

Nguyen Van Thoai

Department of Mathematics, University of Trier, D-54286 Trier, Germany
thoai@uni-trier.de

Abstract. The nonconvex quadratic programming problem with continuous and/or binary variables is a typical NP-hard optimization problem, which has a wide range of applications. This article presents an overview of actual solution methods for solving this interesting and important class of programming problems. Solution methods are discussed in the sense of global optimization.

Keywords: quadratic programming, global optimization, binary quadratic programming, copositive optimization problems.

1 Introduction

In this article, by *'general quadratic programming problem'* we mean an optimization problem, in which all functions involved are quadratic or linear and, in general, local optima can be different from global optima. We also consider the case where some of variables are required to take values in $\{0, 1\}$ (binary variables).

The class of general quadratic programming problems plays a prominent role in the field of nonconvex global optimization because of its theoretical aspects as well as its wide range of applications. On the one hand, many real world problems arising from economies and engineering design can be directly modelled as quadratic programming problems. On the other hand, general quadratic programming includes as special cases the equilalent formulations of many important and well studied optimization problems, e.g., linear zero-one programs, assignment problems, maximum clique problems, linear complementarity problems, bilinear problems, packing problems, etc. Last but not least, some special quadratic programming problems are used as basic subproblems in trust region methods in nonlinear programming. Applications of general quadratic programming problems can be found in almost all of the references given at the end of this article.

The present overview focuses on actual results of solution methods for general quadratic programming problems with continuous and/or binary variables. All methods are discussed in the sense of global optimization.

N.T. Nguyen, T. Van Do, and H.A. Le Thi (Eds.): *ICCSAMA 2013*, SCI 479, pp. 3–17.
DOI: 10.1007/978-3-319-00293-4_1   © Springer International Publishing Switzerland 2013

Section 2 deals with solution methods for problems with continuous variables. Three basic approaches which are successfully used in global optimization and different techniques for the realization of them in general quadratic programming are discussed. In Section 3 we mainly present a new approach for problem with binary variables, namely the class of *copositive optimization problems.*

2   Quadratic Optimization Problems with Continuous Variables

We consider general quadratic optimization problems given by

$$
\begin{aligned}
\min \ &f(x) = \tfrac{1}{2}\langle Ax, x\rangle + \langle b, x\rangle \\
\text{s.t.} \ &g_i(x) = \langle Q_i x, x\rangle + \langle q_i, x\rangle + d_i \leq 0, \ i \in I_1 \\
&g_i(x) = \langle Q_i x, x\rangle + \langle q_i, x\rangle + d_i = 0, \ i \in I_2,
\end{aligned} \tag{1}
$$

where I_1, I_2 are two finite index sets, A, Q_i, $i \in I_1 \cup I_2$, are $n \times n$ symmetric real matrices and $b, q_i \in \mathbb{R}^n$, $d_i \in \mathbb{R}$ for all i. The notation $\langle \cdot, \cdot \rangle$ stands for the standard scalar product in $\mathbb{R}^n$. If $I_2 = \emptyset$ and all functions f, g_i, $i \in I_1$ are convex, then (1) is the well known convex programming problem. Since the subject of this article consists of nonconvex problems, we assume throughout the article that in the case $I_2 = \emptyset$, at least one of the functions f, g_i, $i \in I_1$, is nonconvex. For convenience, we sometimes write the objective function of Problem (1) in the form $f(x) = \langle Ax, x\rangle + 2\langle b, x\rangle$.

Applications of Problem (1) includes subproblems required in trust region methods (cf.,e.g., [28], [44], [45] and references given therein).

The general quadratic programming problem is NP-hard (cf. [51]), [57]). In this section, we present some main solution methods for the global optimization of this NP-hard problem. In general, these methods are developed based on three basic concepts which are successfully used in global optimization. We describe these concepts briefly before presenting different techniques for the realization of them in general quadratic programming. For details of three basic concepts, see, e.g., [30], [31], [34], [35]. It is worth noting that the most techniques to be presented here can be applied to integer and mixed integer quadratic programming problems.

2.1   Basic Concepts

Outer Approximation (OA). To establish this concept, we consider the problem of minimizing a linear function $\langle c, x\rangle$ over a closed subset $F \subset \mathbb{R}^n$. This problem can be replaced by the problem of finding an extreme optimal solution of the problem $\min\{\langle c, x\rangle : x \in \overline{F}\}$, where $\overline{F}$ denotes the convex hull of F. Let C_1 be any closed convex set containing F and assume that x^1 is an optimal solution of problem $\min\{\langle c, x\rangle : x \in C_1\}$. Then x^1 is also an optimal solution of the original problem whenever $x^1 \in F$. The basic idea of the outer approximation concept is to contruct iteratively a sequence of convex subsets $\{C_k\}$, $k = 1, 2, \cdots$

such that $C_1 \supset C_2 \supset \cdots \supset F$ and the corresponding sequence $\{x^k\}$ such that for each k, x^k is an optimal solution of the relaxed problem $\min\{\langle c, x \rangle : x \in C_k\}$. This process is performed until finding $x^k \in F$. An OA procedure is convergent if it holds that $x^k \to x^* \in F$ for $k \to +\infty$.

Branch and Bound Scheme (BB). The BB scheme is developed for the global optimization of problem $f^* = \min\{f(x) : x \in F\}$ with f being a continuous function and F a compact subset of $\mathbb{R}^n$. It begins with a convex compact set $C_1 \supset F$ and proceeds as follows. Compute a lowerbound μ_1 and an upper bound γ_1 for the optimal value of the problem $\min\{f(x) : x \in C_1 \cap F\}$. ($\gamma_1 = f(x^1)$ if some feasible solution $x^1 \in F$ is found, otherwise, $\gamma_1 = +\infty$). At Iteration $k \geq 1$, if $+\infty > \mu_k \geq \gamma_k$ or $\mu_k = +\infty$, then stop, (in the first case, x^k with $f(x^k) = \gamma_k$ is an optimal solution, in the second case, the underlying problem has no solution). Otherwise, divide C_k into finitely many convex sets $C_{k_1}, \ldots, C_{k_r}$ satisfying $\bigcup_{i=1}^{r} C_{k_i} = C_k$ and $C_{k_i} \cap C_{k_j} = \emptyset$ for $i \neq j$, (the sets C_k and C_{k_i} are called 'partition sets'). Compute for each partition set a lower bound and an upper bound. Update the lower bound by choosing the minimum of lower bounds according to all existing partition sets, and update the upper bound by using feasible points found so far. Delete all partition sets such that the corresponding lower bounds are biger than or equal to the actual upper bound. If not all partition sets are deleted, let C_{k+1} be a partition set with the minimum lower bound, and go to Iteration $k + 1$. A BB algorithm is convergent if it holds that $\gamma_k \searrow f^*$ and/or $\mu_k \nearrow f^*$ for $k \to +\infty$.

Combination of BB and OA. In many situations, the use of the BB scheme in combination with an OA in the bounding procedure can lead to efficient algorithms. Such a combination is called branch and cut algorithm, if an OA procedure using convex polyhedral subsets $C_k \; \forall k \geq 1$ is applied.

2.2 Reformulation-Linearization Techniques

Consider quadratic programming problems of the form

$$\min f(x) = \langle c, x \rangle$$
$$\text{s.t. } g_i(x) \leq 0, \; i = 1, \cdots, I \qquad (2)$$
$$\langle a^i, x \rangle - b_i \leq 0, \; i = 1, \cdots, m,$$

where $c \in \mathbb{R}^n$, $a^i \in \mathbb{R}^n$, $b_i \in \mathbb{R} \; \forall i = 1, \cdots, m$, and for each $i = 1, \cdots, I$, the quadratic function g_i is given by

$$g_i(x) = d_i + \sum_{k=1}^{n} q_k^i x_k + \sum_{k=1}^{n} Q_{kk}^i x_k^2 + \sum_{k=1}^{n-1} \sum_{l=k+1}^{n} Q_{kl}^i x_k x_l \qquad (3)$$

with $d_i, q_k^i, Q_{kk}^i, Q_{kl}^i$ being given real numbers for all i, k, l. It is assumed that the polyhedral set $X = \{x \in \mathbb{R}^n : \langle a^i, x \rangle - b_i \leq 0, \; i = 1, \cdots, m\}$ is bounded and contained in $R_+^n = \{x \in \mathbb{R}^n : x \geq 0\}$.

The first linear relaxation of Problem (2) is performed as follows. For each quadratic function of the form

$$h(x) = \sum_{k=1}^{n} q_k x_k + \sum_{k=1}^{n} Q_{kk} x_k^2 + \sum_{k=1}^{n-1} \sum_{l=k+1}^{n} Q_{kl} x_k x_l \qquad (4)$$

define additional variables

$$v_k = x_k^2, \ k = 1, \cdots, n, \ \text{and}$$

$$w_{kl} = x_k x_l, \ k = 1, \cdots, n-1; l = k+1, \cdots, n.$$

From (4), one obtains the following linear function in variables x, v, w:

$$[h(x)]_\ell = \sum_{k=1}^{n} q_k x_k + \sum_{k=1}^{n} Q_{kk} v_k + \sum_{k=1}^{n-1} \sum_{l=k+1}^{n} Q_{kl} w_{kl}. \qquad (5)$$

The linear program (in variables x, v and w)

$$\begin{aligned}
&\min \ f(x) = \langle c, x \rangle \\
&\text{s.t.} \ [g_i(x)]_\ell \le 0, \ i = 1, \cdots, I \\
&\quad [(b_i - \langle a^i, x \rangle)(b_j - \langle a^j, x \rangle)]_\ell \ge 0, \ \forall 1 \le i \le j \le m
\end{aligned} \qquad (6)$$

is then a linear relaxation of (2) in the following sense (cf. [3], [60]):

Let f^* and $\overline{f}$ be the optimal values of problems (2) and (6), respectively, and let $(\overline{x}, \overline{v}, \overline{w})$ be an optimal solution of (6). Then

(a) $f^* \ge \overline{f}$ and
(b) if $\overline{v}_k = \overline{x}_k^2 \ \forall k = 1, \cdots, n, \ \overline{w}_{kl} = \overline{x}_k \overline{x}_l \ \forall k = 1, \cdots, n-1; l = k+1, \cdots, n,$
then $\overline{x}$ is an optimal solution of (2).

Geometrically, the convex hull of the (nonconvex) feasible set of Problem (2) is relaxed by the projection of the polyhedral feasible set of Problem (6) on $\mathbb{R}^n$. As well-known, this projection is polyhedral.

In the case that the condition in (b) is not fulfilled, i.e., either $\overline{v}_k \ne \overline{x}_k^2$ for at least one index k or $\overline{w}_{kl} \ne \overline{x}_k \overline{x}_l$ for at least one index pair (k,l), a family of linear inequalities have to be added to Problem (6) to cut the point $(\overline{x}, \overline{v}, \overline{w})$ off from the feasible set of (6) without cutting off any feasible point of (2). To this purpose, several kinds of cuts are discussed in connection with branch and bound procedures. Resulting branch and cut algorithms can be found, e.g., in [1], [3].

2.3   Lift-and-Project Techniques

The first ideas of lift-and-project techniques were proposed by [58] and [42] for zero-one optimization. These basic ideas can be applied to quadratic programming as follows. The quadratic programming problem to be considered is given in the form

$$\min f(x) = \langle c, x \rangle$$
$$\text{s.t. } g_i(x) \leq 0, \ i = 1, \cdots, m \qquad (7)$$
$$x \in C,$$

where C is a compact convex subset of $\mathbb{R}^n$, $c \in \mathbb{R}^n$, and each function g_i is given by

$$g_i(x) = \langle Q^i x, x \rangle + 2\langle q^i, x \rangle + d_i \qquad (8)$$

with Q^i being $n \times n$ symmetric matrix, $q^i \in \mathbb{R}^n$, and $d_i \in \mathbb{R}$.

To each $x \in \mathbb{R}^n$ a symmetric matrix $X \in \mathbb{R}^{n \times n}$ is assigned. Let $\mathcal{S}^n$ be the set of $n \times n$ symmetric matrices. Then each quadratic function

$$\langle Qx, x \rangle + 2\langle q, x \rangle + d$$

on $\mathbb{R}^n$ is lifted to a function on $\mathbb{R}^n \times \mathcal{S}^n$ defined by

$$X \mapsto \langle Q, X \rangle + 2\langle q, x \rangle + d,$$

where X_{ij}, $(i, j = 1, \cdots, n)$, denote the elements of X, and $\langle Q, X \rangle = \sum\limits_{i=1}^{n} \sum\limits_{j=1}^{n} Q_{ij} X_{ij}$ stands for the inner product of $Q, X \in \mathcal{S}^n$.

Thus, the set

$$\{ x \in \mathbb{R}^n : \langle Qx, x \rangle + 2\langle q, x \rangle + d \leq 0 \}$$

can be approximated by the projection of the set

$$\{ (x, X) \in \mathbb{R}^n \times \mathcal{S}^n : \langle Q, X \rangle + 2\langle q, x \rangle + d \leq 0 \}$$

on $\mathbb{R}^n$. By this way, the feasible set of Problem (7) is approximated by the set

$$\{ x \in \mathbb{R}^n : \langle Q^i, X \rangle + 2\langle q^i, x \rangle + d_i \leq 0 \text{ for some } X \in \mathcal{S}^n,$$
$$i = 1, \cdots, m, \ x \in C \}, \qquad (9)$$

and Problem (7) is then relaxed by the problem

$$\min f(x) = \langle c, x \rangle$$
$$\text{s.t. } \langle Q^i, X \rangle + 2\langle q^i, x \rangle + d_i \leq 0, \ i = 1, \cdots, m \qquad (10)$$
$$x \in C, \ X \in \mathcal{S}^n.$$

Next, notice that for each $x \in \mathbb{R}^n$ the matrix xx^T is an element of $\mathcal{S}^n$, and the matrix

$$\begin{pmatrix} 1 & x^T \\ x & xx^T \end{pmatrix} = \begin{pmatrix} 1 \\ x \end{pmatrix} (1, x^T) \in \mathcal{S}^{n+1}$$

is positive semidefinite. Therefore, Problem (7) can also be relaxed by the problem

$$\min f(x) = \langle c, x \rangle$$
$$\text{s.t. } \langle Q^i, X \rangle + 2\langle q^i, x \rangle + d_i \le 0, \ i = 1, \cdots, m$$
$$x \in C, \ X \in \mathcal{S}^n \tag{11}$$
$$\begin{pmatrix} 1 & x^T \\ x & X \end{pmatrix} \text{ is positive semidefinite.}$$

If the set C is respresented by a finite set of linear inequalities, then Problem (10) is a linear program (LP), which actually corresponds to Problem (6). Let the set C be respresented by a finite set of linear matrix inequalities. Then Problem (11) is a semidefinite programming problem (SDP).

   To improve LP/SDP relaxations of Problem (7) within a OA procedure, convex quadratic inequality constraints have to be added to Problem (7) without cutting off any feasible point of it. Different classes of such additional quadratic inequality constraints are proposed in [23], [36], [37]. Aspects of implementation and parallel computation techniques are given in [64], [65].

2.4   D.C. Decomposition and Convex Envelope Techniques

D.C. Decomposition
The idea of representing a general quadratic function as the difference of two convex quadratic functions is quite simple: for each symmetric matrix $Q \in \mathbb{R}^{n \times n}$, let $\rho(Q)$ be the spectral radius (the maxinal eigenvalue) of Q, then Q can be rewritten as the difference of two positive semidefinite matrices A, B with $A = Q + \lambda I$, $B = \lambda I$, where $\lambda \ge \rho(Q)$ and I is the unit matrix in $\mathcal{S}^n$.

   Consider now the quadratic programming problem (7). Each quadratic fucntion $g_i(x)$ can be represented as the difference of two convex functions as follows.

$$g_i(x) = \langle Q^i x, x \rangle + 2\langle q^i, x \rangle + d_i = a_i(x) - b_i(x) \text{ with}$$
$$a_i(x) = \langle (Q^i + \lambda_i I)x, x \rangle + 2\langle q^i, x \rangle + d_i,$$
$$b_i(x) = \lambda_i \langle x, x \rangle, \text{ and} \tag{12}$$
$$\lambda_i \ge \rho(Q^i).$$

Thus, Problem (7) is rewritten in the form

$$\min f(x) = \langle c, x \rangle$$
$$\text{s.t. } a_i(x) - b_i(x) \le 0, \ i = 1, \cdots, m \tag{13}$$
$$x \in C.$$

Let

$$\lambda = \max\{\lambda_i : i = 1, \cdots, m\},$$
$$a(x, t) = \max\{a_i(x) : i = 1, \cdots, m\} - t,$$
$$b(x, t) = \lambda \langle x, x \rangle - t.$$

Then Problem (13) can be rewritten as a program in $\mathbb{R}^{n+1}$:

$$\begin{aligned}
\min \; & f(x) = \langle c, x \rangle \\
\text{s.t.} \; & b(x, t) \geq 0 \\
& a(x, t) \leq 0 \\
& x \in C,
\end{aligned} \qquad (14)$$

A function is called d.c. if it can be represented as a difference of two convex functions. Each of such representations is called a d.c. decomposition of this function. Problems of the form (13), where all functions $a_i(x), b_i(x)$ are convex, are called d.c. programming problems, and Problem (14) is called canonical d.c. program. Theoretical properties and solution methods for these problems can be found, e.g., in [35], [31], [39], [40], [54], [33].

Convex Envelope Techniques

The concept of convex envelopes of nonconvex functions is a basic tool in theory and algorithms of global optimization, see e.g., [19], [35], [31].

Let $C \subset \mathbb{R}^n$ be nonempty convex, and let $f : C \to \mathbb{R}$ be a lower semicontinuous function on C. The function

$$\begin{aligned}
& \varphi_{C,f} : C \to \mathbb{R}, \\
& x \mapsto \varphi_{C,f}(x) := \sup\{h(x) : h : C \to \mathbb{R} \text{ convex}, h \leq f \text{ on } C\}
\end{aligned}$$

is said to be the convex envelope of f over C. Convex relaxations of quadratic programming problems with inequality constraints can be obtained by replacing all nonconvex quadratic functions by their convex envelopes. Linear relaxations are then simply obtained from convex ralaxations. To construct convex envelopes of quadratic functions, two kinds of convex sets, simplices and rectangles, are chosen for C. Relaxation techniques using convex envelopes in connection with branch and bound procedures for general quadratic problems can be found, e.g., in [50], [2], [32], [35], [55], [31], [41].

2.5 Bilinear Programming Techniques

The problem of minimizing a concave quadratic function

$$f(x) = \langle Qx, x \rangle + \langle q, x \rangle, \text{ with } Q \in \mathbb{R}^{n \times n}, \; q \in \mathbb{R}^n$$

over a polyhedral set $X \in \mathbb{R}^n_+$ can be reduced to an equivalent bilinear programming problem

$$\min\{\phi(x, y) = \langle Qx, y \rangle + \langle q, x \rangle : x \in X, y \in X\}$$

in a sense that from an optimal solution (x^*, y^*) of the latter problem one has two optimal solutions x^*, y^* of the former, and from an optimal solution x^* of the former one obtains an optimal solution (x^*, x^*) of the latter (cf. [38]).

In general, quadratic programming problems of the form

$$\min f(x) = \langle Q^0 x, x \rangle + \langle q^0, x \rangle$$
$$\text{s.t. } g_i(x) = \langle Q^i x, x \rangle + \langle q^i, x \rangle + d_i \leq 0, \ i = 1, \cdots, I$$
$$\langle a^i, x \rangle - b_i \leq 0, \ i = 1, \cdots, m,$$
$$x \in C, \tag{15}$$

where Q^i $(i = 0, \cdots, I)$ are symmetric $n \times n$ matrices, q^i $(i = 0, \cdots, I)$, a^i $(i = 1, \cdots, m)$ vectors of $\mathbb{R}^n$, d_i $(i = 1, \cdots, I)$, b_i $(i = 1, \cdots, m)$ real numbers, and C a polyhedral set with simple structure (e.g., R_+^n, simplex or rectangle) can be reduced to one of the following equivalent bilinear programming problems:

$$\min \langle Q^0 x, y \rangle + \langle q^0, x \rangle$$
$$\text{s.t. } \langle Q^i x, y \rangle + \langle q^i, x \rangle + d_i \leq 0, \ i = 1, \cdots, I$$
$$\langle a^i, x \rangle - b_i \leq 0, \ i = 1, \cdots, m,$$
$$x_i - y_i = 0, \ i = 1, \cdots, n. \tag{16}$$
$$x \in C, \ y \in C$$

and

$$\min \langle x, y^0 \rangle + \langle q^0, x \rangle$$
$$\text{s.t. } \langle x, y^i \rangle + \langle q^i, x \rangle + d_i \leq 0, \ i = 1, \cdots, I$$
$$\langle a^i, x \rangle - b_i \leq 0, \ i = 1, \cdots, m,$$
$$y^i - Q^i x = 0, \ i = 0, \cdots, I \tag{17}$$
$$x \in C,$$
$$y^i \in S^i, \ i = 0, \cdots, I,$$

where, for each $i = 0, \cdots, I$,

$$S^i = \{y \in \mathbb{R}^n : \ell_j \leq y \leq u_j, \ j = 1, \cdots, n\} \text{ with}$$
$$\ell_j = \min\{\langle Q_j^i, x \rangle : \langle a^i, x \rangle - b_i \leq 0, \ i = 1, \cdots, m, \ x \in C\},$$
$$u_j = \max\{\langle Q_j^i, x \rangle : \langle a^i, x \rangle - b_i \leq 0, \ i = 1, \cdots, m, \ x \in C\} \text{ and}$$
$$Q_j^i \text{ is the } j-th \text{ row of the matrix } Q^i.$$

Solutions methods for bilinear programming problems of the types (16) and (17) can be found e.g. in [1], [21], [70], [22], [2] and references given therein. In general, the number of additional variables y_i in Problem (16) is n. However, this number can be reduced based on the structures of the matrices Q^i, $i = 0, \cdots, I$. In [26] and [13] techniques for optimizing the number of additional variables are proposed.

2.6   Duality Bound Techniques

Consider optimization problems given by

$$f^* = \inf\{f(x) : \ x \in C, \ g_i(x) \leq 0, \ i \in I_1, \ g_i(x) = 0, \ i \in I_2\}, \tag{18}$$

where C is a nonempty convex subset of $\mathbb{R}^n$, I_1, I_2 are finite index sets and f, and g_i ($i \in I_1 \cup I_2$) are continuous functions on C.

For $\lambda \in \mathbb{R}^{|I_1|+|I_2|}$, $x \in C$ define the Lagrangian

$$L(x, \lambda) := f(x) + \sum_{i \in I_1 \cup I_2} \lambda_i g_i(x)$$

of Problem (18) and on the set

$$\Lambda = \{\lambda \in \mathbb{R}^{|I_1|+|I_2|} : \lambda_i \geq 0, \ i \in I_1\}$$

define the dual function

$$d : \Lambda \to \mathbb{R}, \ \lambda \mapsto d(\lambda) = \inf_{x \in C} L(x, \lambda).$$

Then the dual of (18) is defined as the problem

$$d^* = \sup\{d(\lambda) : \lambda \in \Lambda\}. \tag{19}$$

It is clear that for each $\lambda \in \Lambda$ it holds $d(\lambda) \leq f^*$. Consequently, we have

$$d^* \leq f^*,$$

which means that a lower bound of the optimal value of Problem (18) can be obtained by solving its dual problem (19).

In general, a duality gap

$$\Delta := f^* - d^* > 0$$

has to be expected.

For some special cases of general quadratic problems described by (18) with $C = \mathbb{R}^n$, dual problems can be formulated equivalently as semidefinite programming problems (cf., e.g., [69], [46], [47], [61], [62]). If we formulate quadratic problems of the form (18) as bilinear programming problems of the form (17), then it can be shown that dual problems of the resulting bilinear problems are equivalent to linear programs, (cf., e.g., [5], [18], [66], [67], [68]).

In general, duality bounds can be improved by adding to the original quadratic problem some suitable redundant quadratic constraints. Examples to this topic can be found in [61], [62].

2.7 Optimizing Quadratic Functions over Special Convex Sets

There are three interesting special cases of quadratic programming problems which can be formulated as

$$\min \{f(x) \ x \in F\}, \tag{20}$$

where f is a general quadratic function and F is respectively a ball, a simplex or a box in $\mathbb{R}^n$.

Optimizing Quadratic Functions over a Ball: When F is a ball given by

$$F = \{x \in \mathbb{R}^n : \|x\| \leq r\}, \tag{21}$$

where $\| \cdot \|$ denotes the Euclidean norm and $r > 0$, (20) is the most important subproblem used within trust region methods for solving nonlinear programming problems. The most procedures for solving this subproblem are developed in connection with trust region methods. A positive semidefinite problem approach for this problem is given in [56]. Branch and bound procedures can be found, e.g., in [53] and refences given therein.

Standard Quadratic Programming Problem: This is the case where $f = x^T Q x$, a quadratic form, and F is the standard simplex defined by

$$F = \{x \in \mathbb{R}^n : x_1 + \cdots + x_n = 1,\ x_i \geq 0\ \forall i = 1, \cdots, n\}. \tag{22}$$

Applications of (23) include the quadratic knapsack problem (cf. [52]), the portfolio section problem and the maximum clique problem. Theoretical results and solution methods for the standard quadratic programming problem can be found in [7], [10], [46], [52].

Optimizing Quadratic Functions over a Box: The third special problem we want to mention here is the case where F is a box of the form

$$F = \{x \in \mathbb{R}^n : \ell \leq x \leq u\} \tag{23}$$

with ℓ and u being given vectors. Methods for solving this special problem include the procedure proposed in [8], branch and bound algorithms (cf., e.g., [50], [27]), [67], [31]), cutting plane techniques (cf., e.g., [71]), and heuristic procedures of interior point type proposed in [25].

3   Mixed-Binary Quadratic Programs

The mixed-binary quadric program is formulated as follows.

$$\begin{aligned}
\min\ & \langle Qx, x \rangle + 2\langle q, x \rangle \\
\text{s.t.}\ & \\
& \langle a^i, x \rangle = b_i,\ i = 1, \cdots, m \\
& x \geq 0 \\
& x_j \in \{0, 1\},\ j \in J \subseteq \{1 \cdots, n\},
\end{aligned} \tag{24}$$

where $Q \in \mathcal{S}^n$, q, $a^i \in \mathbb{R}^n$, $i = 1, \cdots, m$.

Problem (24) contains many classical special cases such as knapsack problem, the clique number problems, quadratic assignment problems, etc.,... For each of these problems, there are efficient algorithms, according to their specific structures, see, e.g., [52], [49], [16] and references given therein.

We mention that global optimization techniques presented in the previous section can be applied to Problem (24). Examples for these techniques are presented in [66].

In this section we only present the approach to formulate Problem (24) as a *copositive optimization problems*. For this, we recall the concept of *copositive matrices*. The set of *copositive matrices*, denoted by $\mathcal{C}^n$, is defined as

$$\mathcal{C}^n = \{A \in \mathcal{S}^n : x^T A x \geq 0 \ \forall x \in \mathbb{R}^n_+\}.$$

It is known, (see e.g., [17], [9]) that $\mathcal{C}^n$ is a convex, full dimensional cone in the space of $n \times n$ real matrices. The dual cone of $\mathcal{C}^n$ is defined by

$$\mathcal{D}^n = \{X : \langle A, X \rangle \geq 0 \ \forall A \in \mathcal{C}^n\},$$

where for two $n \times n$ matrices, A, B, we define as usual

$$\langle A, B \rangle := \text{trace}(B, A) = \sum_{i,j=1}^{n} a_{i,j} b_{i,j}.$$

It can be shown (see e.g. [17], [6]) that $\mathcal{D}^n$ is the cone of so-called *completely positive matrices* given by

$$\mathcal{D}^n = \text{conv}\{xx^T : x \in \mathbb{R}^n_+\},$$

where by $\text{conv}(S)$ we denote the convex hull of the set S.

For Problem (24) we make the following two *Key Assumptions (KA)*:

(KA 1): It holds: $x \in L \implies 0 \leq x_j \leq 1, j = 1, \cdots, n$, where

$$L = \{x \geq 0 : \langle a^i, x \rangle = b_i, \ i = 1, \cdots, m\}.$$

(KA2) $\exists \beta \in \mathbb{R}^m$ such that

$$\sum_{i=1}^{m} \beta_i a^i \geq 0, \ \sum_{i=1}^{m} \beta_i b_i = 1.$$

Note that the assumptions (KA1), (KA2) are fulfilled for many classes of problems, e.g., knapsack problems, clique number problems, quadratic assignment problems,...

Burer [16] showed that under (KA1)-(KA2), by using a vector

$$\alpha = \sum_{i=1}^{m} \beta_i a_i \geq 0, \tag{25}$$

Problem (24) can then be equivalently reformulated as the following completely positive problem:

$$\max \langle Q + 2\alpha q^T, X \rangle$$
$$\text{s.t.}$$
$$\langle \alpha(a^i)^T, Z \rangle = b_i, \ i = 1, \cdots, m$$
$$\langle a^i(a^i)^T, X \rangle = b_i^2, \ i = 1, \cdots, m$$
$$(X\alpha)_j = X_{jj}, \ j = 1, \cdots, n \tag{26}$$
$$\langle \alpha\alpha^T, X \rangle = 1$$
$$X \in \mathcal{D}^n.$$

The equivalence between Problem (24) and Problem (26) is stated as follows (see [16]):

Under (KA1)-(KA2), let α be defined as in (25). Then Problem (24) is equivalent to Problem (26) in the sense that:

(i) The optimal values of both problems are equal;
(ii) If X^* is an optimal solution of Problem (26), then $X^*\alpha$ lies in the convex hull of optimal solutions for Problem (24).

We have seen above that a broad class of NP-hard problems can be transformed into a specific class of well-structured convex minimization problems. However, the difficulty of Problem (24) is transferred in the last constraint of Problem (26), namely the completely positive constraint. Results on theory and solution methods for Problem (26) can be found in [9], [11], [12], [14], [15], [17].

References

1. Al-Khayyal, F.A., Falk, J.E.: Jointly constrained biconvex programming. Math. Oper. Res. 8, 273–286 (1983)
2. Al-Khayyal, F.A., Larsen, C., van Voorhis, T.: A relaxation method for nonconvex quadratically constrained quadratic programs. Journal of Global Optimization 6, 215–230 (1995)
3. Audet, C., Hansen, P., Jaumard, B., Savard, G.: A branch and cut algorithm for nonconvex quadratically constrained quadratic programming. Math. Program. Ser. A 87, 131–152 (2000)
4. Bazaraa, M.S., Sherali, H.D., Shetty, C.M.: Nonlinear Programming: Theory and Algorithms, 2nd edn. John Wiley & Sons, Inc. (1993)
5. Ben-Tal, A., Eiger, G., Gershovitz, V.: Global Minimization by Reducing the Duality Gap. Mathematical Programming 63, 193–212 (1994)
6. Berman, A., Shaked-Monderer, N.: Completely positive matrices. World Scientific (2003)
7. Bomze, I.M.: On standard quadratic optimization problems. Journal of Global Optimization 13, 369–387 (1998)

8. Bomze, I.M., Danninger, G.: A global optimization algorithm for concave quadratic problems. SIAM J. Optimization 3, 826–842 (1993)
9. Bomze, I.M., Duer, M., de Klerk, E., Roos, C., Quist, A.J., Terlaky, T.: On copositive programming and standard quadratic optimization problems. Journal of Global Optimization 18, 301–320 (2000)
10. Bomze, I.M.: Branch and bound approaches to standard quadratic optimization problems. Journal of Global Optimization 22, 17–37 (2002)
11. Bomze, I.M., de Klerk, E.: Solving standard quadratic optimization problems via linear, semidefinite and copositive programming. Journal of Global Optimization 24, 163–185 (2002)
12. Bomze, I.M., Jarre, F., Rendl, F.: Quadratic factorization heuristics for copositive programming. Math. Prog. Comp. 3, 37–57 (2011)
13. Brimberg, J., Hansen, P., Mladenović, N.: A note on reduction of quadratic and bilinear programs with equality constraints. Journal of Global Optimization 22, 39–47 (2002)
14. Bundfuss, S., Dür, M.: An adaptive linear approximation algorithm for copositive programs. SIAM Journal on Optimization 20, 30–53 (2009)
15. Bundfuss, S., Dür, M.: Algorithmic copositivity detection by simplicial partition. Linear Algebra and its Applications 428, 1511–1523 (2008)
16. Burer, S.: On the copositive representation of binary and continuous nonconvex quadratic programs. Mathematical Programming 120, 479–495 (2009)
17. Dür, M.: Copositive programming – a survey. In: Diehl, M., Glineur, F., Jarlebring, E., Michiels, W. (eds.) Recent Advances in Optimization and its Applications in Engineering, pp. 3–20. Springer (2010)
18. Dür, M., Horst, R.: Lagrange Duality and Partioning Techniques in Nonconvex Global Optimization. Journal of Optimization Theory and Applications 95, 347–369 (1997)
19. Falk, J.E., Hoffman, K.L.: A Successive Underestimation Method for Concave Minimization Problems. Mathematics of Operations Research 1, 251–259 (1976)
20. Floudas, C.A., Aggarwal, A., Ciric, A.R.: Global optimum search for nonconvex NLP and MINLP problems. Computers and Chemical Engineering 13, 1117–1132 (1989)
21. Floudas, C.A., Visweswaran, V.: A global optimization algorithm (GOP) for certain classes of nonconvex NLPs- I. Theory. Computers and Chemical Engineering 14, 1397–1417 (1990)
22. Floudas, C.A., Visweswaran, V.: Quadratic optimization. In: Horst, R., Pardalos, P.M. (eds.) Handbook of Global Optimization, pp. 217–269. Kluwer Academic Publihers (1995)
23. Fujie, T., Kojima, M.: Semidefinite programming relaxation for nonconvex quadratic programs. Journal of Global Optimization 10, 367–380 (1997)
24. Geoffrion, A.M.: Duality in Nonlinear Programming: A Simplified Application-Oriented Development. SIAM Review 13, 1–37 (1971)
25. Han, C.G., Pardalos, P.M., Ye, Y.: On the solution of indefinite quadratic problems using an interior point algorithm. Informatica 3, 474–496 (1992)
26. Hansen, P., Jaumard, B.: Reduction of indefinite quadratic programs to bilinear programs. Journal of Global Optimization 2, 41–60 (1992)
27. Hansen, P., Jaumard, B., Ruiz, M., Xiong, J.: Global optimization of indefinite quadratic functions subject to box constraints. Naval Res. Logist. 40, 373–392 (1993)
28. Heinkenschloss, M.: On the solution of a two ball trust region subproblem. Math. Programming 64, 249–276 (1994)

29. Hiriart-Urruty, J.B., Seeger, A.: A variational approach to copositive matrices. SIAM Rev. 52, 593–629 (2010)
30. Horst, R., Pardalos, P.M. (eds.): Handbook of global optimization. Kluwer Academic Publihers (1995)
31. Horst, R., Pardalos, P.M., Thoai, N.V.: Introduction to Global Optimization, 2nd edn. Kluwer, Dordrecht (2000)
32. Horst, R., Thoai, N.V.: A new algorithm for solving the general quadratic programming problem. Computational Optimization and Applications 5, 39–49 (1996)
33. Horst, R., Thoai, N.V.: DC programming: Overview. Journal of Optimization Theory and Applications 103, 1–43 (1999)
34. Horst, R., Thoai, N.V., Benson, H.P.: Concave minimization via conical partitions and polyhedral outer approximation. Mathematical Programming 50, 259–274 (1991)
35. Horst, R., Tuy, H.: Global Optimization: Deterministic Approaches, 3rd edn. Springer, Berlin (1996)
36. Kojima, M., Tunçel, L.: Cones of matrix and successive convex ralaxations of nonconvex sets. SIAM J. Optimization 10, 750–778 (2000)
37. Kojima, M., Tunçel, L.: Discretization and localization in successive convex relaxation methods for nonconvex quadratic optimization. Math. Program. Ser. A 89, 79–111 (2000)
38. Kono, H.: Maximizing a convex quadratic function subject to linear constraints. Mathematical Programming 11, 117–127 (1976)
39. Le Thi, H.A., Pham, D.T.: Solving a class of linearly constrained indefinite quadratic problems by d.c. algorithm. Journal of Global Optimization 11, 253–285 (1997)
40. Le Thi, H.A., Pham, D.T.: A branch and bound method via d.c. optimization algorithms and ellipsoidal technique for box constrained nonconvex quadratic problems. Journal of Global Optimization 13, 171–206 (1998)
41. Locatelli, M., Raber, U.: Packing equal circes into a square: a deterministic global optimization approach. Discrete Apllied Mathematics 122, 139–166 (2002)
42. Lovász, L., Schreijver, A.: Cone of matrices and set functions and 0-1 optimization. SIAM J. Optimization 1, 166–190 (1991)
43. Mangasarian, O.L.: Nonlinear Programming. Robert E. Krieger Publishing Company, Huntington (1979)
44. Martinez, J.M.: Local minimizers of quadratic functions on Euclidean balls and spheres. SIAM J. Optimization 4, 159–176 (1994)
45. Moré, J.: Generalizations of the trust region problem. Optimization Methods & Software 2, 189–209 (1993)
46. Novak, I.: A new semidefinite bound for indefinite quadratic forms over a simplex. Journal of Global Optimization 14, 357–364 (1999)
47. Novak, I.: Dual bounds and optimality cuts for all-quadratic programs with convex constraints. Journal of Global Optimization 18, 337–356 (2000)
48. Pardalos, P.M.: Global optimization algorithms for linearly constrained indefinite quadratic problems. Comput. Math. Appl. 21, 87–97 (1991)
49. Pardalos, P.M., Rendl, F., Wolkowicz, H.: The Quadratic Assignment Problem: A Survey and Recent Developments. DIMACS Series in Discrete Mathematics and Theoretical Computer Science, vol. 16, pp. 1–39 (1994)
50. Pardalos, P.M., Rosen, J.B.: Constrained Global Optimization: Algorithms and Applications. LNCS, vol. 268. Springer, Heidelberg (1987)
51. Pardalos, P.M., Vavasis, S.A.: Quadratic programming with one negative eigenvalue is NP-hard. Journal of Global Optimization 1, 843–855 (1991)

52. Pardalos, P.M., Ye, Y., Han, C.G.: Algorithms for the solution of quadratic knap-sack problems. Linear Algebra & its Applications 152, 69–91 (1991)
53. Pham, D.T., Le Thi, H.A.: D.c. optimization algorithm for solving the trust region subproblem. SIAM J. on Optimization 8, 1–30 (1998)
54. Phong, T.Q., Pham, D.T., Le Thi, H.A.: A method for solving d.c. program-ming problems. Application to fuel mixture nonconvex optimization problem. J. of Global Optimization 6, 87–105 (1995)
55. Raber, U.: A simplicial branch and bound method for solving nonconvex all-quadratic programs. J. of Global Optimization 13, 417–432 (1998)
56. Rendl, F., Wolkowicz, H.: A semidefinite framwork for trust region subproblems with applications to large scale minimization. Mathematical Programming (Serie B) 77, 273–299 (1997)
57. Sahni, S.: Computationally related problems. SIAM J. Comput. 3, 262–279 (1974)
58. Sherali, H.D., Adams, W.P.: A hierarchy of relations between the continuous and convex hull representations of zero-one programming problems. SIAM J. Discrete Mathematics 3, 411–430 (1990)
59. Sherali, H.D., Adams, W.P.: A refomulation-linearization technique for solving discrete and continuous nonconvex problems. Kluwer Academic Publihers (1999)
60. Sherali, H.D., Tuncbilek, C.H.: A reformulation-convexification approach for solv-ing nonconvex quadratic programming problems. J. of Global Optimization 7, 1–31 (1995)
61. Shor, N.Z.: Nondifferentiable optimization and polynomial problems. Kluwer Aca-demic Publihers (1998)
62. Shor, N.Z., Stetsyuk, P.I.: Lagrangian bounds in multiextremal polynomial and discrete optimization problems. J. of Global Optimization 23, 1–41 (2002)
63. Stern, R., Wolkowicz, H.: Indefinite trust region subproblems and nonsymmetric eigenvalue perturbation. SIAM J. Optimization 5, 286–313 (1995)
64. Takeda, A., Dai, Y., Fukuda, M., Kojima, M.: Towards the implementation of suc-cessive convex relaxation method for nonconvex quadratic optimization problems. In: Pardalos, P. (ed.) Approximation and Complexity in Numerical Optimization: Continuous and Discrete Problems. Kluwer Academic Publishers (2000)
65. Takeda, A., Fujisawa, K., Fukaya, Y., Kojima, M.: Parallel implementation of suc-cessive convex relation methods for quadratic optimization problems. J. of Global Optimization 24, 237–260 (2002)
66. Thoai, N.V.: Global Optimization Techniques for Solving the General Quadratic Integer Programming Problem. Computational Optimization and Applications 10, 149–163 (1998)
67. Thoai, N.V.: Duality Bound Method for the General Quadratic Programming Prob-lem with Quadratic Constraints. Journal of Optimization Theory and Applica-tions 107, 331–354 (2000)
68. Thoai, N.V.: Convergence and Application of a Decomposition Method Using Du-ality Bounds for Nonconvex Global Optimization. Journal of Optimization Theory and Applications 113, 165–193 (2002)
69. Vandenberghe, L., Boyd, S.: Semidefinite Programming. SIAM Review 38, 49–95 (1996)
70. Visweswaran, V., Floudas, C.A.: A global optimization algorithm (GOP) for certain classes of nonconvex NLPs- II. Application of theory and test problems. Computers and Chemical Engineering 14, 1419–1434 (1990)
71. Yajima, Y., Fujie, T.: A polyhedral approach for nonconvex quadratic programming problems with box constraints. Journal of Gloabl Optimization 13, 151–170 (1998)

Branch and Bound Algorithm Using Cutting Angle Method for Global Minimization of Increasing Positively Homogeneous Functions

Nguyen Van Thoai

Department of Mathematics, University of Trier, D-54286 Trier, Germany
`thoai@uni-trier.de`

Abstract. We consider the problem of globally minimizing an abstract convex function called increasing positively homogeneous (IPH) function over a compact convex subset of an $n-$dimensional Euclidean space, for short, IPH optimization problem.

A method for solving IPH optimization problems called cutting angle algorithm was proposed by Rubinov and others in 1999. The principle of cutting angle algorithm is a generalization of the cutting plane method for convex programming problems, where the convex objective function is iteratively approximated by the maximum of a family of affine functions defined by its subgradients. In this article, we propose a method for solving IPH optimization problems which is a combination of the cutting angle algorithm with a branch and bound scheme successfully used in global optimization. The lower bounding procedure in the present algorithm is performed by solving ordinary convex (or even linear) programs. From preliminary computational results we hope that the proposed algorithm could work well for some problems with specific structures.

Keywords: IPH optimization, cutting angle method, nonconvex programming, global optimization, branch and bound algorithms.

1 Introduction

A real function f defined on the nonnegative orthant of an $n-$dimensional Euclidean space, $\mathbb{R}_+^n$, is called to be increasing and positively homogeneous of degree one (IPH), if it holds that $f(x^1) \geq f(x^2)$ for arbitrary points $x^1 \geq x^2 \geq 0$ and $f(\lambda x) = \lambda f(x)$ for all $x \in \mathbb{R}_+^n$, $\lambda \geq 0$. Consequently, we call the problem of minimizing an IPH function over a closed convex subset of $\mathbb{R}_+^n$ an IPH optimization problem.

A method for solving IPH optimization problems called cutting angle algorithm was proposed by Rubinov and others in [1], [2], [11], [13], [14], [15]. The main idea of the cutting angle algorithm is to construct iteratively an under-approximation of the IPH objective function by the maximum of a family of so-called min-types functions, and then to solve the relaxed problem of minimizing the under-approximation function over the given feasible set. The algorithm terminates when the optimal value of a relaxed problem is equal to the value

N.T. Nguyen, T. Van Do, and H.A. Le Thi (Eds.): *ICCSAMA 2013*, SCI 479, pp. 19–30.
DOI: 10.1007/978-3-319-00293-4_2   © Springer International Publishing Switzerland 2013

of the original IPH objective function at some feasible point. The principle of cutting angle algorithm is a generalization of the cutting plane method for convex programming problems, where the convex objective function is iteratively approximated by the maximum of a family of affine functions defined by its subgradients, (cf., e.g. [7], [8], [10], [13]).

In this article, we propose a method for solving IPH optimization problems which is a combination of the cutting angle algorithm with a branch and bound scheme successfully used in global optimization (cf., e.g., [5], [6], [9], [17], [18], [19]). As a result, we obtain a branch and bound algorithm, in which the lower bounding procedure is performed by solving ordinary convex (or even linear) programs. Preliminary computational results show that the proposed algorithm could work well for some problems with specific structures.

The article is organized as follows. In the next section, we give briefly some basic concepts and results on abstract convex functions and IPH functions and introduce the standard form of IPH optimization problems. Section 3 deals with some basic operations, which are used to establish the branch and bound algorithm in Section 4. The transformation of a class of problems into the standard form is discussed in Section 5. Section 6 contains preliminary computational test results.

2   Abstract Convex Functions and IPH Optimization Problems in Standard Form

In this section, we first recall briefly some basic concepts and results on abstract convex functions and IPH functions (cf.,e.g. [2], [12] - [13], [16]), and introduce the standard form of IPH optimization problems thereafter.

Let H be a set of real functions on $\mathbb{R}^n_+$. A real function f is said to be abstract convex over $\mathbb{R}^n_+$ with respect to H (or $H-$convex) iff there exists $U \subseteq H$ such that

$$f(x) = \sup\{h(x) : h \in U\} \ \forall x \in \mathbb{R}^n_+. \tag{1}$$

A function $h \in H$ is called a subgradient of an $H-$convex function f at $y \in \mathbb{R}^n_+$, iff it holds that

$$h(x) \leq f(x) \ \forall x \in \mathbb{R}^n_+ \text{ and } h(y) = f(y). \tag{2}$$

For each given vector $p \in \mathbb{R}^n_+ \setminus \{0\}$, a min-type function generated from p is defined by

$$\ell_p(x) = \min\{p_i x_i : p_i > 0\} \ \forall x \in \mathbb{R}^n_+. \tag{3}$$

In particular, if the set H is define by

$$H = \{\ell_p : p \in \mathbb{R}^n_+ \setminus \{0\}\},$$

then it is known that each IPH function $f : \mathbb{R}_+^n \to \mathbb{R}$ is H−convex in the sense of (1) with U being the set of all min-type underestimations of f. In other words, every real finite IPH function f on $\mathbb{R}_+^n$ can be represented by

$$f(x) = \max\{\ell_p(x) : \ell_p \in H, \ \ell_p \leq f\} \ \forall x \in \mathbb{R}_+^n. \tag{4}$$

Moreover, let f be a finite IPH function, and for each $y \in \mathbb{R}_+^n \setminus \{0\}$ satisfying $f(y) > 0$, let $p^y = (p_1^y, \cdots, p_n^y) \in \mathbb{R}_+^n \setminus \{0\}$ be the point constructed from y by

$$p_i^y = \begin{cases} \frac{f(y)}{y_i}, \text{ for } y_i > 0 \\ 0, \text{ for } y_i = 0. \end{cases} \tag{5}$$

Then the min-type function ℓ_{p^y} generated from p^y is a subgradient of the IPH function f at y. More precisely, it holds that

$$\ell_{p^y}(x) \leq f(x) \ \forall x \in \mathbb{R}_+^n \text{ and } \ell_{p^y}(y) = f(y). \tag{6}$$

Properties and examples for IPH functions can be found, e.g., in [2], [13], [14].

For the establishment of the method in next sections, we consider IPH optimization problems in the following standard form.

$$\min\{f(x) : \ x \in S^1 \cap D\}, \tag{7}$$

where $f : \mathbb{R}_+^n \to \mathbb{R}$ is an IPH function, S^1 is the unit simplex of $\mathbb{R}^n$ defined by

$$S^1 = \{x \in \mathbb{R}^n : x \geq 0, \ \sum_{i=1}^n x_i = 1\}, \tag{8}$$

and D is a convex subset of $\mathbb{R}^n$.

We will show in Section 5 that every IPH optimization problem with a compact feasible set can be equivalently transformed into the standard form (7). Moreover, in connection with the transformation of Rubinov and Andramonov in [15], each problem of minimizing a Lipschitz function over a compact set can also be transformed into Problem (7).

In the next two sections, we establish an algorithm for solving IPH optimization problems in the standard form.

3   Basic Operations

As mentioned in the introduction, the method we propose to solve IPH optimization problems of the standard form (7) belongs to the class of branch and bound algorithms, in which two basic operations, branching and bounding, have to be performed. We begin to establish our method with these basic operations.

3.1 Branching Procedure

The unit simplex S^1 defined in (8) is the convex hull of the unit vectors, $e^1, \cdots, e^n$, of $\mathbb{R}^n$. Thus, we also use the notion $S^1 = [e^1, \cdots, e^n]$. The branching procedure begins with this simplex. Throughout the algorithm, the simplex S^1 is iteratively divided into subsimplices with the same structure, i.e., each generated subsimplex $S \subset S^1$ is defined by $S = [u^1, \cdots, u^n]$, where $u^1, \cdots, u^n$ are n linear independent vectors in S^1. Such a simplex subdivision can be performed as follows.

For each simplex $S = [u^1, \cdots, u^n] \subseteq S^1$, choose a point $y \in S \setminus \{u^1, \cdots, u^n\}$. This point y is uniquely represented by

$$y = \sum_{i=1}^{n} \lambda_i u^i, \ \sum_{i=1}^{n} \lambda_i = 1, \ \lambda_i \geq 0 \ \forall i.$$

Let $I = \{i : \lambda_i > 0\}$. Then the simplex S is divided into $|I|$ subsimplices S_i, $i \in I$, where for each $i \in I$, S_i is formed from S by replacing u^i by y. More precisely, let $S = [u^1, \cdots, u^i, \cdots, u^n]$. Then $S_i = [u^1, \cdots, y, \cdots, u^n]$. By this way, it is clear that

$$S = \bigcup_{i \in I} S_i \text{ and int } S_i \cap \text{int } S_j = \emptyset \ \forall i \neq j,$$

(here int S_i denotes the interior of S_i).

If the chosen point y is the midpoint of a longest edge of S, then $|I| = 2$, i.e., S is divided into 2 subsimplices, and this division is called simplex bisection. It is known (cf. e.g., [5], [19]) that the simplex bisection procedure has the following useful property: every infinite nested sequence $\{S^k\}$ such that $S^{k+1} \subset S^k \ \forall k$ shrinks to a unique point, i.e.,

$$\lim_{k \to \infty} S^k = \bigcap_{k=0}^{\infty} S^k = \{s^*\}. \tag{9}$$

In this case, we say that the simplex bisection procedure is exhaustive.

3.2 Lower Bound Estimation

Let $S = [u^1, \cdots, u^n] \subseteq S^1$ be any simplex generated by a simplex subdivision procedure. We compute a lower bound $\mu(S)$ of f over the set $S \cap D$, i.e., a real number $\mu(S)$ satisfying

$$\mu(S) \leq \min\{f(x) : x \in S \cap D\}. \tag{10}$$

First, if $S \cap D = \emptyset$, then, logically, S should be removed from further consideration within any branch and bound algorithm. Therefore, for this case we set

$$\mu(S) = +\infty. \tag{11}$$

In the case that $S \cap D \neq \emptyset$, let $Q(S)$ be any finite subset of $S \cap D$. Note that, since f is an IPH function, it follows that $f(x) \geq 0 \ \forall x \in \mathbb{R}_+^n$. So, if there is a point $\overline{y} \in Q(S)$ with $f(\overline{y}) = 0$, then 'everything is done': $\overline{y}$ is a global optimal solution of Problem (7). Thus, we assume that $f(y) > 0 \ \forall y \in Q(S)$, and our method for computing a lower bound $\mu(S)$ is based on the following.

Proposition 1. *Let U be the $n \times n$ matrix with columns $u^1, \cdots, u^n$. Then a real number, $\nu(S)$, satisfying $\nu(S) \leq \min\{f(x) : x \in S \cap D\}$ can be computed by solving the following optimization problem in variables $\lambda \in \mathbb{R}_+^n$ and $t \in \mathbb{R}$:*

$$\nu(S) = \min \ t$$
$$s.t. \ \sum_{i=1}^{n} \ell_{p^y}(u^i)\lambda_i - t \leq 0, \ y \in Q(S)$$
$$U\lambda \in D \qquad\qquad (12)$$
$$\sum_{i=1}^{n} \lambda_i = 1$$
$$\lambda \geq 0,$$

where for each $y \in Q(S)$, the point p^y is constructed by (5) and the min-type function $\ell_{p^y}(x)$ is defined by (3).

Proof. From (4) and (6) it follows that

$$\min\{f(x) : x \in S \cap D\} \geq \min\{ \max_{y \in Q(S)} \ell_{p^y}(x) : x \in S \cap D\}$$
$$= \min\{t : \max_{y \in Q(S)} \ell_{p^y}(x) - t \leq 0, \ x \in S \cap D\} \qquad (13)$$
$$= \min\{t : \ell_{p^y}(x) - t \leq 0 \ \forall y \in Q(S), \ x \in S \cap D\}.$$

Each point $x \in S$ is uniquely represented by

$$x = \sum_{i=1}^{n} \lambda_i u^i = U\lambda, \ \sum_{i=1}^{n} \lambda_i = 1, \ \lambda_i \geq 0, \ i = 1, \cdots, n. \qquad (14)$$

For each $y \in Q(S)$, the min-type function $\ell_{p^y}(x)$, being the pointwise minimum of a family of linear functions, is concave. Therefore, for each $x \in S$ it holds from (14) that

$$\ell_{p^y}(x) = \ell_{p^y}(\sum_{i=1}^{n} \lambda_i u^i) \geq \sum_{i=1}^{n} \ell_{p^y}(u^i)\lambda_i. \qquad (15)$$

From (14)-(15) it follows that

$$\min\{t : \ell_{p^y}(x) - t \leq 0 \ \forall y \in Q(S), \ x \in S \cap D\} \geq$$
$$\min\{t : \sum_{i=1}^{n} \ell_{p^y}(u^i)\lambda_i - t \leq 0 \ \forall y \in Q(S), \ U\lambda \in D, \ \sum_{i=1}^{n} \lambda_i = 1, \ \lambda \geq 0\},$$

which implies from (13) that

$$\min\{f(x): x \in S \cap D\} \geq$$
$$\min\{t: \ell_{p^y}(x) - t \leq 0 \ \forall y \in Q(S),\ x \in S \cap D\} \geq$$
$$\min\{t: \sum_{i=1}^{n} \ell_{p^y}(u^i)\lambda_i - t \leq 0 \ \forall y \in Q(S),\ U\lambda \in D,\ \sum_{i=1}^{n}\lambda_i = 1,\ \lambda \geq 0\} = \nu(S).$$

$\square$

Remark 1. (a) Since the set D is assumed to be convex, Problem (12) is a convex optimization problem. For the case that D is a polyhedral set given by

$$D = \{x \in \mathbb{R}^n : Cx \leq d\} \tag{16}$$

with C and d being matrix and vector of appropriate sizes, Problem (12) is an ordinary linear program, in which the condition $U\lambda \in D$ becomes $CU\lambda \leq d$.

(b) The quality of lower bound $\nu(S)$ computed above depends on the number of chosen elements of the finite set $Q(S) \subset S \cap D$. Here the principle would be 'the more the better'.

Proposition 1 means that $\nu(S)$ can be used as a lower bound of f over the set $S \cap D$. However, within a branch and bound algorithm, to guarantee the convergence, it is required to construct iteratively a nondecreasing sequence of lower bounds. For this purpose, the lower bound $\nu(S)$ is improved as follows.

Let S' be the simplex, from which the simplex S is directly generated by a simplex division, and let $\mu(S')$ be a known lower bound of f over $S' \cap D$. Then a lower bound $\mu(S)$ is computed by

$$\mu(S) = \begin{cases} +\infty, & \text{if } S \cap D = \emptyset \\ \max\{\mu(S'), \nu(S)\}, & \text{else.} \end{cases} \tag{17}$$

Obviously, the lower bound $\mu(S)$ computed by (17) is nondeceasing in the sense that $\mu(S') \leq \mu(S)$ for $S \subseteq S'$. Moreover, it has the following useful property, which will be used later to establish the convergence of a branch and bound algorithm.

Proposition 2. *For each simplex $S = [u^1, \cdots, u^n]$, let U be the $n \times n$ matrix with columns $u^1, \cdots, u^n$, $Q(S) \subset S \cap D$, and let $\kappa(S)$ be defined by*

$$\kappa(S) = \min\{\ell_{p^y}(u^i): y \in Q(S),\ i = 1, \cdots, n\}. \tag{18}$$

Then the lower bound $\mu(S)$ computed by (17) satisfies that

$$\mu(S) \geq \kappa(S). \tag{19}$$

Proof. From the definition of $\kappa(S)$ in (18), it follows that

$$\{(\lambda, t) \in \mathbb{R}^{n+1} : \sum_{i=1}^{n} \ell_{p^y}(u^i)\lambda_i - t \leq 0 \ \forall y \in Q(S),\ U\lambda \in D,\ \sum_{i=1}^{n}\lambda_i = 1,\ \lambda \geq 0\} \subseteq$$
$$\{(\lambda, t) \in \mathbb{R}^{n+1} : \sum_{i=1}^{n} \kappa(S))\lambda_i - t \leq 0,\ U\lambda \in D,\ \sum_{i=1}^{n}\lambda_i = 1,\ \lambda \geq 0\}.$$

Thus, from the definition of $\nu(S)$ by (12), and $\mu(S)$ by (17), it holds that

$$\mu(S) \geq \nu(S)$$
$$= \min\{t : \sum_{i=1}^{n} \ell_{p^{\nu}}(u^i)\lambda_i - t \leq 0 \; \forall y \in Q(S), \; U\lambda \in D, \; \sum_{i=1}^{n} \lambda_i = 1, \; \lambda \geq 0\}$$
$$\geq \min\{t : \sum_{i=1}^{n} \kappa(S)\lambda_i - t \leq 0, \; U\lambda \in D, \; \sum_{i=1}^{n} \lambda_i = 1, \; \lambda \geq 0\}$$
$$\geq \min\{t : \kappa(S) \sum_{i=1}^{n} \lambda_i - t \leq 0, \; \sum_{i=1}^{n} \lambda_i = 1, \}$$
$$= \min\{t : \kappa(S) \leq t\}$$
$$= \kappa(S).$$

$\square$

3.3 Upper Bounds

For each simplex S generated within the algorithm satisfying $S \cap D \neq \emptyset$, a finite set of feasible points, $Q(S) \subset S \cap D$, is constructed. Throughout the algorithm, more and more feasible points are computed. Upper bounds for the optimal value of Problem (7)) are then iteratively improved using the best feasible point found so far.

4 Algorithm and Convergence Properties

Using the basic operations from the previous section we establish the following branch and bound algorithm, and examine its convergence properties thereafter.

Branch and Bound Algorithm:
Initialization:

Check the set $S^1 \cap D$. If $S^1 \cap D = \emptyset$, then stop, Problem (7) is unsolvable. Otherwise, determine a finite set $Q^1 = Q(S^1) \subset S^1 \cap D$. If there is a point $\bar{y} \in Q^1$ with $f(\bar{y}) = 0$, then stop, $\bar{y}$ is a global optimal solution of Problem (7). Otherwise, compute a lower bound $\mu(S^1) = \nu(S^1)$ by solving Problem (12) according to S^1. Compute

$$\gamma_1 = \min\{f(x) : x \in Q^1\}$$

(γ_1 is an upper bound for the optimal value of the underlying problem). Choose a point $x^1 \in Q^1$ such that $f(x^1) = \gamma_1$. Set $\mu_1 = \mu(S^1)$; $\mathcal{R}^1 = \{S^1\}$; $k = 1$.

Iteration k: Execute the steps (i) to (ix) below.

(i) If $\gamma_k = \mu_k$, then stop, x^k is a global optimal solution and γ_k is the optimal value of Problem (7)).

(ii) If $\gamma_k > \mu_k$, then divide S^k into r_k subsimplices $S_1^k, \ldots, S_{r_k}^k$ (See Subsection 3.1).

(iii) For each $i = 1, \ldots, r_k$, check the set $S_i^k \cap D$. If $S_i^k \cap D = \emptyset$, then set $\mu(S_i^k) = +\infty$. Otherwise, determine a finite set $Q(S_i^k) \subset S_i^k \cap D$. If there is a point $\bar{y} \in Q(S_i^k)$ with $f(\bar{y}) = 0$, then stop, $\bar{y}$ is a global optimal solution of Problem (7). Otherwise, compute $\nu(S_i^k)$ by solving Problem (12) according to S_i^k, and set

$$\mu(S_i^k) = \max\{\mu(S^k), \nu(S_i^k)\}. \qquad (20)$$

(iv) Set $Q^{k+1} = (\overset{r_k}{\underset{i=1}{\bigcup}} Q(S_i^k)) \cup \{x^k\}$. Compute $\gamma_{k+1} = \min\{f(x) : x \in Q^{k+1}\}$.

(v) Choose $x^{k+1} \in Q^{k+1}$ such that $f(x^{k+1}) = \gamma_{k+1}$.

(vi) Set $\mathcal{R}^{k+1} = \left(\mathcal{R}^k \setminus \{S^k\}\right) \cup \left(\overset{r_k}{\underset{i=1}{\bigcup}} S_i^k\right)$.

(vii) Delete all $S \in \mathcal{R}^{k+1}$ such that $\mu(S) \geq \gamma_{k+1}$.

(viii) If $\mathcal{R}^{k+1} = \emptyset$, then set $\mu_{k+1} = \gamma_{k+1}$, otherwise, set

$$\mu_{k+1} = \min\{\mu(S) : S \in \mathcal{R}^{k+1}\}$$

and choose $S^{k+1} \in \mathcal{R}^{k+1}$ such that $\mu(S^{k+1}) = \mu_{k+1}$.

(ix) Go to iteration $k + 1$.

Convergence

If the algorithm terminates at some iteration k, then obviously the point x^k is a global optimal solution and γ_k is the optimal value of Problem (7). For the case that the algorithm does not terminate after a finite number of iterations, it generates an infinite sequence of feasible points, $\{x^k\}$, and accordingly the sequences $\{\mu_k\}$ and $\{\gamma_k\}$ of lower bounds and upper bounds, respectively. Convergence properties of the algorithm are discussed below.

Proposition 3. *Assume that the algorithm does not terminate after a finite number of iterations and the branching procedure is exhaustive, i.e., each nested subsequence of simplices, $\{S^q\}$ with $S^{q+1} \subset S^q \, \forall q$, satisfies that*

$$\lim_{q \to \infty} S^q = \bigcap_{q=1}^{\infty} S^q = \{s^*\} \text{ for some } s^* \in S^1 \cap D.$$

Then it holds that

(a) $\lim_{k \to \infty} \mu_k = \lim_{k \to \infty} f(x^k) = \lim_{k \to \infty} \gamma_k$, *and*

(b) Every accumulation point x^* of the sequence $\{x^k\}$ is a global optimal solution of Problem (7).

Proof. (a) Let x^* be an accumulation point of the sequence $\{x^k\} \subset S^1 \cap D$, and let f^* be the optimal value of Problem (7). Since the sequence $\{\mu_k\}$ of lower bounds is computed by (20), it is monotonically nondecreasing and bounded from above by f^*. Moreover, the sequence $\{\gamma_k\}$ of upper bounds is nonincreasing and bounded from below by f^*. Therefore, $\mu^* = \lim\limits_{k\to\infty} \mu_k$ and $\gamma^* = \lim\limits_{k\to\infty} \gamma_k$ exist, and, because of $\gamma_k = f(x^k)$ and continuity of $f(x)$, we obtain

$$\lim_{k\to\infty} \mu_k = \mu^* \le f^* \le \gamma^* = \lim_{k\to\infty} \gamma_k = \lim_{k\to\infty} f(x^k) = f(x^*). \tag{21}$$

Now, let $\{x^q\}$ be a subsequence of $\{x^k\}$ which converges to x^*. By passing to further subsequences if necessary, we can assume that the corresponding sequence $\{S^q\}$ satisfies $S^{q+1} \subset S^q$ and $\mu_q = \mu(S_q) \forall q$.

For each q, let $y^q \in Q(S^q) \subset S^q \cap D$. Since $\lim\limits_{q\to\infty} S^q = \{s^*\}$ for some $s^* \in S^1 \cap D$ and f is continuous, it follows that

$$\gamma^* \le \lim_{q\to\infty} f(x^q) \le \lim_{q\to\infty} f(y^q) = f(s^*). \tag{22}$$

On the other hand, from (19) in Proposition (?), it holds that $\mu(S^q) \ge \kappa(S^q) \,\forall q$. Letting $q \to \infty$, we obtain, by continuity of the functions f, ℓ_{p^y}, $y \in Q(S^q)$, by Property (6), and because of $\lim\limits_{q\to\infty} S^q = \{s^*\}$, that

$$\mu^* = \lim_{q\to\infty} \mu_q = \lim_{q\to\infty} \mu(S^q) \ge \lim_{q\to\infty} \kappa(S^q) = f(s^*). \tag{23}$$

Finally, from (21), (22) and (23), it follows that

$$\lim_{k\to\infty} \mu_k = \lim_{k\to\infty} f(x^k) = \lim_{k\to\infty} \gamma_k.$$

(b) From (23) it follows that s^* is a global optimal solution of Problem (7). Since $f(x^q) \le f(y^q) \,\forall q$, it follows that $f(x^*) = \lim\limits_{q\to\infty} f(x^q) \le \lim\limits_{q\to\infty} f(y^q) = f(s^*)$, which implies that x^* is also a global optimal solution of Problem (7). $\square$

To obtain a finiteness property of the algorithm, we use the concept of approximate optimal solutions in following sense.

Let ε be a positive number. A point $\overline{x} \in S \cap D$ is called an $\varepsilon-$ optimal solution of Problem (7) if it satisfies that $f(x) \le f(\overline{x}) + \varepsilon$ for all $x \in S \cap D$.

From Proposition 3, we obtain immediately the following finiteness result.

Proposition 4. *Let ε be a pre-chosen positive number. Assume that the branching procedure used in Algorithm BB is exhaustive, and at iteration k, steps (i) and (vii) are replaced by following steps (i)' and (vii)', respectively:*

(i)' If $\gamma_k - \mu_k \le \varepsilon$, then stop, x^k is an $\varepsilon-$ optimal solution of Problem (7),
(vii)' Delete all $S \in \mathcal{R}^{k+1}$ such that $\mu(S) \ge \gamma_{k+1} - \varepsilon$.
Then the algorithm always terminates after finitely many iterations.

5   Transformation into Standard Form

In this section, we present a way to transform an arbitrary IPH optimization problem with a compact feasible set into the standard form. This transformation can also be applied to Lipschitz optimization problems so that, in connection with the transformation of Rubinov and Andramonov in [15], problems of minimizing a Lipschitz function over a compact set can in principle be transformed into Problem (7).

Consider the optimization problem

$$\min\{g(z) : z \in Z\} \tag{24}$$

where $g : \mathbb{R}_+^q \to \mathbb{R}$ is an IPH (or Lipschitz) function, and Z is a convex compact subset of $\mathbb{R}_+^q$.

From the compactness of the feasible set Z, we can pack it into a q–simplex

$$T = [v^1, \cdots, v^{q+1}] \subset \mathbb{R}_+^q, \tag{25}$$

which is the convex hull of $q + 1$ affine independent points $v^1, \cdots, v^{q+1} \in \mathbb{R}_+^q$. The construction of such a simplex T is discussed, e.g., in [5].

Let $n = q + 1$, and let $V = (v^1, \cdots, v^n)$ be the $q \times n$ matrix with columns $v^1, \cdots, v^n$. Further, denoting by C the cone generated by vectors $v^1, \cdots, v^n$, i.e.,

$$C = \{z \in \mathbb{R}^q : z = \sum_{j=1}^n x_j v^j, \ x_j \geq 0, \ j = 1, \cdots, n\},$$

we define a mapping

$$t : \mathbb{R}_+^n \to C, \quad x \mapsto z = Vx. \tag{26}$$

Next, let $e^T = (1, \cdots, 1) \in \mathbb{R}^n$ and let S be the unit simplex of $\mathbb{R}^n$ defined by

$$S = \{x \in \mathbb{R}^n : e^T x = 1, \ x \geq 0\}. \tag{27}$$

Then the restriction of t to S is a bijective mapping from S to T satisfying

$$t(e^j) = v^j, \ j = 1, \cdots, n,$$

where e^j, $j = 1, \cdots, n$, are unit vectors of $\mathbb{R}^n$.

Finally, define a function $f : \mathbb{R}_+^n \to \mathbb{R}$ by

$$f(x) = g(Vx) \ \forall x \in \mathbb{R}_+^n. \tag{28}$$

Then Problem (24) is equivalently transformed into problem

$$\min\{f(x) : x \in S \cap D\} \text{ with } D = \{x \in \mathbb{R}^n : Vx \in Z\} \tag{29}$$

in the sense that a point $\overline{x} \in S \cap D$ is a global optimal solution of Problem (29) if and only if $\overline{z} = V\overline{x}$ is a global optimal solution of Problem (24).

The following result shows some useful properties of the function f defined by (28).

Proposition 5. *The function f defined by (28) has following properties:*

(a) If g is an IPH function on the cone C, then f is an IPH function on $\mathbb{R}^n_+$.

(b) If g is a Lipschitz function on the cone C, then f is a Lipschitz function on $\mathbb{R}^n_+$.

Proof. (a) It is obvious that $f(\lambda x) = g(V\lambda x) = g(\lambda Vx) = \lambda g(Vx) = \lambda f(x) \; \forall \lambda \geq 0$, and from (25), it follows that $x^1 \leq x^2 \Rightarrow Vx^1 \leq Vx^2 \Rightarrow f(x^1) = g(Vx^1) \leq g(Vx^2) = f(x^2)$.

(b) Let L be the Lipschitz constant for g in C and $\|\cdot\|_z$ a chosen norm in $\mathbb{R}^q$. Then, for arbitrary $x^1, x^2 \in \mathbb{R}^n_+$, it follows that

$$|f(x^1) - f(x^2)| = |g(Vx^1) - g(Vx^2)| \leq L\|Vx^1 - Vx^2)\|_z.$$

Thus, by choosing a suitable matrix norm, $\|\cdot\|_V$, which is consistent with given vector norms $\|\cdot\|_x$ and $\|\cdot\|_z$ on $\mathbb{R}^n$ and $\mathbb{R}^q$, respectively, in the sense that

$$\|Vx\|_z \leq \|V\|_V \cdot \|x\|_x \; \forall \, V \in \mathbb{R}^{q \times n}, \; x \in \mathbb{R}^n, \tag{30}$$

we obtain the Lipschitz property of the function f with the Lipschitz constant $L \cdot \|V\|_V$. □

6 Preliminary Computational Experiments

The algorithm is tested on problems of the standard form (7), where the IPH function f has some forms given in [2], [11], and the convex set D is given by (16) with C being an $m \times n$ matrix, and $d \in \mathbb{R}^m$.

For each pair (m, n) ($10 \leq m \leq 50$ and $20 \leq n \leq 100$), the elements of matrix C and vector d are randomly generated by using a pseudo-random number from an uniform distribution on $(0, 1)$. For all test problems, the stopping criterion in Proposition 4 is used with $\varepsilon = 10^{-3}$.

Test problems were run on a Sun SPARC station 10 Modell 20 workstation. We note that linear problems were solved by our own version of the simplex algorithm. For each given pair (m, n) the algorithm was run on 20 randomly generated test problems. Test results (in average values) are summarized in Table 1.

Table 1. Computational results

m	n	ITER	SMAX	TIME
10	20	100	50	5.05
20	20	323	160	15.98
10	50	825	412	28.45
20	50	635	346	18.93
10	100	3799	1656	800.27
50	100	1788	868	727.27

ITER: Average number of iterations,
SMAX: Maximal number of simplices stored at an iteration,
TIME: Average CPU-Time in seconds.

References

1. Andramonov, M.Y., Rubinov, A.M., Glover, B.M.: Cutting angle method in global optimization. Applied Mathematics Letters 12, 95–100 (1999)
2. Bagirov, A.M., Rubinov, A.M.: Global Minimization of Increasing Positively Homogeneous Functions over the Unit Simplex. Annals of Operations Research 98, 171–187 (2000)
3. Bartels, S.G., Kuntz, L., Sholtes, S.: Continuous selections of linear functions and nonsmooth critical point theory. Nonlinear Analysis, TMA 24, 385–407 (1995)
4. Demyanov, V.F., Rubinov, A.M.: Constructive Nonsmooth Analysis, Peter Lang, Frankfurt am Main (1995)
5. Horst, R., Pardalos, P.M., Thoai, N.V.: Introduction to Global Optimization, 2nd edn. Kluwer, Dordrecht (2000)
6. Horst, R., Thoai, N.V.: DC Programming: Overview. Journal of Optimization Theory and Applications 103, 1–43 (1999)
7. Horst, R., Thoai, N.V., Tuy, H.: Outer aproximation by polyhedral convex sets. O.R. Spektrum 9, 153–159 (1987)
8. Horst, R., Thoai, N.V., Tuy, H.: On an outer approximation concept in global optimization. Optimization 20, 255–264 (1989)
9. Horst, R., Thoai, N.V., Benson, H.P.: Concave minimization via conical partitions and polyhedral outer approximation. Mathematical Programming 50, 259–274 (1991)
10. Kelley, J.: The cutting plane method for solving convex programs. SIAM Journal 8, 703–712 (1960)
11. Ordin, B.: The Modified Cutting Angle Method for Global Minimization of Increasing Positively Homogeneous Functions over the Unit Simplex. Journal of Industrial and Management Optimization 5, 825–834 (2009)
12. Pallaschke, D., Rolewicz, S.: Foundations of Mathematical Optimization (Convex Analysis without Linearity). Kluwer Academic, Dordrecht (1997)
13. Rubinov, A.M.: Abstract Convexity and Global Optimization. Kluwer, Dordrecht (2000)
14. Rubinov, A.M., Andramonov, M.Y.: Minimizing increasing star-shaped functions based on abstract convexity. Journal of Global Optimization 15, 19–39 (1999)
15. Rubinov, A., Andramonov, M.Y.: Lipschitz programming via increasing convex-along-rays functions. Optimization Methods and Software 10, 763–781 (1999)
16. Singer, I.: Abstract Convex Analysis. Wilsey & Sons (1997)
17. Thoai, N.V.: Convergence and Application of a Decomposition Method Using Duality Bounds for Nonconvex Global Optimization. Journal of Optimization Theory and Applications 113, 165–193 (2002)
18. Thoai, N.V.: Decomposition Branch and Bound Algorithm for Optimization Problems over Efficient Sets. Journal of Industrial and Management Optimization 4, 647–660 (2008)
19. Thoai, N.V., Tuy, H.: Convergent algoritms for minimizing a concave function. Mathematics of Operations Research 5, 556–566 (1980)

A DC Programming Framework for Portfolio Selection by Minimizing the Transaction Costs

Pham Viet-Nga[1], Hoai An Le Thi[2,3], and Pham Dinh Tao[1]

[1] Laboratory of Mathematics, National Institute for Applied Sciences - Rouen, 76801 Saint Etienne du Rouvray, France
[2] Laboratory of Theorical and Applied computer Science LITA EA 3097, University of Lorraine, Ile du Saulcy-Metz 57045, France
[3] Lorraine Research Laboratory in Computer Science and its Applications LORIA CNRS UMR 7503, University of Lorraine, 54506 Nancy, France
{viet.pham,pham}@insa-rouen.fr, hoai-an.le-thi@univ-lorraine.fr

Abstract. We consider a single-period portfolio selection problem which consists of minimizing the total transaction cost subject to different types of constraints on feasible portfolios. The transaction cost function is separable, i.e., it is the sum of the transaction cost associated with each trade, but discontinuous. This optimization problem is nonconvex and very hard to solve. We investigate in this work a DC (Difference of Convex functions) programming framework for the solution methods. First, the objective function is approximated by a DC function. Then a DC formulation for the resulting problem is proposed for which two approaches are developed: DCA (DC Algorithm) and a hybridization of Branch and Bound and DCA.

Keywords: portfolio selection, separable transaction cost, DC programming, DCA, Branch and Bound.

1 Introduction

The mean-variance's model proposed by Markowitz [9] in 1952 is known as a basic for the development of various portfolio selection techniques. While the Markowitz' model is a convex program, extended models considering some factors like transaction costs, cardinality constraints, shortselling, buy-in threshold constraints, etc,... are, in most of cases, nonconvex and very difficult to solve. The portfolio optimization problems including transaction costs have been studied by many researchers [1–4].

In [8], the authors studied two alternative models for the problem of single-period portfolio optimization. The first consists of maximizing the expected return, taking transaction costs into account, and subject to different type of constraints on the feasible portfolios. They proposed a heuristic method for solving this model where the transaction cost is separable and discontinuous. The second model deals with minimizing the total nonconvex transaction cost subject to feasible portfolio constraints. The authors claimed that their heuristic method for solving the former model can be adapted to solve the later.

N.T. Nguyen, T. Van Do, and H.A. Le Thi (Eds.): *ICCSAMA 2013*, SCI 479, pp. 31–40.
DOI: 10.1007/978-3-319-00293-4_3　　　ⓒ Springer International Publishing Switzerland 2013

The starting point of our work is the second model introduced in [8]. We consider a little modified model where the constraints include shortselling constraints, limit on expected return, limit on variance, and diversification constraints. The considered transaction cost is assumed to be separable, say the sum of the transaction cost associated with each trade. It is a discontinuous function that results to a difficult nonconvex program.

We investigate DC programming and DCA for designing solution methods to this problem. DC programming and DCA were first introduced by Pham Dinh Tao in 1985 and have been extensively developed since 1994 by Le Thi Hoai An and Pham Dinh Tao in their common works. DCA has been successfully applied to many large-scale nonconvex programs in various domains of applied sciences, to become now classic and popular (see e.g. [5, 7, 6] and references therein). We first approximate the discontinuous nonconvex objective function by a DC function and then develop DCA for tackling the resulting DC problem. For globally solving the original problem, we propose a hybrid algorithm that combines DCA and a Branch-and-Bound (B&B) scheme. DCA is used for solving the DC approximation problem to compute good upper bounds in the B&B algorithm. Lower bounds are obtained by solving relaxation problems which consist of minimizing a linear function under linear and convex quadratic constraints.

The rest of this paper is organized as follows. In the next section, we describe the considered portfolio problem and its mathematical formulation. Section 3 is concerned with the DC approximation of the considered problem and the description of DCA for solving it. The hybrid Branch and Bound - DCA algorithm is presented in Section 4 while some conclusions are included in the last section.

2   Problem Description and Mathematical Formulation

Consider an investment portfolio that consists of holdings in some or all of n assets.

The current holdings in each asset are $w = (w_i, \ldots, w_n)^T$. The total current wealth is then $\mathbf{1}^T w$, where $\mathbf{1}$ is a vector with all entries equal to one. The amount transacted in asset i is x_i, with $x_i > 0$ for buying, $x_i < 0$ for selling and $x = (x_1, \ldots, x_n)^T$ is a portfolio selection. After transactions, the adjusted portfolio is $w + x$.

The adjusted portfolio $w + x$ is held for a fixed period time. At the end of that period, the return on asset i is the random variable a_i. We assume knowledge of the first and the second moments of the joint distribution of $a = (a_1, \ldots, a_n)$,

$$\mathbf{E}(a) = \bar{a}, \qquad \mathbf{E}(a - \bar{a})(a - \bar{a})^T = \Sigma.$$

A riskless asset can be included, in which case the corresponding $\bar{a}_i$ equal to its return and the i-th row and column of Σ are zero.

The wealth at the end of the period is a random variable, $W = a^T(w + x)$ with expected value and variance given by

$$\mathbf{E}W = \bar{a}^T(w + x), \qquad \mathbf{E}(W - \mathbf{E}W)^2 = (w + x)^T \Sigma (w + x). \tag{1}$$

We consider the problem of minimizing the total transaction costs subject to portfolio constraints:

$$\begin{cases} \min \ \phi(x) \\ \text{s.t. } \bar{a}(w + x) \geq r_{\min}, \\ \quad\quad w + x \in \mathcal{S}, \end{cases} \tag{2}$$

where $r_{\min}$ is the desired lower bound on the expected return and $\mathcal{S} \subseteq \mathbb{R}^n$ is the portfolio constraint set.

The portfolio constraint set $\mathcal{S}$ can be defined from the following convex constraints:

1. *Shortselling constraints:* Individual bounds s_i on the maximum amount of shortselling allowed on asset i are

$$w_i + x_i \geq -s_i, \quad i = 1, \ldots, n. \tag{3}$$

    If shortselling is not permitted, the s_i are set to zero. Otherwise, $s_i > 0$.
2. *Variance:* The standard deviation of the end period wealth W is constrained to be less than $\sigma_{\max}$ by the convex quadratic inequality

$$(w + x)^T \Sigma (w + x) \leq \sigma_{\max}^2. \tag{4}$$

    ((4) is a *second-order cone constraint*).
3. *Diversification constraints:* Constraints on portfolio diversification can be expressed in terms of linear inequalities and therefore are readily handled by convex optimization. Individual diversification constraints limit the amount invested in each asset i to a maximum of p_i,

$$w_i + x_i \leq p_i, \quad i = 1, \ldots, n. \tag{5}$$

    Alternatively, we can limit the fraction of the total wealth held in each asset,

$$w_i + x_i \leq \lambda_i \mathbf{1}^T (w + x), \quad i = 1, \ldots, n. \tag{6}$$

    They are convex inequality constraints on x.

Transaction costs can be used to model a number of costs, such as brokerage fee, bid-ask spread, taxes or even fund loads. In this paper, the transaction costs $\phi(x)$ is defined by

$$\phi(x) = \sum_{i=1}^n \phi_i(x_i), \tag{7}$$

where ϕ_i is the transaction cost function for asset i. We will consider a simple model that includes fixed plus linear costs. Let β_i be the fixed costs common associated with buying and selling asset i. The fixed plus linear transaction cost function is given by

$$\phi_i(x_i) = \begin{cases} 0 & \text{if } x_i = 0, \\ \beta_i - \alpha_i^1 x_i & \text{if } x_i < 0, \\ \beta_i + \alpha_i^2 x_i & \text{if } x_i > 0. \end{cases} \tag{8}$$

The function ϕ is nonconvex, unless the fixed costs are zero.

We develop below two approaches based on DC programming and DCA for solving the problem (2) with S being defined in (3) - (6) and ϕ being given in (7), (8).

3   DC Programming and DCA for Solving (2)

3.1   DC Approximation Problem

Let C be the feasible set of (2). Since ϕ is discontinuous, we will construct a DC approximation of ϕ. We first compute upper bounds u_i^0 and lower bounds l_i^0 for variables x_i by solving $2n$ convex problems:

$$\min\{x_i \ : \ x \in C\} \quad (LB_i), \qquad \max\{x_i \ : \ x \in C\} \quad (UB_i). \qquad (9)$$

Let $R_0 = \prod_{i=1}^{n} [l_i^0, u_i^0]$. The problem (2) can be rewritten as

$$\omega = \min\left\{\phi(x) = \sum_{i=1}^{n} \phi_i(x_i) \ : \ x \in C \cap R_0\right\}. \qquad (P)$$

For each $i = 1, \ldots, n$, let $\epsilon_i > 0$ be a sufficiently small number chosen as follows:

$$\begin{cases} \epsilon_i < \min\{-l_i^0, u_i^0\} & \text{if } l_i^0 < 0 < u_i^0, \\ \epsilon_i < u_i^0 & \text{if } l_i^0 = 0 < u_i^0, \\ \epsilon_i < -l_i^0 & \text{if } l_i^0 < u_i^0 = 0. \end{cases}$$

Consider the functions $\overline{\phi_i}, \psi_i : \mathbb{R} \longrightarrow \mathbb{R}$ given by

$$\overline{\phi_i}(x_i) = \begin{cases} \beta_i - \alpha_i^1 x_i, & x_i \leq 0 \\ \beta_i + \alpha_i^2 x_i, & x_i \geq 0 \end{cases}, \qquad \psi_i(x_i) = \begin{cases} -c_i^1 x_i, & x_i \leq 0 \\ c_i^2 x_i, & x_i \geq 0 \end{cases},$$

where $c_i^j = \left(\frac{\beta_i}{\epsilon_i} + \alpha_i^j\right)$, $j = 1, 2$. By definition, $\overline{\phi_i}, \psi_i$ are convex functions. Then, a DC approximation function f of ϕ can be

$$f(x) = \sum_{i=1}^{n} f_i(x_i), \qquad (10)$$

where $f_i(x_i) = g_i(x_i) - h_i(x_i)$ with g_i, h_i being determined by

- $g_i(x_i) = 0, h_i(x_i) = -\beta_i + \alpha_i^1 x_i$ if $l_i^0 < u_i^0 < 0$;
- $g_i(x_i) = 0, h_i(x_i) = -\beta_i - \alpha_i^2 x_i$ if $0 < l_i^0 < u_i^0$;
- $g_i(x_i) = 0, h_i(x_i) = -\min\{-c_i^1 x_i, \beta_i - \alpha_i^1 x_i\}$ if $l_i^0 < u_i^0 = 0$;
- $g_i(x_i) = 0, h_i(x_i) = -\min\{c_i^2 x_i, \beta_i + \alpha_i^2 x_i\}$ if $0 = l_i^0 < u_i^0$;
- and if $l_i < 0 < u_i$:

$$g_i(x_i) = \overline{\phi_i}(x_i) + \psi_i(x_i) = \begin{cases} \beta_i - (\alpha_i^1 + c_i^1)x_i & \text{if } x_i \leq 0 \\ \beta_i + (\alpha_i^2 + c_i^2)x_i & \text{if } x_i \geq 0 \end{cases},$$

$$h_i(x_i) = \max\{\overline{\phi_i}(x_i), \psi_i(x_i)\} = \begin{cases} -c_i^1 x_i & \text{if } x_i \leq -\epsilon_i \\ \beta_i - \alpha_i^1 x_i & \text{if } -\epsilon_i \leq x_i \leq 0 \\ \beta_i + \alpha_i^2 x_i & \text{if } 0 \leq x_i \leq \epsilon_i \\ c_i^2 x_i & \text{if } x_i \geq \epsilon_i. \end{cases}$$

It is easy to show that for all cases, g_i, h_i are convex polyhedral functions over $\mathbb{R}$. Therefore, with $g(x) = \sum_{i=1}^{n} g_i(x_i)$ and $h(x) = \sum_{i=1}^{n} h_i(x_i)$, $g - h$ is a DC decomposition of f. In addition,

- $\min\{f(x) : x \in C \cap R_0\} \leq \min\left\{\phi(x) = \sum_{i=1}^{n} \phi_i(x_i) : x \in C \cap R_0\right\}$.
- For each i, the smaller value of ϵ_i, the better approximation of f_i to ϕ_i over $[l_i^0, u_i^0]$.

The problem (P) with ϕ being replaced by f,

$$\mu = \min\{f(x) = g(x) - h(x) : x \in C \cap R_0\} \qquad (P_{dc})$$

is a DC approximation problem of (P). We will investigate a DCA scheme for solving this problem.

3.2   DCA for Solving (P_{dc})

DC Programming and DCA. For a convex function θ, the subdifferential of θ at $x_0 \in \text{dom}\theta := \{x \in \mathbb{R}^n : \theta(x_0) < +\infty\}$, denoted by $\partial\theta(x_0)$, is defined by

$$\partial\theta(x_0) := \{y \in \mathbb{R}^n : \theta(x) \geq \theta(x_0) + \langle x - x_0, y \rangle, \forall x \in \mathbb{R}^n\},$$

and the conjugate θ^* of θ is

$$\theta^*(y) := \sup\{\langle x, y \rangle - \theta(x) : x \in \mathbb{R}^n\}, \quad y \in \mathbb{R}^n.$$

A general DC program is that of the form:

$$\alpha = \inf\{F(x) := G(x) - H(x) \,|\, x \in \mathbb{R}^n\}, \qquad (11)$$

where G, H are lower semi-continuous proper convex functions on $\mathbb{R}^n$. Such a function F is called a DC function, and $G - H$ a DC decomposition of F while G and H are the DC components of F. Note that, the closed convex constraint $x \in C$ can be incorporated in the objective function of (11) by using the indicator function on C denoted by χ_C which is defined by $\chi_C(x) = 0$ if $x \in C$, and $+\infty$ otherwise.

A point x^* is called a *critical point* of $G - H$, or a generalized Karush-Kuhn-Tucker point (KKT) of (P_{dc})) if

$$\partial H(x^*) \cap \partial G(x^*) \neq \emptyset. \tag{12}$$

Based on local optimality conditions and duality in DC programming, the DCA consists in constructing two sequences $\{x^k\}$ and $\{y^k\}$ (candidates to be solutions of (11) and its dual problem respectively). More precisely, each iteration k of DCA approximates the concave part $-H$ in (11) by its affine majorization (that corresponds to taking $y^k \in \partial H(x^k)$) and minimizes the resulting convex program.

Generic DCA Scheme

Initialization: Let $x^0 \in \mathbb{R}^n$ be an initial guess, $0 \leftarrow k$.

Repeat

- Calculate $y^k \in \partial H(x^k)$
- Calculate $x^{k+1} \in \arg\min\{G(x) - \langle x, y^k\rangle : x \in \mathbb{R}^n\}$ (P_k)
- $k + 1 \leftarrow k$

Until convergence of $\{x^k\}$.

It is worth noting that DCA works with the convex DC components G and H but not the DC function F itself (see [5, 6, 10, 11]). Moreover, a DC function F *has infinitely many DC decompositions* which have crucial impacts on the performance (speed of convergence, robustness, efficiency, globality of computed solutions,...) of DCA.

Convergence properties of DCA and its theoretical basis can be found in [5, 6, 10]. For instant, it is important to mention that (for simplify we omit here the dual part)

- DCA is a descent method (the sequences $\{G(x^k) - H(x^k)\}$ is decreasing) without linesearch.
- If the optimal value α of problem (11) is finite and the infinite sequence $\{x^k\}$ is bounded then every limit point $\tilde{x}$ of the sequence $\{x^k\}$ is a critical point of $G - H$.
- DCA has a linear convergence for general DC programs.
- DCA has a finite convergence for polyhedral DC programs.

The next subsection is devoted to the development of DCA applied on (P_{dc}).

DC Algorithm for Solving the Problem (P_{dc}). According to the generic DCA scheme, at each iteration k, we have to compute a subgradient $y^k \in \partial h(x^k)$ and then solve the convex program of the form (P_k)

$$\min\{g(x) - \langle y^k, x\rangle : x \in C \cap R_0\} \tag{13}$$

which is equivalent to

$$\min_{x,t} \left\{ \sum_{i=1}^n t_i - \langle y^k, x\rangle : g_i(x_i) \leq t_i, \forall i = 1, \ldots, n, \ x \in C \cap R_0 \right\}. \tag{14}$$

A subgradient $y^k \in \partial h(x^k)$ is computed by

- if $l_i^0 < u_i^0 < 0 : y_i^k = \alpha_i^1$;
- if $0 < l_i^0 < u_i^0 : y_i^k = -\alpha_i^2$;
- if $l_i^0 < u_i^0 = 0 : y_i^k = \alpha_i^1$ if $x_i^k < -\epsilon i, c_i^1$ if $x_i^k > -\epsilon_i, \in [\alpha_i^1, c_i^1]$ if $x_i^k = -\epsilon_i$;
- if $0 = l_i^0 < u_i^0 : y_i^k = -c_i^2$ if $x_i^k < \epsilon_i, -\alpha_i^2$ if $x_i^k > \epsilon_i, \in [-c_i^2, -\alpha_i^2]$ if $x_i^k = \epsilon_i$;
- if $l_i^0 < 0 < u_i^0$:

$$
y_i^k = \begin{cases}
-c_i^1, & \text{if } x_i^k < -\epsilon_i, \\
\in [-c_i^1, -\alpha_i^1], & \text{if } x_i^k = -\epsilon_i, \\
-\alpha_i^1, & \text{if } -\epsilon_i < x_i^k < 0, \\
\in [-\alpha_i^1, \alpha_i^2], & \text{if } x_i^k = 0, \\
\alpha_i^2, & \text{if } 0 < x_i^k < \epsilon_i, \\
\in [\alpha_i^2, c_i^2], & \text{if } x_i^k = \epsilon_i, \\
c_i^2, & \text{if } x_i^k > \epsilon_i.
\end{cases}
$$

Hence, DCA applied on (P_{dc}) can be described as follows.

Algorithm 1 (DCA applied on (P_{dc})):

- **Initialization:** Let $x^0 \in \mathbb{R}^n$ and ε be a sufficiently small positive number; iteration $k \longleftarrow 0$.
- **Repeat:**
  ◇ Compute $y^k \in \partial h(x^k)$ as indicated above.
  ◇ Solving the convex program (14) to obtain x^{k+1}.
  ◇ $k \longleftarrow k + 1$
- **Until:** $|f(x^{k+1}) - f(x^k)| \le \varepsilon$.

4   A Hybrid Branch and Bound-DCA Algorithm

In this section we propose a combined B&B-DCA algorithm to globally solve the problem (P).

   As DCA is a descent and efficient method for nonconvex programming, DCA will be used to improving upper bounds for ω in B&B scheme while lower bounds will be provided by solving relaxation problems constructed over the rectangle $R = \prod_{i=1}^{n} [l_i, u_i]$, subsets of R_0, at each iteration.

4.1   Lower Bounding

A lower bound for ϕ on $C \cap R := \prod_{i=1}^{n} [l_i, u_i] \subset R_0$ can be determined by the following way. Let $B_i = [l_i, u_i]$, $i = 1, \dots, n$. A convex underestimator of the objective function ϕ over the domain $C \cap R$ can be chosen as follows (since ϕ is separable):

$$
\widetilde{\phi}_R(x) = \sum_{i=1}^{n} \widetilde{\phi}_{B_i}(x_i) \tag{15}
$$

where $\widetilde{\phi}_{B_i}(x_i)$ is defined by the following way:

- if $l_i < u_i < 0$, let $\widetilde{\phi}_{B_i}(x_i) = \beta_i - \alpha_i^1 x_i$;
- if $0 < l_i < u_i$, let $\widetilde{\phi}_{B_i}(x_i) = \beta_i + \alpha_i^2 x_i$;
- if $l_i < u_i = 0$, $\widetilde{\phi}_{B_i}(x_i) = \left(\frac{\beta_i}{l_i} - \alpha_i^1\right) x_i$;
- if $0 = l_i < u_i$, let $\widetilde{\phi}_{B_i}(x_i) = \left(\frac{\beta_i}{u_i} + \alpha_i^2\right) x_i$;
- if $l_i < 0 < u_i$,

$$\widetilde{\phi}_{B_i}(x_i) = \begin{cases} \left(\frac{\beta_i}{l_i} - \alpha_i^1\right) x_i, & x_i \leq 0 \\ \left(\frac{\beta_i}{u_i} + \alpha_i^2\right) x_i, & x_i \geq 0. \end{cases}$$

Hence, solving the convex program

$$\eta(R) = \min\{\widetilde{\phi}_R(x) \ : \ x \in C \cap R\} \tag{16}$$

provides a point $x^R \in C$ satisfying

$$\eta(R) = \widetilde{\phi}_R(x^R) \leq \min\{\phi(x) \ : \ x \in C \cap R\},$$

i.e. $\eta(R)$ is a lower bound for ϕ over $C \cap R$.

4.2   Upper Bounding

Since x^R is a feasible solution to (P), $\phi(x^R)$ is an upper bound for the global optimal value ω of (P). To use DCA for finding a better upper bound for ω, we will construct a DC approximation problem $\min\{f(x) \ : \ x \in C \cap R\}$ of (P) over $C \cap R$ by the same way mentioned in section 3.1 and launch DCA from x^R for solving the corresponding DC approximation problem. Note that we does not restart DCA at every iteration of B&B scheme but only when $\phi(x^R)$ is smaller than the current upper bound.

4.3   Subdivision Process

Let R_k be the rectangle to be subdivided at iteration k of the B&B algorithm and x^{R_k} be an optimal solution of the corresponding relaxation problem of (P) over $C \cap R_k$. We adopt the following rule of bisection of R_k: Choose an index i_k^* satisfying

$$i_k^* \in \arg\max_i\{\phi_i(x_i^{R_k}) - \widetilde{\phi}_i(x_i^{R_k})\}$$

and subdivide R_k into two subsets:

$$R_{k_1} = \{v \in R_k \ : \ v_{i_k^*} \leq x_{i_k^*}^{R_k}\}, \quad R_{k_2} = \{v \in R_k \ : \ v_{i_k^*} \geq x_{i_k^*}^{R_k}\}.$$

We are now in a position to describe our hybrid algorithm for solving (P).

4.4 Hybrid Algorithm

Algorithm 2 (BB-DCA):

- **Initialization:** Compute the first bounds l_i^0, u_i^0 for variables x_i and the first rectangle $R_0 = \prod_{i=1}^{n} [l_i^0, u_i^0]$. Construct the convex underestimator function ϕ_{R_0} of ϕ over R_0 then solve the convex program

$$\min\{\widetilde{\phi}_{R_0}(x) \ : \ x \in C \cap R_0\} \qquad (R_0 cp)$$

  to obtain an optimal solution x^{R_0} and the optimal value $\eta(R_0)$.
  Launch DCA from x^{R_0} for solving the corresponding DC approximation problem (P_{dc}). Let $\overline{x}^{R_0}$ be a solution obtained by DCA.
  Set $\mathcal{R}_0 := \{R_0\}$, $\eta_0 := \eta(R_0)$, $\omega_0 := \phi(\overline{x}^{R_0})$.
  Set $x^* := \overline{x}^{R_0}$.
- **Iteration $k = 0, 1, 2, \dots$:**
  - $k.1$ Delete all $R \in \mathcal{R}_k$ with $\eta(R) \geq \omega_k$. Let $\mathcal{P}_k$ be the set of remaining rectangles. If $\mathcal{P}_k = \emptyset$ then STOP: x^* is a global optimal solution.
  - $k.2$ Otherwise, select $R_k \in \mathcal{P}_k$ such that

$$\eta_k := \eta(R_k) = \min\{\eta(R) \ : \ R \in \mathcal{P}_k\}$$

    and subdivide R_k into R_{k_1}, R_{k_2} according to the subdivision process.
  - $k.3$ For each R_{k_j}, $j = 1, 2$, construct relaxation function $\widetilde{\phi}_{R_{k_j}}$, and solve

$$\min\{\widetilde{\phi}_{R_{k_j}}(x) \ : \ x \in C \cap R_{k_j}\} \qquad (R_{k_j} cp)$$

    to obtain $x^{R_{k_j}}$ and $\eta(R_{k_j})$.
    If $\phi(x^{R_{k_j}}) < \omega_k$, i.e., the current upper bound is improved on rectangle R_{k_j} then construct a DC approximation problem for (P) over $C \cap R_{k_j}$ by replacing ϕ with DC function $f_{R_{k_j}}$ and launch DCA from $x^{R_{k_j}}$ for solving

$$\min\{f_{R_{k_j}}(x) = g_{R_{k_j}}(x) - h_{R_{k_j}}(x) \ : \ x \in C \cap R_{k_j}\}. \qquad (R_{k_j} DC)$$

    Let $\overline{x}^{R_{k_j}}$ be a solution obtained by DCA. Let

$$\gamma_k = \min\{\phi(x^{R_{k_j}}), \phi(\overline{x}^{R_{k_j}})\}.$$

  - $k.4$ Update ω_{k+1} and the best feasible solution known so far x^*.
  - $k.5$ Set $\mathcal{R}_{k+1} = (\mathcal{P}_k \setminus R_k) \bigcup \{R_{k_1}, R_{k_2}\}$ and go to the next iteration.

5   Conclusion

We have rigorously studied the model and solution methods for solving a hard portfolio selection problem where the total transaction cost function is nonconvex. Attempting to use DC programming and DCA, an efficient approach in nonconvex programming, we construct an appropriate DC approximation of the objective function, and then investigate a DCA scheme for solving the resulting DC program. The DCA based algorithm is quite simple: each iteration we have to minimize a linear function under linear and convex quadratic constraints for which the powerful CPLEX solver can be used. To get a global minimizer of the original problem we combine DCA with a Branch and Bound scheme. We propose an interesting way to compute lower bounds that leads to the same type of convex subproblems in DCA, say linear program with additional convex quadratic constraints. In the next step we will implement the algorithms and study the computational aspects of the proposed approaches.

References

1. Kellerer, H., Mansini, R., Speranza, M.G.: Selecting Portfolios with Fixed Costs and Minimum Transaction Lots. Annals of Operations Research 99, 287–304 (2000)
2. Konno, H., Wijayanayake, A.: Mean-absolute deviation portfolio optimization model under transaction costs. Journal of the Operation Research Society of Japan 42(4), 422–435 (1999)
3. Konno, H., Wijayanayake, A.: Portfolio optimization problems under concave transaction costs and minimal transaction unit constraints. Mathematical Programming 89(B), 233–250 (2001)
4. Konno, H., Yamamoto, R.: Global Optimization Versus Integer Programming in Portfolio Optimization under Nonconvex Transaction Costs. Journal of Global Optimization 32, 207–219 (2005)
5. Le Thi, H.A.: Contribution à l'optimisation non convexe et l'optimisation globale: théorie, algorithmes et applications. Habilitation à Diriger de Recherches. Université de Rouen, France (1997)
6. Le Thi, H.A., Pham, D.T.: The DC (Difference of convex functions) Programming and DCA Revisited with DC Models of Real World Nonconvex Optimization Problems. Annals of Operations Research 133, 23–46 (2005)
7. Le Thi, H.A.: DC Programming and DCA, http://lita.sciences.univ-metz.fr/~lethi/DCA.html
8. Lobo, M.S., Fazel, M., Boyd, S.: Portfolio optimization with linear and fixed transaction costs. Annals of Operations Research 157, 341–365 (2007)
9. Markowitz, H.: Portfolio selection. The Journal of Finance 7(1), 77–91 (1952)
10. Pham, D.T., Le Thi, H.A.: Convex analysis approach to d.c. programming: Theory, Algorithms and Applications. Acta Mathematica Vietnamica (dedicated to Professor Hoang Tuy on the occasion of his 70th birthday) 22(1), 289–355 (1997)
11. Pham, D.T., Le Thi, H.A.: A d.c. optimazation algorithm for solving the trust region subproblem. SIAM Journal of Optimization 8(2), 476–505 (1998)

Efficient Algorithms for Feature Selection in Multi-class Support Vector Machine*

Hoai An Le Thi[1,2] and Manh Cuong Nguyen[1]

[1] Laboratory of Theoretical and Applied Computer Science EA 3097
University of Lorraine, Ile de Saulcy, 57045 Metz, France
[2] Lorraine Research Laboratory in Computer Science and its Applications
CNRS UMR 7503, University of Lorraine, 54506 Nancy, France
hoai-an.le-thi@univ-lorraine.fr,
manh-cuong.nguyen9@etu.univ-lorraine.fr

Abstract. This paper addresses the problem of feature selection for Multi-class Support Vector Machines (MSVM). Basing on the l_0 and the l_2-l_0 regularization we consider two models for this problem. The l_0-norm is approximated by a suitable way such that the resulting optimization problems can be expressed as DC (Difference of Convex functions) programs for which DC programming and DC Algorithms (DCA) are investigated. The preliminary numerical experiments on real-world datasets show the efficiency and the superiority of our methods versus one of the best standard algorithms on booth feature selection and classification.

Keywords: Feature selection, MSVM, DC programming, DCA.

1 Introduction

One of challenges of Machine Learning is the handling of the input datasets with very large number of features. Many techniques are proposed to address this challenge. Its goals are to remove the irrelevant and redundant features, reduce store space and execution time, and avoid the course of dimensionality to improve prediction performance [11].

We are interested in this paper the feature selection task for Multi-class Support Vector Machine (MSVM). The objective is to simultaneously select a subset of features (representative features) and construct a good classifier. Whereas several feature-selection methods for SVM have been proposed in the literature (see e.g. [1], [4], [7], [10], [11], [16], [21], [24]), there exist a few works on feature selection for MSVM. We first consider the model of MSVM proposed by Weston and Watkins [22], known to be appropriate to capture correlations between the different classes, which can be described as follows.

Let $\mathcal{X}$ be a set of vectors in $\mathbb{R}^d$ and $\mathcal{Y} = \{1, ..., Q\}$ be a set of class labels. Given a training dataset $X = \{(x_1, y_1), (x_2, y_2), .., (x_n, y_n)\} \in \mathbb{R}^{n*(d+1)}$, where

---

* This research has been supported by "Fonds Européens de Développement Régional" (FEDER) Lorraine via the project InnoMaD (Innovations techniques d'optimisation pour le traitement Massif de Données).

N.T. Nguyen, T. Van Do, and H.A. Le Thi (Eds.): *ICCSAMA 2013*, SCI 479, pp. 41–52.
DOI: 10.1007/978-3-319-00293-4_4   © Springer International Publishing Switzerland 2013

$x_i \in \mathcal{X}$, $y_i \in \mathcal{Y}$, $i = \{1, ..., n\}$. Denote by $\mathcal{F}$ the class of functions $f : \mathcal{X} \mapsto \mathcal{Y}$ with $f = \{f_1, f_2, ..., f_Q\}$, $f_i : \mathcal{X} \mapsto \mathbb{R}$, $i = 1, .., Q$.

A (multi-class) classifier is a function $f : \mathcal{X} \mapsto \mathcal{Y}$ that maps an element x to a category $y \in \mathcal{Y}$. Let $w = (w_1, w_2, ..., w_Q)$ be the vector in $\mathbb{R}^{Q \cdot d}$ consisting of Q vector $w_i \in \mathbb{R}^d$, $i \in \{1, ..., Q\}$ and let b be a vector in $\mathbb{R}^Q$. We consider the function f of the form

$$f(x) = arg \max_{1 \leq i \leq Q} f_i(x), \tag{1}$$

where $f_i(x) = \langle w_i, x \rangle + b_i$, $i \in [1..Q]$ and $\langle ., . \rangle$ is the scalar product in $\mathbb{R}^d$ space.

The goal is to determine the most appropriate hyperplanes $f_i(x)$, $i \in \{1, ..., Q\}$, that separate the training dataset in the best way. So, the MSVM model [22] is defined by:

$$\min_{w, b, \xi} C \sum_{i=1}^{n} \sum_{k \neq y_i} \xi_{ik} + \sum_{k=1}^{Q} \|w_k\|_2^2 \tag{2}$$

subject to

$$\Omega : \begin{cases} \langle w_{y_i} - w_k, x_i \rangle + b_{y_i} - b_k \geq 1 - \xi_{ik}, & (1 \leq i \leq n), (1 \leq k \neq y_i \leq Q) \\ \xi_{ik} \geq 0, & (1 \leq i \leq n), (1 \leq k \neq y_i \leq Q), \end{cases}$$

where $\xi_{ik} \in \mathbb{R}^{n*Q}$ are slack variables. In the objective function, $C \sum_{i=1}^{n} \sum_{k \neq y_i} \xi_{ik}$ is the hinge loss term which presents the training classification errors. The remaining term is known as a regularization. C is a parameter that presents the trade-off between the hinge loss and the regularizer term.

For the feature selection purpose, we study two models obtained from (2) by replacing the second term in the objective function by the l_0 (the zero norm) and the l_2-l_0 regularization. Naturally, using the zero norm (the zero norm of a vector is defined as the number of its nonzero components) is the best way to feature selection. However, the resulting optimizations problems are very hard. Minimizing a function involving the zero norm is a challenge of the community of researchers in optimization and in machine learning.

Our solution method is based on Difference of Convex functions (DC) programming and DC Algorithms (DCA) that were introduced by Pham Dinh Tao in their preliminary form in 1985 and have been extensively developed since 1994 by Le Thi Hoai An and Pham Dinh Tao and become now classic and more and more popular (see, e.g. [12, 13, 17, 18], and references therein), in particular in machine learning for which they provide quite often a global solution and proved to be more robust and efficient than standard methods. Basing on the concave approximation [1] of the zero norm we reformulate the resulting optimization problems as DC programs and then design DCA schemes for solving them.

The remainder of the paper is organized as follows. Section 2 is devoted to the description of the two models of feature selection for MSVM with l_0 and l_2-l_0 regularization, and the resulting optimization problems via the concave approximation of the zero norm. In Section 3 we show how to investigate DC programming and DCA for solving the approximate optimization problems. Finally, the computational experiments are reported in Section 4.

2   The l_0 and l_2-l_0 Regularization Models of Feature Selection for MSVM

First, let us describe some notations and background materials that will be used in the sequel. The base of the natural logarithm will be denoted by ε and for a vector $x \in \mathbb{R}^n$, ε^{-x} will denote a vector in $\mathbb{R}^n$ with components ε^{-x_i}, $i = 1,\dots n$. For $w_i \in \mathbb{R}^d$, let w_i^* be the step vector defined by

$$w_{ij}^* = 1 \ \text{ if } w_{ij} \neq 0; \ 0 \text{ otherwise}, j = 1, 2, ..., d.$$

Hence, the zero-norm $\|w_i\|_0$ can be written as $e^T w_i^*$.

For the purpose of feature selection, the l_2-norm in (2) is respectively replaced by l_0-norm and l_2-l_0-norm. So, we have the l_0-MSVM problem

$$\min_{(w,b,\xi) \in \Omega} C \sum_{i=1}^{n} \sum_{k \neq y_i} \xi_{ik} + \sum_{k=1}^{Q} \|w_k\|_0, \tag{3}$$

and the l_2-l_0-MSVM problem

$$\min_{(w,b,\xi) \in \Omega} C \sum_{i=1}^{n} \sum_{k \neq y_i} \xi_{ik} + \beta \sum_{k=1}^{Q} \|w_k\|_2^2 + \sum_{k=1}^{Q} \|w_k\|_0, \tag{4}$$

which are nonsmooth and nonconvex.

For solving these problems by DCA we first approximate the zero norm by a concave function.

For $x \in \mathbb{R}^n$ and a given $\alpha > 0$, let η be the function defined by

$$\eta(x) = \begin{cases} 1 - \varepsilon^{-\alpha x} & \text{if } x \geq 0, \\ 1 - \varepsilon^{\alpha x} & \text{if } x \leq 0. \end{cases}$$

The step vector w^*, for all $w \in \mathbb{R}^n$ can be approximated by

$$w_i^* \simeq \eta(w_i),$$

and then, an approximation of the zero-norm $\|w\|_0$ would be

$$\|w\|_0 \simeq \sum_{i=1}^{n} \eta(w_i).$$

By the way, the l_0-norm in (3) and (4) is approximated by

$$\|w_i\|_0 \simeq \sum_{j=1}^{d} \eta(w_{ij}).$$

Hence, the resulting optimization problem of (3) takes the form:

$$\min_{(w,b,\xi)\in\Omega} \left\{ C\sum_{i=1}^{n}\sum_{k\neq y_i}\xi_{ik} + \sum_{i=1}^{Q}\sum_{j=1}^{d}\eta(w_{ij}) \right\}, \qquad (5)$$

and that of (4) can be written as:

$$\min_{(w,b,\xi)\in\Omega} \left\{ C\sum_{i=1}^{n}\sum_{k\neq y_i}\xi_{ik} + \beta\sum_{k=1}^{Q}\|w_k\|_2^2 + \sum_{i=1}^{Q}\sum_{j=1}^{d}\eta(w_{ij}) \right\}. \qquad (6)$$

In the sequel we will investigate DC programming and DCA for solving the two problems (5) and (6).

3   DC Programming and DCA for Solving Problems (5) and (6)

3.1   A Brief Presentation of DC Programming and DCA

DC programming and DCA constitute the backbone of smooth/nonsmooth nonconvex programming and global optimization [17]. A general DC program takes the form:

$$\inf\{F(x) := G(x) - H(x) : x \in \mathbb{R}^p\}, \qquad (P_{dc})$$

where G and H are lower semicontinuous proper convex functions on $\mathbb{R}^p$. Such a function F is called DC function, and $G-H$, DC decomposition of F while G and H are DC components of F. The convex constraint $x \in C$ can be incorporated in the objective function of (P_{dc}) by using the indicator function on C denoted χ_C which is defined by $\chi_C(x) = 0$ if $x \in C$; $+\infty$ otherwise:

$$\inf\{f(x) := G(x) - H(x) : x \in C\} = \inf\{\chi_C(x) + G(x) - H(x) : x \in \mathbb{R}^p\}.$$

A convex function θ is called *convex polyhedral* if it is the maximum of a finite family of affine functions, i.e.

$$\theta(x) = \max\{\langle a_i, x\rangle + b : i = 1, ...m\}, a_i \in \mathbb{R}^p.$$

Polyhedral DC optimization occurs when either G or H is polyhedral convex. This class of DC optimization problems, which is frequently encountered in practice, enjoys interesting properties (from both theoretical and practical viewpoints) concerning local optimality and the convergence of DCA ([12]).

A point x^* is said to be *a local minimizer* of $G - H$ if $G(x^*) - H(x^*)$ is finite and there exists a neighbourhood $\mathcal{U}$ of x^* such that

$$G(x^*) - H(x^*) \leq G(x) - H(x), \quad \forall x \in \mathcal{U}. \qquad (7)$$

The necessary local optimality condition for (primal) DC program (P_{dc}) is given by

$$\emptyset \neq \partial H(x^*) \subset \partial G(x^*). \qquad (8)$$

The condition (8) is also sufficient (for local optimality) in many important classes of DC programs, for example, when (P_{dc}) is a polyhedral DC program with H being polyhedral convex function, or when f is locally convex at x^* (see [12]).

A point x^* is said to be *a critical point* of $G - H$ if

$$\partial H(x^*) \cap \partial g(x^*) \neq \emptyset. \tag{9}$$

The relation (9) is in fact the generalized KKT condition for (P_{dc}) and x^* is also called a generalized KKT point.

DCA is based on local optimality conditions and duality in DC programming. The main idea of DCA is simple: each iteration of DCA approximates the concave part $-H$ by its affine majorization (that corresponds to taking $y^k \in \partial H(x^k)$) and minimizes the resulting convex function.

The generic DCA scheme can be described as follows:

Initialization: Let $x^0 \in \mathbb{R}^p$ be a best guess, $l \leftarrow 0$.
Repeat
    - Calculate $y^l \in \partial H(x^l)$.
    - Calculate $x^{l+1} \in arg\min\{G(x) - H(x^l) - \langle x - x^l, y^l \rangle\} : x \in \mathbb{R}^p$.
    - $l \leftarrow l + 1$.
Until convergence of $\{x^l\}$.

Convergences properties of DCA and its theoretical basic can be found in [17, 12]. It is worth mentioning that (for simplify we omit here the dual part of DCA)

- DCA is a descent method (*without line search*): the sequences $\{G(x^l) - H(x^l)\}$ is decreasing.
- If $G(x^{l+1}) - H(x^{l+1}) = G(x^l) - H(x^l)$, then x^l is a critical point of $G - H$. In such a case, DCA terminates at l-th iteration.
- If the optimal value α of problem (P_{dc}) is finite and the infinite sequences $\{x^l\}$ is bounded then every limit point x^* of the sequences $\{x^l\}$ is a critical point of $G - H$.
- DCA has a *linear convergence* for general DC programs, and has a finite convergence for polyhedral DC programs.

A deeper insight into DCA has been described in [12]. For instant it is crucial to note the main features of DCA: DCA is constructed from DC components and their conjugates but not the DC function f itself which has infinitely many DC decompositions, and there are as many DCA as there are DC decompositions. Such decompositions play a critical role in determining the speed of convergence, stability, robustness, and globality of sought solutions. It is important to study various equivalent DC forms of a DC problem. This flexibility of DC programming and DCA is of particular interest from both a theoretical and an algorithmic point of view.

For a complete study of DC programming and DCA the reader is referred to [12, 17, 18] and the references therein.

It should be noted that the convex concave procedure (CCCP) for constructing discrete time dynamical systems mentioned in [20] is a special case of DCA
applied to smooth optimization. Likewise, the SLA (Successive Linear Approximation) algorithm developed in [1] is a version of DCA for concave minimization
program.

In the last decade, a variety of works in Machine Learning based on DCA have
been developed. The efficiency and the scalability of DCA have been proved in
a lot of works (see e.g. [2, 8–11, 14–16, 19] and the list of reference in [13]).

3.2   DCA for Solving Problem (5)

First, we find a DC decomposition for the objective function of (5). We express
η as a DC function

$$\eta(x) = g_1(x) - h_1(x),$$

where

$$g_1(x) = \begin{cases} \alpha x & \text{if } x \geq 0 \\ -\alpha x & \text{if } x \leq 0 \end{cases} ; \quad h_1(x) = \begin{cases} \alpha x - 1 + \varepsilon^{-\alpha x} & \text{if } x \geq 0 \\ -\alpha x - 1 + \varepsilon^{\alpha x} & \text{if } x \leq 0. \end{cases}$$

The objective function of (5) can be now written as:

$$C \sum_{i=1}^{n} \sum_{k \neq y_i} \xi_{ik} + \sum_{i=1}^{Q} \sum_{j=1}^{d} g_1(w_{ij}) - \sum_{i=1}^{Q} \sum_{j=1}^{d} h_1(w_{ij}).$$

Set

$$G_1(w, b, \xi) = C \sum_{i=1}^{n} \sum_{k \neq y_i} \xi_{ik} + \sum_{i=1}^{Q} \sum_{j=1}^{d} g_1(w_{ij}), \qquad (10)$$

and

$$H_1(w, b, \xi) = \sum_{i=1}^{Q} \sum_{j=1}^{d} h_1(w_{ij}). \qquad (11)$$

Then Problem (5) can be expressed as:

$$\min \{ G_1(w, b, \xi) - H_1(w, b, \xi) : (w, b, \xi) \in \Omega \}. \qquad (12)$$

Obviously, g_1 and h_1 are convex functions, and so are G_1 and H_1. Therefore
(12) is a DC program. Note that since $g_1(x) = \max(\alpha x, -\alpha x)$, the function g_1
is polyhedral convex and so is G_1. Therefore (12) is a polyhedral DC program.

We are now developing a DCA scheme to solve the DC program (12). According to the generic DCA scheme, applying DCA on (12) amounts to computing,
at each iteration k, a subgradient $Y^k = (\overline{w}^k, \overline{b}^k, \overline{\xi}^k)$ of H_1 at $X^k = (w^k, b^k, \xi^k)$
and then, solve the convex program:

$$\min \{ G_1(w, b, \xi) - \langle Y^k, X \rangle : X = (w, b, \xi) \in \Omega \}. \qquad (13)$$

Clearly, H_1 is differentiable and $Y^k = \nabla H_1(X^k)$ is computed as

$$Y^k = (\overline{w}^k, 0, 0), \quad \overline{w}_{ij}^k = \begin{cases} \alpha(1 - \varepsilon^{-\alpha w_{ij}^k}) & \text{if } w_{ij} \geq 0 \\ -\alpha(1 - \varepsilon^{\alpha w_{ij}^k}) & \text{if } w_{ij} < 0, \\ i = 1, ..., Q, \quad j = 1, ..., d, \end{cases} \tag{14}$$

and the convex program (13) is defined as

$$\min\left\{ G_1(X) - \sum_{i=1}^{Q} \sum_{j=1}^{d} \overline{w}_{ij}^k w_{ij} : (w, b, \xi) \in \Omega \right\} \tag{15}$$

$$\Leftrightarrow \min\left\{ C\sum_{i=1}^{n} \sum_{k \neq y_i} \xi_{ik} + \sum_{i=1}^{Q} \sum_{j=1}^{d} \max(\alpha w_{ij}, -\alpha w_{ij}) - \sum_{i=1}^{Q} \sum_{j=1}^{d} \overline{w}_{ij}^k w_{ij}, (w, b, \xi) \in \Omega \right\}. \tag{16}$$

The last problem is equivalent to the next linear program

$$\min_{w,b,\xi,t}\left\{ \begin{array}{l} C\sum_{i=1}^{n} \sum_{k \neq y_i} \xi_{ik} + \sum_{i=1}^{Q} \sum_{j=1}^{d} t_{ij} - \sum_{i=1}^{Q} \sum_{j=1}^{d} \overline{w}_{ij}^k w_{ij} \\ s.t. \ (w, b, \xi) \in \Omega, t_{ij} \geq \alpha w_{ij}, t_{ij} \geq -\alpha w_{ij}, i = 1, ..., Q, j = 1, ..., d \end{array} \right. \tag{17}$$

Finally, DCA applied on the problem (12) is described as follows:

Algorithm 1. l_0-DCA

Initialization:
Let $\epsilon > 0$ be given and $X^0 = (w^0, \xi^0, b^0)$ be an initial point. Select α and set $k = 0$;
  Repeat:
    1. Calculate Y^k via (14).
    2. Calculate X^{k+1}, an optimal solution of the linear program (17).
    3. $k = k + 1$.
  Until $\| X^{k+1} - X^k \| \leq \epsilon \| X^k \|$.

Remark. Algorithm 1 (l_0-**DCA**) enjoys interesting convergence properties: since G_1 is polyhedral convex, (12) is a polyhedral DC program, and consequently l_0-**DCA** has a finite convergence. Moreover, thank to the differentiability of H_1, the solution furnished by DCA satisfies the necessary local optimality condition (8) which is almost alway sufficient condition in our case, say DC polyhedral program with G_1 being polyhedral convex function (see [12, 17]).

3.3   DCA for Solving Problem (6)

We observe that the objective function of (6) is the one of (5) with addition the l_2-norm term. Hence, in a similar way to (5), we get the following DC formulation of (6)

$$\min\left\{ G_2(w, b, \xi) - H_1(w, b, \xi) : (w, b, \xi) \in \Omega \right\}, \tag{18}$$

where G_2 is defined by

$$G_2(w, b, \xi) = G_1(w, b, \xi) + \beta \sum_{i=1}^{Q} \|w_i\|_2^2. \qquad (19)$$

Therefore DCA applied on (18) is similar to l_0-**DCA**, but at each iteration k, instead of solving the linear problem (17), we solve the following convex quadratic program

$$\Leftrightarrow \min_{w,b,\xi,t} \left\{ \begin{array}{l} \beta \sum_{i=1}^{Q} \|w_i\|_2^2 + C \sum_{i=1}^{n} \sum_{k \neq y_i} \xi_{ik} + \sum_{i=1}^{Q} \sum_{j=1}^{d} t_{ij} - \sum_{i=1}^{Q} \sum_{j=1}^{d} \overline{w}_{ij}^k w_{ij} \\ s.t. \ (w, b, \xi) \in \Omega, t_{ij} \geq \alpha w_{ij}, t_{ij} \geq -\alpha w_{ij}, i = 1, ..., Q, j = 1, ..., d \end{array} \right.$$
$$(20)$$

Algorithm 2. $l_2 - l_0$-**DCA**

Initialization:
Let $\epsilon > 0$ be given, $X^0 = (w^0, \xi^0, b^0)$ be an initial point. Set $k = 0$;
    Repeat:
        1. Calculate Y^k via (14).
        2. Calculate X^{k+1}, an optimal solution of the convex quadratic program (20)
        3. $k = k + 1$.
    Until $\| X^{k+1} - X^k \| \leq \epsilon \| X^k \|$.

---

4   Numerical Experiments

The environment used for the experiments is Intel Core^{TM} I7 (2×2.2 Ghz) processor, 4 GB RAM. We have tested the algorithms on the seven popular datasets often used for feature selection. The Lung Cancer (LUN), Libras Movement (MOV), CNAE-9 (CNA), Hill-Valley (HIL), Spambase (SPA) and Internet Advertisement (ADV) are the datasets taken from UCI Machine Learning Repository. The ADN dataset (ADN) can be found at ftp://genbank.bio.net. Each of dataset is divided into two parts: the training set and the test set. These datasets are described in details in the table 1.

For the l_0-**DCA** and l_2-l_0-**DCA**, we set $\alpha = 0.9$. The parameter β associated with l_2 term of (6) is set to 0.1. For all methods, the most appropriate values of the parameter C are chosen by a five-fold cross-validation. The CPLEX 11.2 solver is used to solve linear or convex quadratic problems.

After solving the Problem (5) or (6) by DCA, to select relevant features, we first compute the feature ranking score $c_j, j = 1, ..., d$ for each feature ([3]). More precisely, we normalize the vectors w_i

$$w_i = \frac{w_i}{\|w_i\|}, i = 1, ..., Q$$

and set the ranking score c_j as $c_j = \frac{1}{Q} \sum_{i=1}^{Q} w_{ij}^2$. Then we remove the features j for which c_j is smaller than a given threshold ρ. After removing features, for

Table 1. The description of the datasets

Dataset	#train	#test	#feature	#class
LUN	16	16	56	3
MOV	255	105	90	15
CNA	500	580	856	9
HIL	606	606	100	2
SPA	2301	2300	57	2
ADV	1000	2279	1558	2
ADN	1064	2122	60	3

Table 2. The number and the percentage of selected features and the corresponding accuracy of classifiers

Datasets	Selected features			Accuracy of classifiers (%)		
	l_2-l_0-**DCA**	l_0-**DCA**	ASN	l_2-l_0-**DCA**	l_0-**DCA**	ASN
LUN	**7** (12.50%)	8 (18.49%)	19 (33.93%)	**56.25**	**56.25**	37.50
MOV	33 (36.67%)	31 (34.44%)	**28** (31.11%)	75.32	**76.19**	64.76
CNA	54 (6.31%)	48 (5.61%)	**40** (4.67%)	**83.79**	80.89	60.86
HIL	6 (6.00%)	**4** (4.00%)	12 (12.00%)	99.51	**100.0**	**100.0**
SPA	20 (35.09%)	**15** (26.32%)	22 (38.36%)	87.61	**89.57**	89.30
ADV	**15** (0.96%)	**15** (0.96%)	51 (3.27%)	**93.55**	90.52	93.20
ADN	**4** (6.67%)	16 (26.67%)	27 (45.00%)	80.87	**85.01**	79.36
Average	**14.88(%)**	16.04(%)	24.08(%)	82.41	**82.63**	75.00

computing the accuracy of classification, we apply again l_2-MSVM (2) on the new training datasets and calculate the classification's accuracy on the new test sets.

We compare our methods with one of the best algorithms to feature selection in MSVM, called the Adaptive Sub-Norm (ASN) method (see [24] for more details). We are interested in the efficiency (the sparsity and the classification error) and the rapidity of the algorithms. Comparative computational results are reported in the tables 2, 3 and the figures 1, 2. CPU time is computed as the sum of running time of all learning steps, say feature selection and classification by solving the optimization problem (5) or (6) as well as the ranking procedure, and classification with selected feature by standard l_2-MSVM (2).

We observe from computational results that:

- The two DCA based algorithms reduce considerably the number of selected features (from 65% and 99%) while the classifiers are quite good (more than 80% for 5/7 datasets).

- Our DCA based algorithms perform better than ASN on both feature selection and classification. On average, l_2-l_0-**DCA** and l_0-**DCA** selected, respectively, 14.88% and 16.68% of features while the corresponding result of ASN is 24.08%. The accuracy classifiers given by our methods are greater than 82%, and the one of ASN is 75%.

Table 3. CPU time (in seconds) of each method

Datasets	l_2-l_0-**DCA**	l_0-**DCA**	ASN
LUN	**0.20**	0.23	0.26
MOV	249.96	220.18	**116.19**
CNA	32.34	**31.67**	59.78
HIL	7.89	**6.32**	6.93
SPA	3.85	**3.84**	18.10
ADV	5.33	**5.01**	36.70
ADN	29.68	**19.68**	52.97
AVERAGE	47.04	**40.99**	41.56

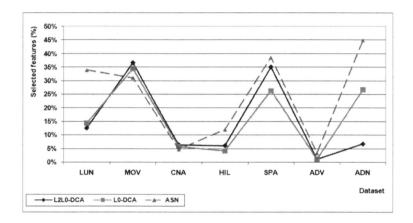

Fig. 1. The feature selection performance of each algorithm

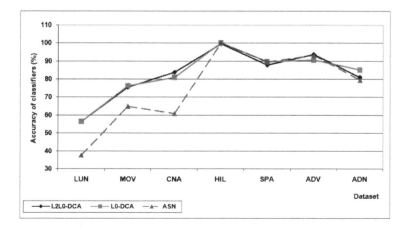

Fig. 2. The classification accuracy of each algorithm

- The performance of two versions of DCA on the feature selection and classification are comparable: on average, 14.88% of selected feature (resp. accuracy classification 82.41%) in l_2-l_0-**DCA** versus 16.04% (resp. 82.63%) in l_0-**DCA**.
- As for CPU time, l_0-**DCA** is the fastest algorithm. Comparing the two versions of DCA, this result is not surprise since the convex subproblems in $l_2 - l_0$-DCA are quadratic programs while that in l_0-**DCA** are linear programs.

5   Conclusion

We have developed an efficient approach based on DC programming and DCA for feature selection in multi-class support vector machine. Using an appropriate approximation function of zero-norm, we get two DC programs for the l_0 and $l_2 - l_0$ regularizer formulation of MSVM. It fortunately turns out that the corresponding DCA consists in solving, at each iteration, one linear program (in l_0 regularization) and/or one convex quadratic program (in $l_2 - l_0$ regularization). Moreover, l_0-**DCA** converges, after a finite number of iteration, almost always to a local solution. Numerical results on several real datasets showed the robustness, the effectiveness of the DCA based schemes. We are convinced that DCA is a promising approach for feature selection in MSVM. We plan to investigate other approximations of the zero norm and design corresponding DCAs for this important topic.

References

1. Bradley, P.S., Mangasarian, O.L.: Eature selection via concave minimization and support vector machines. In: Shavlik, J. (ed.) Proceedings of the Fifteenth International Conferences Machine Learning (ICML1998), pp. 82–90. Morgan Kaufmann, San Francisco (1998)
2. Collobert, R., Sinz, F., Weston, J., Bottou, L.: Large scale transductive SVMs. J. Machine Learn. 7, 1687–1712 (2006)
3. Duan, K.B., Rajapakse, J.C., Wang, H., Azuaje, F.: Multiple SVM-RFE for Genne Selection in Cancer Classification With Expression Data. IEEE Transactions on NANOBIOSCIENCE 4, 228–234 (2005)
4. Fan, J., Li, R.: Variable selection via nonconcave penalized likelihood and its Oracle Properties. Journal of the American Statistical Association 96, 1348–1360 (2001)
5. Guyon, I., Elisseeff, A.: An Introduction to Variable and Feature Selection. Journal of Machine Learning Research 3, 1157–1182 (2003)
6. Hui, Z.: The Adaptive Lasso and Its oracle Properties. Journal of the American Statistical Association 101(476), 1418–1429 (2006)
7. Huang, J., Ma, S., Zhang, C.H.: Adaptive Lasso for sparse high-dimentional regression models. Statistica Sinica 18, 1603–1618 (2008)
8. Le Thi, H.A., Belghiti, T., Pham, D.T.: A new efficient algorithm based on DC programming and DCA for Clustering. Journal of Global Optimization 37, 593–608 (2006)
9. Le Thi, H.A., Le Hoai, M., Pham, D.T.: Optimization based DC programming and DCA for Hierarchical Clustering. European Journal of Operational Research 183, 1067–1085 (2007)

10. Le Thi, H.A., Le Hoai, M., Nguyen, N.V., Pham, D.T.: A DC Programming approach for Feature Selection in Support Vector Machines learning. Journal of Advances in Data Analysis and Classification 2(3), 259–278 (2008)
11. Le Thi, H.A., Van Nguyen, V., Ouchani, S.: Gene selection for cancer classification using DCA. In: Tang, C., Ling, C.X., Zhou, X., Cercone, N.J., Li, X. (eds.) ADMA 2008. LNCS (LNAI), vol. 5139, pp. 62–72. Springer, Heidelberg (2008)
12. Le Thi, H.A., Pham, D.T.: The DC (Difference of convex functions) programming and DCA revisited with DC models of real world nonconvex optimization problems. Annals of Operations Research 133, 23–46 (2005)
13. Le Thi, H.A.: DC Programming and DCA, http://lita.sciences.univ-metz.fr/~lethi/DCA.html
14. Liu, Y., Shen, X., Doss, H.: Multicategory ψ-Learning and Support Vector Machine: Computational Tools. Journal of Computational and Graphical Statistics 14, 219–236 (2005)
15. Liu, Y., Shen, X.: Multicategory ψ-Learning. Journal of the American Statistical Association 101, 500–509 (2006)
16. Neumann, J., Schnörr, C., Steidl, G.: SVM-based Feature Selection by Direct Objective Minimisation. In: Proc. of 26th DAGM Symposium Pattern Recognition, pp. 212–219 (2004)
17. Pham, D.T., Le Thi, H.A.: Convex analysis approach to d.c. programming: Theory, Algorithm and Applications. Acta Mathematica Vietnamica 22, 289–355 (1997)
18. Pham, D.T., Le Thi, H.: Optimization algorithms for solving the trust region subproblem. SIAMJ. Optimization 2, 476–505 (1998)
19. Ronan, C., Fabian, S., Jason, W., Lé, B.: Trading Convexity for Scalability. In: Proceedings of the 23rd International Conference on Machine Learning, ICML 2006, Pittsburgh, Pennsylvania, pp. 201–208 (2006)
20. Yuille, A.L., Rangarajan, A.: The Convex Concave Procedure. Neural Computation 15(4), 915–936 (2003)
21. Wang, L., Shen, X.: On l_1-norm multi-class support vector machine: methodology and theory. Journal of the American Statistical Association 102, 583–594 (2007)
22. Weston, J., Watkins, C.: Support Vector Machines for Multi-Class Pattern Recognition. In: Proceedings - European Symposium on Artificial Neural Networks, ESANN 1999, pp. 219–224. D-Facto public (1999)
23. Weston, J., Elisseeff, A., Schölkopf, B.: Use of Zero-Norm with Linear Models and Kernel Methods. Journal of Machine Learning Research 3, 1439–1461 (2003)
24. Zhang, H.H., Liu, Y., Wu, Y., Zhu, J.: Variable selection for the multicategory SVM via adaptive sup-norm regularization. Journal of Statistics 2, 149–167 (2008)

Image Segmentation via Feature Weighted Fuzzy Clustering by a DCA Based Algorithm*

Hoai Minh Le, Bich Thuy Nguyen Thi, Minh Thuy Ta, and Hoai An Le Thi

Laboratory of Theoretical and Applied Computer Science (LITA)
UFR MIM, University of Lorraine,
Ile du Saulcy, 57045 Metz, France
{minh.le,hoai-an.le-thi}@univ-lorraine.fr,
{thi-bich-thuy.nguyen9,minh-thuy.ta5}@etu.univ-lorraine.fr

Abstract. Image segmentation plays an important role in a variety of applications such as robot vision, object recognition and medical imaging,... Fuzzy clustering is undoubtedly one of the most widely used methods for image segmentation. In many cases, it happens that some characteristics of image are more significant than the others. Therefore, the introduction of a weight for each feature which defines its relevance is a natural way in image segmentation.

In this paper, we develop an efficient method for image segmentation via feature weighted fuzzy clustering model. Firstly, we formulate the feature weighted fuzzy clustering problem as a DC (Difference of Convex functions) program. DCA (DC Algorithm), an innovative approach in nonconvex programming, is then developed to solve the resulting problem. Experimental results on synthetic and real color images have illustrated the effectiveness of the proposed algorithm and its superiority with respect to the standard feature weighted fuzzy clustering algorithm in both running-time and quality of solutions.

Keywords: Image Segmentation, Feature Weighted, Fuzzy Clustering, DC programming, DCA.

1   Introduction

Image segmentation is an important processing step in many image, video and computer vision applications. It is a critical step towards content analysis and image understanding. The aim of image segmentation is to partition an image into a set of non-overlapped, consistent regions with respect to some characteristics such as colors/gray values or textures. Image segmentation is an important research field and many segmentation methods have been proposed in the literature. For a more complete review on image segmentation methods, the reader is

---
* This research has been supported by "Fonds Européens de Développement Régional" (FEDER) Lorraine via the project InnoMaD (Innovations techniques d'optimisation pour le traitement Massif de Données).

N.T. Nguyen, T. Van Do, and H.A. Le Thi (Eds.): *ICCSAMA 2013*, SCI 479, pp. 53–63.
DOI: 10.1007/978-3-319-00293-4_5   © Springer International Publishing Switzerland 2013

referred to [4,14,16,17] and the references therein. We can classify image segmentation methods into four categories ([17]): methods based on pixels, on areas, on contours and on physical model for image formation. Pixel based methods, which consist in regrouping in different regions the pixels contained in an image, are the simplest approach for image segmentation. Pixel based methods are the easiest to understand and to implement. There are three main classes of techniques in pixel based methods:

- Histogram-based technique: firstly, a histogram is computed from all of the pixels in the image. Then, image pixels are classified as belonging to one of those classes thus formed by using the peaks and valleys in the histogram.
- Clustering techniques: pixels are grouped, using a hard clustering method, by means of their color values/textures.
- Fuzzy clustering techniques: instead of using hard clustering, fuzzy clustering is used for pixel classification task. A popular choice is the Fuzzy C-Means algorithm ([1]).

Fuzzy C-Means (FCM) clustering, introduced by Bezdek in 1981 ([1]), is a most widely used fuzzy clustering method. The FCM problem is formulated as a non convex optimization problem for which only heuristic algorithms are available before the work of Le Thi et al. 2007 ([10]). In this work, the authors reformulated FCM model as DC (Difference of Convex function) programs and then developed three DCA (DC Algorithm) schemes to solve the three resulting DC programs. DC programming and DCA, an innovative approach in nonconvex programming, were introduced by Pham Dinh Tao in a preliminary form in 1985. They have been extensively developed since 1994 by Le Thi Hoai An and Pham Dinh Tao and become now classic and increasingly popular (see e.g. [8,12] and the list of references in [7]). The numerical results on several real data sets show that the proposed DCA is an efficient approach for fuzzy clustering in large data sets of high dimension and it is superior to the FCM algorithm in both running-time and quality of solutions. Later, in ([11]), Le Thi et al. have successfully applied the DCA based algorithm for FCM in noisy image segmentation problems.

On another hand, usually in classification, the distance measure involves all attributes of the data set. It is applicable if most attributes are important to every cluster. However, the performance of clustering algorithms can be significantly degraded if many irrelevant attributes are used. In the literature, various approaches have been proposed to address this problem. The first strategy is feature selection that finds irrelevant features and removes them from the feature set before constructing a classifier. Feature weighting is an extension of the feature selection where the features are assigned continuous weights. Relevant features correspond to high weight values, whereas weight values close to zero represent irrelevant features. Clustering using weighted dissimilarity measures attracts more and more attention in recent years ([3,5]). In [5], the authors investigated the FCM problem using weighted features for segmentation image.

The problem FCM using features weighted can be stated as follows. Let $\mathcal{X} :=$ $\{x_1, x_2, ..., x_n\}$ be a data set of n entities with m attributes and the known number of clusters k ($2 \leq k \leq n$). Denote by Λ a $k \times m$ matrix defined as

$\Lambda = (\lambda_{l,i})$ where $\lambda_{l,i}$ defines the relevance of i-th feature to the cluster C_l. $W = (w_{j,l}) \in \mathbb{R}^{n \times k}$ with $j = 1, \ldots, n$ and $l = 1, \ldots, k$ called the *fuzzy partition* matrix in which each element $w_{j,l}$ indicates the membership degree of each point x_j in the cluster C_l (the probability that a point x_j belongs to the cluster C_l).

We are to regrouping the set $\mathcal{X}$ into k clusters in order to minimize the sum of squared distances from the entities to the centroid of their cluster. The dissimilarity measure is defined by m weighted attributes. Then a straightforward formulation of the clustering using weighted dissimilarity measures is (μ, β are exponents greater than 1):

$$
\begin{cases}
\min F(W, Z, \Lambda) := \sum_{l=1}^{k} \sum_{j=1}^{n} \sum_{i=1}^{m} w_{jl}^{\mu} \lambda_{li}^{\beta} (z_{li} - x_{ji})^2 \\
s.t : \sum_{l=1}^{k} w_{jl} = 1, j = 1..n, \\
\quad \sum_{i=1}^{m} \lambda_{li} = 1, l = 1..k, \\
\quad w_{jl} \in [0,1], j = 1..n, l = 1..k, \\
\quad \lambda_{li} \in [0,1], l = 1..k, i = 1..m.
\end{cases}
\tag{1}
$$

Problem (1) is difficult due to the nonconvexity of the objective function. Moreover, in real applications this is a very large scale problem (high dimension and large data set, i.e. m and n are very large), that is why global optimization approaches such as Branch & Bound, Cutting plane algorithms etc. cannot be used. In [5], the authors proposed a FCM type algorithm, called **SCAD** (Simultaneous Clustering and Attribute Discrimination), to solve the problem (1). At first, **SCAD** fixes Z, Λ and finds W to minimize $F(W, ., .)$. Then W, Λ are fixed for finding Z minimizing $F(., Z, .)$. Finally, Λ is obtained by minimizing $F(., ., \Lambda)$ with W and Z fixed. The process is repeated until no more improvement in the objective function can be made.

We investigate in this work, for solving the problem (1), an efficient nonconvex programming approach based on DC Programming and DCA. Our work is motivated by the fact that DCA has been successfully applied to many (smooth or nonsmooth) large-scale nonconvex programs in various domains of applied sciences, in particular in Machine Learning (see e.g. the list of references in [7]).

We will develop the solution method based on DC programming and DCA for solving the problem (1) in Section 2 and then present the computational results in Section 3.

2   DCA for Solving Problem (1)

2.1   Outline of DC Programming and DCA

DC programming and DCA constitute the backbone of smooth/nonsmooth nonconvex programming and global optimization. They address the problem of minimizing a function f which is the difference of two convex functions on the whole space $\mathbb{R}^d$ or on a convex set $C \subset \mathbb{R}^d$. Generally speaking, a DC program is an optimisation problem of the form :

$$\alpha = \inf\{f(x) := g(x) - h(x) : x \in \mathbb{R}^d\} \qquad (P_{dc})$$

where g, h are lower semi-continuous proper convex functions on $\mathbb{R}^d$. The idea of DCA is simple: each iteration l of DCA approximates the concave part $-h$ by its affine majorization (that corresponds to taking $y^l \in \partial h(x^l)$) and minimizes the resulting convex function (that is equivalent to determining a point $x^{l+1} \in \partial g^*(y^l)$ with g^* is the conjugate function of the convex function g).

The generic DCA scheme is shown below.

DCA scheme
Initialization:
Let $x^0 \in \mathbb{R}^d$ be a best guess, $r = 0$.
Repeat

- Calculate $y^r \in \partial h(x^r)$
- Calculate $x^{r+1} \in \arg\min\{g(x) - h(x^r) - \langle x - x^r, y^r \rangle : x \in \mathbb{R}^d\}$ $\quad (P_l)$
- $r = r + 1$

Until convergence of $\{x^r\}$.

For a complete study of DC programming and DCA the reader is referred to [8,12], and the references therein. For instant, it is worth to note that the construction of DCA involves the convex DC components g and h but not the DC function f itself. Moreover, a DC function f has infinitely many DC decompositions $g - h$ which have a crucial impact on the qualities (speed of convergence, robustness, efficiency, globality of computed solutions,...) of DCA. The solution of a nonconvex program by DCA must be composed of two stages: the search of an *appropriate* DC decomposition and that of a *good* initial point.

2.2   A DC Formulation of the Problem (1)

In the problem (1), the variables W and Λ are a priori bounded. One can also find a constraint for bound the variable Z. Indeed, let $\alpha_i := \min_{j=1,...,n} x_{j,i}$, $\gamma_i := \max_{j=1,...,n} x_{j,i}$. Hence $z_l \in \mathcal{T}_l := \Pi_{i=1}^m [\alpha_i, \gamma_i]$ for all $l = 1, ..., k$, and $Z \in \mathcal{T} := \Pi_{l=1}^k \mathcal{T}_l$.

Let Δ_l (resp. $\mathcal{C}_j$) be the $(m-1)$-simplex in $\mathbb{R}^m$(resp. $(k-1)$-simplex in $\mathbb{R}^k$), for each $l \in \{1, ..., k\}$ (resp. for each $j \in \{1, ..., n\}$), defined by:

$$\Delta_l := \left\{ \Lambda_l := (\lambda_{l,i})_l \in [0,1]^m : \sum_{i=1}^m \lambda_{l,i} = 1 \right\};$$
$$\mathcal{C}_j := \left\{ W_j := (w_{j,l})_j \in [0,1]^k : \sum_{l=1}^k w_{j,l} = 1 \right\},$$

and $\mathcal{C} := \Pi_{j=1}^n \mathcal{C}_j, \mathcal{T} := \Pi_{l=1}^k \mathcal{T}_l, \Delta := \Pi_{l=1}^k \Delta_l$.

The problem (1) can be rewritten as:

$$\min \left\{ F(W, Z, \Lambda) : (W, Z, \Lambda) \in (\mathcal{C} \times \mathcal{T} \times \Delta) \right\}. \tag{2}$$

Our DC decomposition of F is based on the following result.

Proposition 1. *There exists $\rho > 0$ such that the function*

$$h(u, v, y) := \frac{\rho}{2} \left(u^2 + v^2 + y^2 \right) - u^\mu y^\beta (v - a)^2$$

is convex on $(u, v, y) \in [0, 1] \times [\alpha, \gamma] \times [0, 1]$.

Proof: Let us consider the function $f : \mathbb{R} \times \mathbb{R} \to \mathbb{R}$ defined by:

$$f(u, v, y) = u^\mu y^\beta (v - a)^2. \tag{3}$$

The Hessian of f, denoted $J(u, v, y)$, is given by:

$$\begin{pmatrix} \mu(\mu - 1)u^{\mu-2}y^\beta(v-a)^2 & 2\mu u^{\mu-1}y^\beta(v-a) & \mu u^{\mu-1}\beta y^{\beta-1}(v-a)^2 \\ 2\mu u^{\mu-1}y^\beta(v-a) & 2u^\mu y^\beta & 2\beta u^\mu y^{\beta-1}(v-a) \\ \beta y^{\beta-1}\mu u^{\mu-1}(v-a)^2 & 2\beta u^\mu y^{\beta-1}(v-a) & \beta(\beta-1)u^\mu y^{\beta-2}(v-a)^2 \end{pmatrix}. \tag{4}$$

For all $(u, v, y) \in [0, 1] \times [\alpha, \gamma] \times [0, 1]$, the determinant $|J(u, v, y)|_1$ of $J(u, v, y)$ is computed as follows:

$$\max \left\{ |\mu(\mu - 1)u^{\mu-2}y^\beta(v - a)^2| + |2\mu u^{\mu-1}y^\beta(v - a)| + |\mu u^{\mu-1}\beta y^{\beta-1}(v - a)^2|; \right.$$
$$|2\mu u^{\mu-1}y^\beta(v - a)| + |2u^\mu y^\beta| + |2\beta u^\mu y^{\beta-1}(v - a)|;$$
$$\left. |\beta y^{\beta-1}\mu u^{\mu-1}(v - a)^2| + |2\beta u^\mu y^{\beta-1}(v - a)| + |\beta(\beta - 1)u^\mu y^{\beta-2}(v - a)^2| \right\}. \tag{5}$$

For all $(u, v, y) : u \in [0, 1], v \in [\alpha, \gamma], y \in [0, 1], \mu > 1, \beta > 1$ we have

$$|J(u, v, y)|_1 < \rho := \max\{\mu(\mu - 1)\delta^2 + 2\mu\delta + \beta\mu\delta^2; 2\mu\delta + 2 + 2\beta\delta;$$
$$\beta\mu\delta^2 + 2\beta\delta + \beta(\beta - 1)\delta^2\}, \tag{6}$$

where $\delta = \gamma - \alpha$. As a consequence, with ρ defined above, the function

$$h(u, v, y) = \frac{\rho}{2} \left(u^2 + v^2 + y^2 \right) - u^\mu y^\beta (v - a)^2 \tag{7}$$

is convex on $\{u \in [0, 1], v \in [\alpha, \gamma], y \in [0, 1]\}$. ∎

Using the above proposition, for $u \leftarrow w_{jl}, v \leftarrow z_{li}, y \leftarrow \lambda_{li}$, the function

$$h_{lij}(w_{jl}, z_{li}, \lambda_{li}) = \frac{\rho}{2} \left(w_{jl}^2 + z_{li}^2 + \lambda_{li}^2 \right) - w_{jl}^\mu \lambda_{li}^\beta (z_{li} - x_{ji})^2 \tag{8}$$

is convex on $([0, 1] \times [\alpha_i, \gamma_i] \times [0, 1])$.

As a consequence, the function $H(W, Z, \Lambda)$ defined by:

$$H(W, Z, \Lambda) := \sum_{l=1}^{k} \sum_{j=1}^{n} \sum_{i=1}^{m} \left[\frac{\rho}{2} \left(w_{jl}^2 + z_{li}^2 + \lambda_{li}^2 \right) - w_{jl}^{\mu} \lambda_{li}^{\beta} (z_{li} - x_{ji})^2 \right] \qquad (9)$$

is convex on $(\mathcal{C} \times \mathcal{T} \times \Delta)$.

Finally, we can express F as follows:

$$F(W, Z, \Lambda) := G(W, Z, \Lambda) - H(W, Z, \Lambda), \qquad (10)$$

where

$$G(W, Z, \Lambda) := \frac{\rho}{2} \sum_{l=1}^{k} \sum_{j=1}^{n} \sum_{i=1}^{m} \left(w_{jl}^2 + z_{li}^2 + \lambda_{li}^2 \right);$$

and $H(W, Z, \Lambda)$ as (9) are clearly convex functions. Therefore, we get the following DC formulation of (1):

$$\min \left\{ F(W, Z, \Lambda) := G(W, Z, \Lambda) - H(W, Z, \Lambda) : (W, Z, \Lambda) \in (\mathcal{C} \times \mathcal{T} \times \Delta) \right\}. \qquad (11)$$

2.3   DCA Applied to (11)

For designing a DCA applied to (11), we first need to compute $(\bar{W}^r, \bar{Z}^r, \bar{\Lambda}^r) \in \partial H(W^r, Z^r, \Lambda^r)$ and then solve the convex program

$$\min \left\{ \frac{\rho}{2} \sum_{l=1}^{k} \sum_{j=1}^{n} \sum_{i=1}^{m} \left(w_{jl}^2 + z_{li}^2 + \lambda_{li}^2 \right) - \langle (W, Z, \Lambda), (\bar{W}^r, \bar{Z}^r, \bar{\Lambda}^r) \rangle : \right.$$

$$\left. (W, Z, \Lambda) \in (\mathcal{C} \times \mathcal{T} \times \Delta) \right\}. \qquad (12)$$

The function H is differentiable and its gradient at the point (W^r, Z^r, Λ^r) is given by:

$$\bar{W}^r = \nabla_W H(W, Z, \Lambda) = \left(m\rho w_{jl} - \sum_{i=1}^{m} \mu w_{jl}^{\mu-1} \lambda_{li}^{\beta} (z_{li} - x_{ji})^2 \right)_{\substack{l=1..k \\ j=1..n}},$$

$$\bar{Z}^r = \nabla_Z H(W, Z, \Lambda) = \left(n\rho z_{li} - \sum_{j=1}^{n} 2 w_{jl}^{\mu} \lambda_{li}^{\beta} (z_{li} - x_{ji}) \right)_{\substack{i=1..m \\ l=1..k}}, \qquad (13)$$

$$\bar{\Lambda}^r = \nabla_\Lambda H(W, Z, \Lambda) = \left(n\rho \lambda_{li} - \sum_{j=1}^{n} \beta w_{jl}^{\mu} \lambda_{li}^{\beta-1} (z_{li} - x_{ji})^2 \right)_{\substack{i=1..m \\ l=1..k}}.$$

The solution of the auxiliary problem (12) is explicitly computed as (Proj stands for the projection)

$$(W^{r+1})_j = \text{Proj}_{\mathcal{C}_j} \left(\frac{1}{m\rho} (\bar{W}^r)_j \right) \; j = 1, ...n;$$

$$(Z^{r+1})_{li} = \text{Proj}_{[\alpha_i, \gamma_i]} \left(\frac{1}{n\rho} (\bar{Z}^r)_{li} \right) \; l = 1, .., k, i = 1, ...m; \qquad (14)$$

$$(\Lambda^{r+1})_l = \text{Proj}_{\Delta_l} \left(\frac{1}{n\rho} (\bar{\Lambda}^r)_l \right) \; l = 1, ...k.$$

Finally, DCA applied to (11) can be described as follows.

DCA-SI: DCA applied to (11)

- **Initialization:** Choose W^0, Z^0 and Λ^0. Let $\epsilon > 0$ be sufficiently small, $r = 0$.
- **Repeat**
  - Compute $(\bar{W}^r, \bar{Z}^r, \bar{\Lambda}^r)$ via (13).
  - Compute $(W^{r+1}, Z^{r+1}, \Lambda^{r+1})$ via (14).
  - $r = r + 1$
- **Until** $\|(W^{r+1}, Z^{r+1}, \Lambda^{r+1}) - (W^r, Z^r, \Lambda^r)\| \leq \epsilon$ or $|F(W^{r+1}, Z^{r+1}, \Lambda^{r+1}) - F(W^r, Z^r, \Lambda^r)| \leq \epsilon$.

Theorem 1. *(Convergence properties of DCA-SI)*

(i) **DCA-SI** *generates a sequence* $\{W^r, Z^r, \Lambda^r\}$ *such that the sequence* $\{F(W^r, Z^r, \Lambda^r)\}$ *is monotonously decreasing.*

(ii) **DCA-SI** *has a linear convergence.*

(iii) *The sequence* $\{W^r, Z^r, \Lambda^r\}$ *generated by* **DCA-SI** *converges to a critical point of* $F = G - H$.

Proof: (i) - (iii) are direct consequences of the convergence properties of general DC programs.

2.4   Finding a Good Starting Point of DCA

Finding a good starting point is an important question while designing DCA schemes. The research of such a point depends on the structure of the problem being considered and can be done by, for example, a heuristic procedure. Generally speaking a good starting point for DCA must not be a local minimizer, because DCA is stationary from such a point. As proposed in ([11]), we use an alternative SCAD - DCA-SI procedure for (11) which is described as follows.

SCAD - DCA-SI procedure

- **Initialization:** Choose randomly W^0, Z^0 and Λ^0. Let *maxiter* > 0 be a given integer. Set $s = 0$.
- **Repeat**
  - Perform one iteration of **SCAD** from (W^s, Z^s, Λ^s).
  - Perform one iteration of **DCA-SI** from the solution given by **SCAD** to obtain $(W^{s+1}, Z^{s+1}, \Lambda^{s+1})$.
  - $s = s + 1$.
- **Until** $s = maxiter$.

In our experiments, we use $maxiter = 2$.

3 Computational Experiments and Results

All clustering algorithms were implemented in the Visual C++ 2008, and performed on a PC Intel i5 CPU650, 3.2 GHz of 4GB RAM. Images for experiment are taken from *Berkeley segmentation dataset*. As the same way in [5], we map each pixel to an 8-dimensional feature vector consisting of three colors, three texture features and the two positions of pixels. The three color features are L^*a^*b coordinates of the color image. The three texture features (polarity, anisotropy and contrast (cf. [2,5]) are computed as follows. First, the image $I(x, y)$ is convolved with Gaussian smoothing kernels $G_\delta(x, y)$ of several scales δ: $M_\delta(x, y) = G_\delta(x, y) \otimes (\Delta I(x, y))(\Delta I(x, y))^t$.

- The polarity is defined by $p = |E_+ - E_-|/(E_+ - E_-)$, where E_+ and E_- represent, respectively, the number of gradient vectors in the matrix Gauss kernels $G_\delta(x, y)$ of scale δ at the pixel (x, y) on the positive and negative sides of the dominant orientation. For each pixel, an optimal scale value is selected such that it corresponds to the value where polarity stabilizes with respect to scale.
- The anisotropy is computed by $a = 1 - \lambda_2/\lambda_1$, where λ_1, λ_2 are the eigenvalues of $M_\delta(x, y)$ at the selected scale.
- The texture contrast is defined as $c = 2(\sqrt{\lambda_1 + \lambda_2})^3$.

We compare our algorithm with **SCAD** ([5]). For both algorithms, the parameter β is chosen in the interval $[1.0, 4.0]$ while the parameter of μ is taken in $[1.5, 2.5]$. We stop all algorithm with the tolerance $\epsilon = 10^{-4}$. The computational results are reported in the figures below. In these figures

- (a) corresponds to original image;
- (b1) represents the resulting image given by **DCA-SI** while (c1), (d1),... correspond to each segment detected by **DCA-SI**;
- (b2) represents the resulting image given by **SCAD** while (c2), (d2),... correspond to each segment detected by **SCAD**.

First Experiment (Figure 1): Figure 1.(a) contains 5 regions based on their shapes and colors: the background, the green circle, the orange square, the yellow square, and the pink region. Figure 1 shows that the segmentation obtained by **DCA-SI** is quite good. **DCA-SI** detects well 3 out of 5 regions (the pink region 1.(c1), the green circle 1.(d1) and the background 1.(f1)). However, **DCA-SI** puts the yellow square and the orange square in same segment, since the two colors are close together. Concerning **SCAD**, only the yellow square is well segmented.

Second Experiment (Figure 2): In the second experiment, we deal with the case where texture features are the most important, in contrast to the first experiment where the colors are the most significant. We see that, in Figure 2, the color of the sun flower and the butterfly are closed. Then, to obtain a good segmentation, the weight of texture attributes should be greater than the weight of other attributes. We observe that, **DCA-SI** furnishes better results

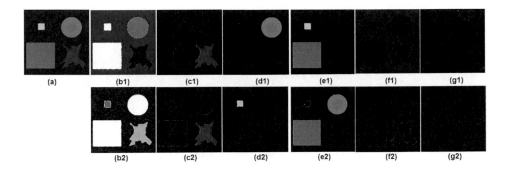

Fig. 1. Image segmentation by DCA-SI and SCAD

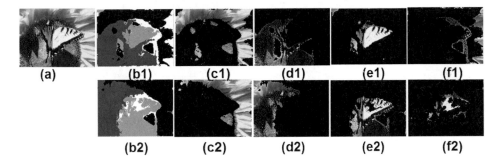

Fig. 2. Image segmentation by DCA-SI and SCAD

than **SCAD**. **DCA-SI** can separate the butterfly, the petals and the stamens with fewer errors than **SCAD**.

Third Experiment (Figure 3): In this experiment, we study the influence of parameters β and μ. The Figure 3.(a) contains two regions: the red peppers and the green peppers. We observe that, with a small value of β ($\beta \in [1.1, 1.5]$), the boundaries of objects are detected. Whereas the regions can be better detected with a greater value of β ($\beta \in [2.5, 4.0]$). Concerning μ, we get the best results with μ in the interval $[2.0, 2.5]$.

Finally, in Table. 1, we report the computation CPU time of each algorithm. We observe that **DCA-SI** is faster than **SCAD** in all experiment. The ratio of gain varies from 1.5 to 5 times.

Table 1. CPU Time in seconds

Image	Size	N^0.Classes	DCA-SI	SCAD
Figure 1	256×256	5	11.13	29.00
Figure 2	192×144	4	3.38	16.93
Figure 3	512×512	2	21.01	31.34

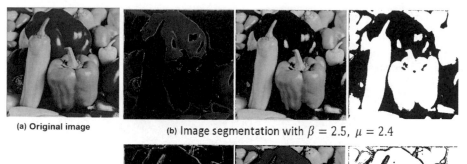

(a) Original image   (b) Image segmentation with $\beta = 2.5$, $\mu = 2.4$

(c) Image segmentation with $\beta = 1.5$, $\mu = 2.1$

Fig. 3. Image segmentation by DCA-SI with different values of μ and β

4   Conclusion

We have rigorously studied DC programming and DCA for image segmentation via feature weighted fuzzy clustering. First, the optimization model has been formulated as a DC program. It fortunately turns out that the corresponding DCA consists in computing, at each iteration, the projection of points onto a simplex and/or a rectangle, that all are given in the explicit form. Computational experiments show the efficiency and the superiority of DCA with respect to the standard SCAD algorithm ([5]). We are convinced that our approach is promising for weighted fuzzy clustering. In future works we will investigate this approach for other applications of clustering.

References

1. Bezdek, J.C.: Pattern recognition with fuzzy objective function algorithm. Plenum Press, New York (1981)
2. Carson, C., Belongie, S., Greenspan, H., Malik, J.: Color and Texture-Based Image Segmentation Using EM and Its Application to Content-Based Image Retrieval. In: Proceedings of the Sixth International Conference on Computer Vision, January 4-7, pp. 675–682 (1998)
3. Chan, E.Y., Ching, W.K., Michael, K.N., Huang, Z.J.: An optimizationalgorithm for clustering using weighted dissimilarity measures. Pattern Recognition 37(5), 943–952 (2004)
4. Haralick, R., Shapiro, L.: Image segmentation techniques. Computer Vision. Graphics and Image Processing 29, 100–132 (1985)

5. Hichem, F., Olfa, N.: Unsupervised learning of prototypes and attribute weights. Pattern Recognition 37(3), 567–581 (2004)
6. Hung, W.L., Yang, M.S., Chen, D.H.: Parameter selection for suppressed fuzzy c-means with an application to MRI segmentation. Pattern Recognition Letters 27, 424–438 (2006)
7. Le Thi, H.A.: DC Programming and DCA, http://lita.sciences.univ-metz.fr/$\sim$lethi
8. Le Thi, H.A.: Contribution à l'optimisation non convexe et l'optimisation globale: Théorie, Algoritmes et Applications. Habilitation à Diriger des Recherches, Uni. Rouen (1997)
9. Le Thi, H.A., Belghiti, M.T., Pham, D.T.: A new efficient algorithm based on dc programming and dca for clustering. J. of Global Optimization 37(4), 593–608 (2007)
10. Le Thi, H.A., Le Hoai, M., Pham, D.T.: Fuzzy clustering based on nonconvex optimisation approaches using difference of convex (DC) functions algorithms. Journal of Advances in Data Analysis and Classification 2, 1–20 (2007)
11. Le Hoai An, T., Le Minh, H., Phuc, N.T., Dinh Tao, P.: Noisy image segmentation by a robust clustering algorithm based on DC programming and DCA. In: Perner, P. (ed.) ICDM 2008. LNCS (LNAI), vol. 5077, pp. 72–86. Springer, Heidelberg (2008)
12. Le Thi, H.A., Pham, D.T.: The DC (Difference of Convex functions) Programming and DCA revisited with DC models of real world nonconvex optimization problems. Annals of Operations Research 133, 23–46 (2005)
13. Le Thi, H.A., Pham, D.T., Huynh, V.N.: Exact penalty techniques in DC programming. Journal of Global Optimization, 1–27 (2011), doi:10.1007/s10898-011-9765-3
14. Pal, N., Pal, S.: A review on image segmentation techniques. Pattern Recognition 26, 1277–1294 (1993)
15. Pham, D.T., Le Thi, H.A.: Convex analysis approach to d.c. programming: Theory, algorithms and applications. Acta Mathematica Vietnamica, dedicated to Professor Hoang Tuy on the occasion of his 70th birthday 22(1), 289–355 (1997)
16. Skarbek, W., Koschan, A.: Colour Image Segmentation: A Survey, Leiter der Fachbibliothek Informatik, Sekretariat FR, 5–4 (1994)
17. Verge, L.J.: Color Constancy and Image Segmentation Techniques for Applications to Mobile Robotics Universitat Politécnica de Catalunya, Thesis (2005)

Clustering Data Streams over Sliding Windows by DCA*

Ta Minh Thuy[1], Le Thi Hoai An[1,2], and Lydia Boudjeloud-Assala[1]

[1] Laboratory of Theoretical and Applied Computer Science
LITA EA 3097, University of Lorraine, Ile du Saulcy, 57045 Metz, France
[2] Lorraine Research Laboratory in Computer Science and its Applications,
LORIA CNRS UMR 7503, University of Lorraine, 54506 Nancy, France
minh-thuy.ta5@etu.univ-lorraine.fr
{hoai-an.le-thi,lydia.boudjeloud-assala}@univ-lorraine.fr

Abstract. Mining data stream is a challenging research area in data mining, and concerns many applications. In stream models, the data is massive and evolving continuously, it can be read only once or a small number of times. Due to the limited memory availability, it is impossible to load the entire data set into memory. Traditional data mining techniques are not suitable for this kind of model and applications, and it is required to develop new approaches meeting these new paradigms. In this paper, we are interested in clustering data stream over sliding window. We investigate an efficient clustering algorithm based on DCA (Difference of Convex functions Algorithm). Comparative experiments with clustering using the standard K-means algorithm on some real-data sets are presented.

Keywords: Clustering, Data streams, Sliding windows, clustering, DCA.

1 Introduction

Data streams concern many applications involving large and temporal data sets such as telephone records, banking, multimedia, network traffic monitoring, and sensor network data processing, etc. In these applications, the data patterns may evolve continuously and depend on time or depend on the events. In stream models, the data elements can only be accessed in the order in which they arrive; random access to the data is not allowed; and memory is limited in relative to the number of data elements. Due to the limited memory availability, it is impossible to load the entire data set into memory. Mining data streams poses great challenges for researchers and recently, several researches are devoted to this topic.

We focus in this paper the clustering over data streams. As a common data mining task, clustering is widely studied to reveal similar features among data records. In general, clustering over data streams is dominated by the outdated

* This research has been supported by "Fonds Européens de Développement Régional" (FEDER) Lorraine via the project InnoMaD (Innovations techniques d'optimisation pour le traitement Massif de Données).

N.T. Nguyen, T. Van Do, and H.A. Le Thi (Eds.): *ICCSAMA 2013*, SCI 479, pp. 65–75.
DOI: 10.1007/978-3-319-00293-4_6   © Springer International Publishing Switzerland 2013

historic information of the stream, since the most recent N records are considered to be more critical and preferable in many applications. Hence, clustering data streams over sliding windows is a natural choice and becomes one of the most popular models. In the sliding windows model, data elements arrive continually, and only the most recent elements are considered. In this context, at the period t - called window t, we consider N recent elements that contain some elements of window $t - 1$.

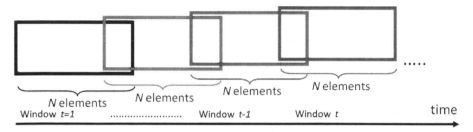

Fig. 1. Analysis on sliding-windows

Several subjects should be studied in clustering data streams over sliding windows: the design of effective and scalable streaming clustering algorithms, the analysis of characteristics of a cluster (e.g., the number of objects, the center and radius of the cluster) and of the evolving behaviors of different clusters as well as the evolution of the individual clusters, etc. In this work, assuming that each window contains N records and k clusters (N and k are fixed), we attempt to investigate efficiency clustering algorithms over sliding windows.

The data in one window can formally be written as $(x_1, x_2, ..., x_N)$, where a single observation x_i has d-dimensional. In our sliding windows model, the evolution of data items can be presented as follows:

$$Data^{window\ t} = \{x_1, x_2, ..., x_M, x_{M+1}, x_{M+2}, ..., x_{N-1}, x_N\}$$
$$Data^{window\ t+1} = \qquad \{x_{M+1}, x_{M+2}, ..., x_{N-1}, x_N, x_{N+1}, ..., x_{N+M-1}, x_{N+M}\}.$$

This means that M elements oldest will be remove and M new elements are inserted. Then the speed of evolution is defined as $(M/N)\%$. Since the new streams will differ from the current ones but slightly, it is important to study the initialization strategy of clustering algorithm performing on new windows. This is the main purpose of our work.

We investigate an efficient clustering algorithm in the nonconvex programming framework called DCA. Basing on an efficient and scalable DCA scheme to solve the MSSC (Minimum Sum-of-Squares Clustering) problem, we propose an initialization strategy for clustering on the new window and design a streaming clustering algorithms for data streams over sliding windows. To evaluate the efficiency of this strategy, we compare it with the same version of DCA using random initial points. We also study the effect of these strategies toward the speed of evolution of data, as well as the performance of DCA based approaches versus the standard K-means algorithm.

The rest of the article is organized as follows. The streaming clustering algorithm based on DCA is discussed in Section 2 while computational experiments are presented in Section 3. Finally, Section 4 concludes the paper.

2   A DCA Based Algorithm for Clustering Data Streams

DC programming and DCA which constitute the backbone of smooth/nonsmooth nonconvex programming and global optimization were introduced by Pham Dinh Tao in a preliminary form in 1985. These tools have been extensively developed since 1994 by Le Thi Hoai An and Pham Dinh Tao (see e.g. [5,8] and the references therein) and become now classic and increasingly popular (see the list of references in [6]). They address the problem of minimizing a function f which is the difference of convex functions on the whole space $\mathbb{R}^d$ or on a convex set $C \subset \mathbb{R}^d$. Generally speaking, a DC program is an optimisation problem of the form :

$$\alpha = \inf\{f(x) := g(x) - h(x) : x \in \mathbb{R}^d\} \qquad (P_{dc})$$

where g, h are lower semi-continuous proper convex functions on $\mathbb{R}^d$. The convex constraint $x \in C$ can be incorporated in the objective function of (P_{dc}) by using the indicator function on C denoted by χ_C which is defined by $\chi_C(x) = 0$ if $x \in C$, and $+\infty$ otherwise. The construction of DCA involves the convex DC components g and h but not the DC function f itself. Moreover, a DC function f has infinitely many DC decompositions $g - h$ which have a crucial impact on the qualities (speed of convergence, robustness, efficiency, globality of computed solutions,...) of DCA. The solution of a nonconvex program by DCA must be composed of two stages: the search of an *appropriate* DC decomposition and that of a *good* initial point.

The DCA has been successfully applied to real world non convex programs in various fields of applied sciences, in particular in machine learning (see e.g. the list of references in [6]). It is one of the rare efficient algorithms for non smooth non convex programming which allows solving large-scale DC programs.

2.1   DCA for Solving the MSSC Problem

An instance of the partition clustering problem consists of a data set $\mathcal{A} := \{a^1, ...a^m\}$ of m points in $\mathbb{R}^n$, a measured distance, and an integer k; we are to choose k members x^ℓ ($l = 1, ...k$) in $\mathbb{R}^n$) as "centroid" and assign each member of $\mathcal{A}$ to its closest centroid. The assignment distance of a point $a \in \mathcal{A}$ is the distance from a to the centroid to which it is assigned, and the objective function, which is to be minimized, is the sum of assignment distances. If the squared Euclidean distance is used, then the corresponding optimization formulation is expressed as: ($\|.\|$ denotes the Euclidean norm)

$$\min \quad \left\{\sum_{i=1}^m \min_{\ell=1,...,k} \left\|x^\ell - a^i\right\|^2 : x^\ell \in \mathbb{R}^n, \ell = 1, \ldots, k\right\}. \qquad (1)$$

The DCA applied to problem (1) has been developed in [7]. For the reader's convenience we will give below a brief description of this method.

To simplify related computations in DCA for solving problem (1) we will work on the vector space $\mathbb{R}^{k \times n}$ of $(k \times n)$ real matrices. The variables are then $X \in \mathbb{R}^{k \times n}$ whose i^{th} row X_i is equal to x^i for $i = 1, ..., k$. The Euclidean structure of $\mathbb{R}^{k \times n}$ is defined with the help of the usual scalar product

$$\mathbb{R}^{k \times n} \ni X \longleftrightarrow (X_1, X_2, \ldots, X_k) \in (\mathbb{R}^n)^k, \ X_i \in \mathbb{R}^n, (i = 1, .., k),$$

$$\langle X, Y \rangle : = Tr(X^T Y) = \sum_{i=1}^{k} \langle X_i, Y_i \rangle$$

and its Euclidean norm $\|X\|^2 := \sum_{i=1}^{k} \langle X_i, X_i \rangle = \sum_{i=1}^{k} \|X_i\|^2$ (Tr denotes the trace of a square matrix).

An interesting DC formulation of (1) is minimizing the difference of the simplest convex quadratic function G and the nonsmooth convex function H defined, respectively, in (3) and (4):

$$(1) \Leftrightarrow \min \ \{F(X) := G(X) - H(X) : \ X \in \mathbb{R}^{k \times n}\}, \tag{2}$$

$$G(X) := \frac{m}{2} \|X\|^2 - \langle B, X \rangle + \frac{k}{2} \|A\|^2, \tag{3}$$

$$H(X) := \sum_{i=1}^{m} \max_{j=1,\ldots,k} \sum_{\ell=1, \ell \neq j}^{k} \frac{1}{2} \|X_\ell - a^i\|^2. \tag{4}$$

Here $A \in \mathbb{R}^{m \times n}, B \in \mathbb{R}^{k \times n}$ are defined as

$$A_i := a^i \quad \text{for} \quad i = 1, \ldots, m; \quad B_\ell := a = \sum_{i=1}^{m} a^i \text{ for } \ell = 1, \ldots, k.$$

Denote by $A^{[i]} \in \mathbb{R}^{k \times n}$ the matrix whose rows are all equal to a^i, and by $\{e_j^{[k]} : j = 1, ..., k\}$ the canonical basis of $\mathbb{R}^k$. DCA applied on (2) has an explicit form that is described as follows.

DCA-MSSC (DCA to solve the problem (2)-(4))
Initialization: Let $\epsilon > 0$ be given, let $X^{(0)}$ be an initial point in $\mathbb{R}^{k \times n}$, set $p := 0$;

Repeat:

Calculate $Y^{(p)} \in \partial H(X^{(p)})$ via the next equation

$$Y^{(p)} = mX^{(p)} - B - \sum_{i=1}^{m} e_{j(i)}^{[k]} (X_{j(i)}^{(p)} - a^i)$$

and set
$$X^{(p+1)} := \frac{1}{m}(B + Y^{(p)}).$$

Set $p + 1 \leftarrow p$

Until: $\| X^{(p+1)} - X^{(p)} \| \leq \epsilon(\| X^{(p)} \| + 1)$
or $\left| F(X^{(p+1)}) - F(X^{(p)}) \right| \leq \epsilon(\left| F(X^{(p)}) \right| + 1)$.

Remark 1. The DC decomposition (2) gives birth to a very simple DCA. It requires only elementary operations on matrices (matrix addition and matrix scalar multiplication) and can so handle large-scale clustering problems.

2.2   A DCA Streaming Algorithm

For clustering data stream over sliding window we propose the following DCA based scheme, denoted by **DCA-stream** . The main idea is to start DCA in the window t from the optimal solution computed in the window $t - 1$. In other words, the clustering structure of the current streams is taken as an initialization for the clustering structure of the new streams. This initialization will usually be good since the new streams will differ from the current ones but slightly.

DCA-stream :

Initialization:
   At the first window, $t = 0$.

- Let $X^{(0,t)}$ be an initial point randomly chosen from the data of the first window.
- Apply DCA-MSSC from the initial point $X^{(0,t)}$ to get an optimal solution $X^{(t,*)}$.

Repeat

- Set $t \leftarrow t + 1$.
- Set $X^{(0,t)} = X^{(t,*)}$.
- Apply DCA-MSSC from the initial point $X^{(0,t)}$ to get an optimal solution $X^{(t+1,*)}$ at window $t + 1$.

Until All of windows are performed.

To study the performance of **DCA-stream** we consider another version, called **DCA-random**, which consists of applying DCA-MSSC at each window t from an initial points randomly chosen among the data set of this window.

DCA-random :

   Set $t = 0$.
Repeat

- Let $X^{(0)}$ be an initial point randomly chosen from the data of the window t.
- Apply DCA-MSSC from $X^{(0)}$.
- Set $t \leftarrow t + 1$.

Until All of windows are performed.

3   Computational Experiments

Data

We execute experiments with 3 real-world datasets: *KDD99* from [12], *KDD98* from [4] and *SEA* from [14].

The data *KDD-CUP'99* Network Intrusion Detection is an important problem of automatic and real-time detection of cyber attacks. This is a challenging problem for dynamic stream clustering in its own right ([3]). This dataset contains totally 494.021 elements, and each element contains 42 attributes, including 34 continuous attributes, 7 categorical attributes and 1 class attribute. As in [3], all 34 continuous attributes will be used for clustering, and the number of clusters is 5.

KDD98 contains 95.412 records of information about people who have made charitable donations in response to direct mailing request, and clustering can be used to group donors showing similar donation behavior. We use 56 fields which can be extracted from the total 481 fields of each record. The number of clusters is 10 (as in [3], [4]). This dataset is converted into a data stream by taking the data input order as the order of streaming and assuming that they flow-in with a uniform speed (as in [3], [4]).

SEA [14] is proposed by Street and Kim [10], with 60.000 elements, 3 attributes and 2 clusters.

Evaluation

To evaluate the clustering quality, we are using the CH value introduced by Calinski and Harabasz ([2], [11]). The CH value is expressed as follows:

$$CH(k) := \frac{[traceB/(k-1)]}{[traceW/(n-k)]},$$

where

$$traceB := \sum_{i=1}^{k} |C_i| \, \|\overline{C_i} - \overline{x}\|^2; \quad traceW := \sum_{i=1}^{k} \sum_{j \in C_i} \|x_j - \overline{C_i}\|^2.$$

Here $|C_i|$ is the number of objects assigned to the cluster C_i ($i = 1, \ldots, k$); $\overline{C_i}$ is the center of cluster i and $\overline{x} = \frac{1}{n} \sum_{i=1}^{n} x_i$ is the center of whole data set. The best clustering is achieved when CH is maximal.

Numerical Results

For all data sets, the number of windows is 10, and all of data are normalized before perform clustering. The size of each window in the data set SEA is $N = 5.000$, and in other data sets $N = 10.000$. All algorithms have been implemented in the Visual C++ 2008, and run on a PC Intel i5 CPU650, 3.2 GHz of 4GB RAM.

Comparison between DCA-Stream and DCA-Random

We perform clustering on each window by two algorithms: **DCA-stream** and **DCA-random** and execute experiment with three values of the speed of evolution data: 75%, 50%, 25% (remember that the speed of evolution is defined as $(M/N)\%$).

In the first window, the two algorithms start from the same solution randomly chosen from the data points.

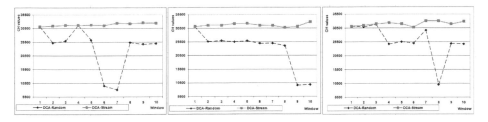

Fig. 2. The CH values given by **DCA-random** & **DCA-stream** versus the speeds of evolution data: 25% (left) 50% (center) 75% (right) on KDD99 dataset

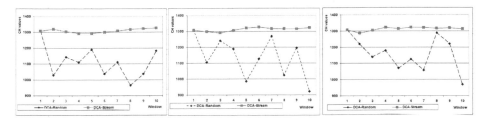

Fig. 3. The CH values given by **DCA-random** & **DCA-stream** versus the speeds of evolution data: 25% (left) 50% (center) 75% (right) on KDD98 dataset

From the numerical results (Figures 2,3,4 and Tables 1, 2, 3), we observe that:

i) With an appropriate starting point in the first window, the **DCA-stream** algorithm is more efficient than **DCA-random** in all cases, on both the clustering quality (the value CH) and the rapidity (the number of iterations and average running time).

ii) The behavior of **DCA-stream** is quite stable: after the first one or two windows, the algorithm converges with a small number of iterations and in a short time.

Comparison with K-Means Stream

In this experiment we compare **DCA-stream** with the classical algorithm K-means [9] with the same initialization strategy at each window. The initial points in the first window are the same for both **DCA-stream** and **K-means stream**.

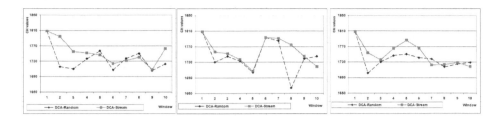

Fig. 4. The CH values given by **DCA-random** & **DCA-stream** versus the speeds of evolution data: 25% (left) 50% (center) 75% (right) on SEA dataset

Table 1. Number of iterations and running time in seconds of **DCA-random** (1) & **DCA-stream** (2) on KDD99 dataset

KDD99	Speed 25%				Speed 50%				Speed 75%			
	No.Iter		Times		No.Iter		Times		No.Iter		Times	
Window	(1)	(2)	(1)	(2)	(1)	(2)	(1)	(2)	(1)	(2)	(1)	(2)
1	29	29	0,843	0,798	29	29	0,796	0,885	29	29	0,796	0,870
2	7	3	0,141	0,125	7	4	0,156	0,156	24	2	0,656	0,078
3	7	2	0,125	0,075	7	2	0,140	0,077	7	2	0,125	0,080
4	7	2	0,140	0,096	21	2	0,546	0,083	7	21	0,140	0,660
5	7	2	0,125	0,073	7	2	0,125	0,080	8	2	0,202	0,083
6	7	2	0,140	0,077	7	4	0,125	0,159	7	2	0,125	0,106
7	7	2	0,141	0,073	8	3	0,171	0,125	36	2	0,967	0,095
8	19	2	0,500	0,087	7	5	0,141	0,164	7	1	0,124	0,055
9	20	2	0,531	0,085	7	2	0,141	0,108	7	1	0,125	0,048
10	7	10	0,125	0,302	7	2	0,171	0,084	36	1	0,983	0,074
Avg.	11,7	**5,6**	0,2811	**0,1791**	10,7	**5,5**	0,2512	**0,1921**	16,8	**6,3**	0,4243	**0,2149**

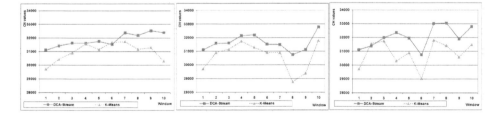

Fig. 5. The CH values giben by **DCA-stream** & K-means versus the speeds of evolution data: 25% (left) 50% (center) 75% (right) on KDD99 dataset

From numerical results (Fig. 5,6,7 and Table 4), we see that:

i) With the same appropriate starting point in the first window, **DCA-stream** is better than K-means on the quality of clustering (the CH value).

Table 2. Number of iterations and running time in seconds of **DCA-random** (1) & **DCA-stream** (2) on KDD98 dataset

KDD98	Speed 25%				Speed 50%				Speed 75%			
	No.Iter		Times		No.Iter		Times		No.Iter		Times	
Window	(1)	(2)	(1)	(2)	(1)	(2)	(1)	(2)	(1)	(2)	(1)	(2)
1	50	50	4,131	3,939	50	50	3,851	4,095	50	50	4,418	4,061
2	16	22	1,106	1,831	25	22	1,733	1,905	52	23	4,060	1,953
3	39	4	3,166	0,361	7	7	0,419	0,594	6	7	0,363	0,629
4	15	6	1,106	0,526	37	6	2,609	0,500	8	7	0,493	0,606
5	24	3	1,664	0,319	8	4	0,497	0,339	33	5	2,357	0,442
6	15	4	1,048	0,354	22	6	1,613	0,519	15	5	1,059	0,451
7	45	3	3,418	0,304	27	4	1,936	0,340	6	5	0,340	0,448
8	6	2	0,358	0,197	18	4	1,250	0,369	45	5	3,369	0,430
9	12	3	0,775	0,270	6	4	0,344	0,377	37	5	2,780	0,443
10	18	4	1,341	0,398	7	5	0,408	0,407	9	4	0,574	0,364
Avg.	24	**10,1**	1,8113	**0,8499**	20,7	**11,2**	1,4660	**0,9445**	26,1	**11,6**	1,9813	**0,9827**

Table 3. Number of iterations and running time of **DCA-random** (1) & **DCA-stream** (2) on SEA dataset

SEA	Speed 25%				Speed 50%				Speed 75%			
	No.Iter		Times		No.Iter		Times		No.Iter		Times	
Window	(1)	(2)	(1)	(2)	(1)	(2)	(1)	(2)	(1)	(2)	(1)	(2)
1	10	10	0,039	0,033	10	10	0,038	0,032	10	10	0,043	0,031
2	14	2	0,055	0,008	25	2	0,102	0,008	15	3	0,057	0,011
3	22	2	0,087	0,008	30	2	0,118	0,008	30	4	0,120	0,020
4	30	2	0,109	0,009	30	3	0,105	0,013	20	6	0,082	0,021
5	29	2	0,114	0,010	14	3	0,053	0,010	19	2	0,073	0,011
6	12	3	0,044	0,013	22	6	0,091	0,022	12	6	0,050	0,019
7	23	2	0,096	0,012	22	2	0,082	0,008	20	3	0,077	0,012
8	14	2	0,052	0,009	17	2	0,071	0,009	26	2	0,111	0,008
9	20	3	0,078	0,011	13	2	0,048	0,008	19	3	0,081	0,013
10	26	4	0,094	0,013	12	3	0,042	0,011	16	3	0,073	0,014
Avg.	20	**3,2**	0,0768	**0,0126**	19,5	**3,5**	0,0750	**0,0129**	18,7	**4,2**	0,0767	**0,0160**

ii) **K-means stream** is slightly less expensive than **DCA-stream** on running time in 2/3 data sets, whereas both algorithms are fast.

The effect of streaming algorithms toward the speed of evolution of data

Not surprisingly, for all experiments, in both **DCA-stream** and **K-means stream**, the smaller speed of evolution is, the better algorithms will be (here we compare the number of iterations and running-time but not the CH value because the data sets change).

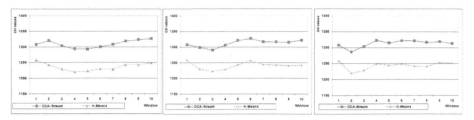

Fig. 6. The CH values given by **DCA-stream** & K-means versus the speeds of evolution data: 25% (left) 50% (center) 75% (right) on KDD98 datasets

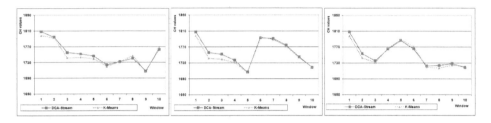

Fig. 7. The CH values given by **DCA-stream** & K-means versus the speeds of evolution data: 25% (left) 50% (center) 75% (right) on SEA dataset

Table 4. Average of number of iterations and running time over 10 windows of **DCA-stream** & **K-means stream**

Data	Algorithm	No.Iteration			Times (s)		
		Speed25%	Speed50%	Speed75%	Speed25%	Speed50%	Speed75%
KDD99	**K-means stream**	2,5	2,8	3,1	0,0618	0,0710	0,0805
	DCA-stream	5,6	5,5	6,3	0,1791	0,1921	0,2149
KDD98	**K-means stream**	4,0	4,9	6,2	0,2204	0,2588	0,3122
	DCA-stream	10,1	11,2	11,6	0,8499	0,9445	0,9827
SEA	**K-means stream**	8,0	8,8	9,9	0,0187	0,0199	0,0243
	DCA-stream	3,2	3,5	4,2	0,0126	0,0129	0,0160

4   Conclusion

We have studied the efficiency of DCA clustering algorithm in the context of data stream over sliding windows. We have improved the computational aspects of DCA clustering algorithm by investigating a good initialization strategy for performing clustering on new windows. Preliminary numerical experiments show the advantage of this strategy and the efficiency of **DCA-stream** algorithm for clustering over sliding windows. The performance of DCA suggests us to develop this approach for other tasks of mining data streams. On the other hand, a deeper study on other subjects of clustering data streams could be interesting and useful for designing efficient streaming algorithms. Works in these directions are in progress.

References

1. Zhou, A., Cao, F., Qian, W., Jin, C.: Tracking clusters in evolving data streams over sliding windows. Knowl. Inf. Syst. 15(2), 181–184 (2008)
2. Calinski, T., Harabasz, J.: A dendrite method for cluster analysis. Communications in Statistics Simulation and Computation 3(1), 1–27 (1974)
3. Aggarwal, C.C., Han, J., Wang, J., Yu, P.S.: A framework for clustering evolving data streams. In: Proceedings of the 29th International Conference on Very Large Data Bases, vol. 29, pp. 81–92 (2003)
4. Farnstrom, F., Lewis, J., Elkan, C.: Scalability for clustering algorithms revisited. SIGKDD Explor. Newsl. 2(1), 51–57 (2000)
5. Le Thi, H.A.: Contribution à l'optimisation non convexe et l'optimisation globale: Théorie, Algoritmes et Applications, Habilitation à Diriger des Recherches, Université de Rouen (1997)
6. Le Thi, H.A.: DC programming and DCA,
   `http://lita.sciences.univ-metz.fr/~lethi/english/DCA.html`
7. Le Thi, H.A., Belghiti, M.T., Pham, D.T.: A new efficient algorithm based on dc programming and dca for clustering. J. of Global Optimization 37(4), 593–608 (2007)
8. Le Thi, H.A., Pham, D.T.: The DC (difference of convex functions) Programming and DCA revisited with DC models of real world nonconvex optimization problems. Annals of Operations Research 133, 23–46 (2005)
9. MacQueen, J.B.: Some Methods for classification and analysis of multivariate observations. In: Proceedings of 5th Berkeley Symposium on Mathematical Statistics and Probability, vol. 1, pp. 281–288. University of California Press, Berkeley (1967)
10. Street, W.N., Kim, Y.S.: A streaming ensemble algorithm (sea) for largescale classification. In: Proceedings of the Seventh ACM SIGKDD International Conference on Knowledge Discovery and Data Mining, KDD 2001, pp. 377–382. ACM, New York (2001)
11. Vendramin, L., Campello, R.J.G.B., Hruschka, E.R.: On the comparison of relative clustering validity criteria. In: Proceedings of the Ninth SIAM International Conference on Data Mining, Nevada, pp. 733–744 (April 2009)
12. Zhu, X.: Stream data mining repository, `http://cse.fau.edu/xqzhu/stream.html` (accessed on September 2012)
13. `http://cseweb.ucsd.edu/users/elkan/skm.html` (accessed on September 2012)
14. `http://www.liaad.up.pt/kdus/kdus_5.html` (accessed on September 2012)

A Column Generation Based Label Correcting Approach for the Sensor Management in an Information Collection Process

Duc Manh Nguyen*, Frédéric Dambreville, Abdelmalek Toumi,
Jean-Christophe Cexus, and Ali Khenchaf

Lab-STICC UMR CNRS 6285, ENSTA Bretagne
2 rue François Verny, 29806 Brest Cedex 9, France
{nguyendu,dambrefr,toumiab,cexusje,khenchal}@ensta-bretagne.fr

Abstract. This paper deals with problems of sensor management in a human driven information collection process. This applicative context results in complex sensor-to-task assignment problems, which encompass several difficulties. First of all, the tasks take the form of several information requirements, which are linked together by logical connections and priority rankings. Second, the assignment problem is correlated by many constraint paradigms. Our problem is a variant of Vehicle Routing Problem with Time Windows (VRPTW), and it also implements resource constraints including refuelling issues. For solving this problem, we propose a column generation approach, where the label correcting method is used to treat the sub-problem. The efficiency of our approach is evaluated by comparing with solution given by CPLEX on different scenarios.

Keywords: Sensor management, Information collection, Vehicle Routing Problem, Column generation, Mixed integer linear programming.

1 Introduction

Sensor planning is a research domain that treats the problem of how to manage or coordinate the usage of a suite of sensors or measurement devices in a dynamic, uncertain environment, to improve the performance of data fusion and ultimately that of perception [24]. It is also beneficial to avoid overwhelming storage and computational requirements in a sensor and data rich environment by controlling the data gathering process such that only the truly necessary data are collected and stored [18]. The literature on sensor planning closely followed the appearance of the first significant sensor capacity, and its history tracks back to the seminal works of Koopman during World War II [13, 23]. Nowadays, because sensors are becoming more complex with the advances in sensor technology and also due to the perplexing nature of the environment to be sensed, sensor planning has

---

* Corresponding author.

N.T. Nguyen, T. Van Do, and H.A. Le Thi (Eds.): *ICCSAMA 2013*, SCI 479, pp. 77–89.
DOI: 10.1007/978-3-319-00293-4_7 © Springer International Publishing Switzerland 2013

evolved out of the need for some form of assigning and scheduling tasks to the sensors [16].

Sensor planning has been studied extensively and is becoming increasingly important due to its practical implementations and applications. Besides several military applications, sensor planning currently deals with the general domain of search and surveillance [9, 10], and also is one of the key points to optimize the performance of a sensor network [4, 12]. In sensor planning, the global issue is to optimize an implementation of sensors in order to maximize the positive effect of subsequent data processing in regards to mission objectives. Therefore, we have to deal with both the optimization of implementation of sensors and the information processing (typically data fusion). From this point of view, sensor planning is also related to difficult topics in robotic - *e.g.* Partially Observable Markov Decision Process [5, 22].

In this paper, we will consider the planing of sensors, which are monitored by human teams. This problem is reduced to a generalization of Vehicle Routing Problem with Time Windows (VRPTW). Therefore, it is a NP-complete problem. For solving this problem, we introduce an approach based on the column generation method [6, 7], which is one of the most famous methods in the literature for solving VRPTW. In order to successfully apply the column generation method, we propose an suitable integer programming formulation of this problem, and then develop a label correcting method [8] for treating the sub-problem. The numerical results will show the efficiency of our approach.

The rest of paper is organized as follows. In Section 2, we introduce the considered sensor planing problem and its formulation. Our column generation based label correcting approach for solving this problem is presented in Section 3. Numerical experiments are reported in Section 4 while some conclusions and perspectives are discussed in Section 5.

2   Problem Formulation

When sensors are planned by human teams, the planning process is typically divided into two stages: the first is purely human driven, and results in the definition of an assignment problem with time and travel constraints; the second is based on optimization processes and results from the formalization of the first step. In such case, the human interaction with the optimization process is fundamental. Therefore, the human operators should be highly skilled in their domain, and may provide useful information to the optimization processes. Moreover, the human operators need to know, to understand and to interact with the optimization processes. These requirements quite often lead to the intricate sensor planning problems, for instance, the sensor-to-task assignment problems [14, 19], or the variants of the vehicle routing problem with time constraint satisfaction [11], etc.

In this work, we are interested in the second step of the planning. Our problem is to design the trajectories for a set of sensors in order to perform a set of missions with maximum performance. Besides taking into account several constraints (trajectory constraints, time windows constraints) like those in the Vehicle Routing Problem with Time Windows (VRPTW), we have to deal with both refuelling steps and a plan evaluation doctrine. Our problem could be considered as a generalization of the VRPTW.

This sensor planning problem is characterized by the following objects:

Formal Information Requirements: we have a set F of formal information requirements (FIR) needed to be satisfied. For each requirement $u \in F$, we have a set of missions corresponding $\mu(u)$ to satisfy this requirement. Here, we suppose that $\mu(u) \cap \mu(v) = \emptyset$ if $u \neq v$, and denote $M = \bigcup_{u \in F} \mu(u)$ the set of all missions for all requirements.

Sensors: A sensor is a resource unit which may be used for some FIR acquisition. We denote K the set of all sensors.

Starting points: A starting point is a possible state from which a sensor have to start. S is the set of all starting points.

Refuelling centres: A refuelling centre is a possible state where a sensor will reset its autonomy levels. R is the set of all refuelling centres.

Arrival points: An arrival point is a possible state where a sensor have to end. E is the set of all arrival points (endpoints).

Sensor states: In our model, starting points, refuelling centres and ending points could be considered as particular cases of missions, and represent a possible *state* of the sensor. For this reason, we denote $N = S \cup M \cup R \cup E$ the set of all possible states (also called tasks, or points).

Some states may be incompatible with some sensors. Thus, for each state $i \in N$, we define $K(i) \subset K$ the set of all sensors being compatible with state i. Also, the following definitions will be useful:

$S(k) \subset S$ is the set of all starting points for sensor $k \in K$;

$E(k) \subset E$ is the set of all arrival points for sensor $k \in K$.

Variables for trajectories and affectations: The boolean variables x and y are used for modelling edges and vertices of the sensors trajectories.

$$y_{ik} = \begin{cases} 1 & \text{if sensor } k \text{ performs task } i, \\ 0 & \text{otherwise,} \end{cases}$$

$$x_{ijk} = \begin{cases} 1 & \text{if sensor } k \text{ performs task } j \text{ after task } i, \\ 0 & \text{otherwise.} \end{cases}$$

Moreover, the following instrumental variable will be used in order to prevent any cyclic trajectory:

$\omega_{ik} \in \mathbb{R}$ is a counting variable for the passed states of the trajectories.

Constant parameters for performances and costs: Performances are evaluated by means of the degrees of importance of the FIR and by means of precomputed evaluations of the efficiency of any sensor in performing a mission:

- p_u is the weight of requirement $u \in F$ with respect to its priority;
- g_{ik} is the efficiency of sensor $k \in K$ in performing mission $i \in M$;

- c_{ijk} evaluates the resources expended by sensor $k \in K$ while performing state $j \in N$ after state $i \in N$;
- d_{ijk} evaluates the distance travelled by sensor $k \in K$ while performing state $j \in N$ after state $i \in N$.

The following corrected cost is defined by weighting the actual cost and the distance:

- $\widetilde{c}_{ijk} = \epsilon_1(c_{ijk} + \epsilon_2 d_{ijk})$ is the corrected cost for $i, j \in N$ and $k \in K$.

Here $\epsilon_1, \epsilon_2 \in \mathbb{R}_+$ are small positive numbers.

Variables and constant parameters for resources: Depending on the nature of the state (*e.g.* is it a refuelling centre or not?), the resources of each sensor may be replenished or not after each state. We consider the following variables:

- α_{ik} is the level of autonomy of sensor $k \in K$ after performing state $i \in N$ and before a possible refuelling;
- β_{ik} is the level of autonomy of sensor $k \in K$ after a possible refuelling at state $i \in N$.

By the way, the levels of supply, after possible refuelling, are also defined as constant parameters:

- A_{ik} is the level of supply of sensor $k \in K$ after leaving state $i \in S \cup R$.

Variables and constant parameters for time:
- $[a_i, b_i]$ is the time windows related to the state $i \in N$;
- o_{ik} is the starting time of state $i \in N$ for sensor $k \in K$;
- Δ_{ik} is the necessary time period for sensor $k \in K$ to perform state $i \in N$; (execution time)
- t_{ijk} is the necessary time period for sensor $k \in K$ to move from state $i \in N$ to state $j \in N$. (transition time)

As a conclusion: the variables of the problem are x, y, ω, α and β. Next paragraphs will present the relationship between these parameters and variables, under the form of constraints and optimization criterion.

Trajectory constraints: We consider the constraints which link variables x, y and ω, and which state that the sensors perform non cyclic states trajectories, from starting points to arrival points:

$$y_{ik} + y_{jk} \geq 2x_{ijk}, \forall i, j \in N, k \in K, \tag{1}$$

$$1 + \sum_{i,j \in N} x_{ijk} = \sum_{i \in N} y_{ik}, \forall k \in K, \tag{2}$$

$$\sum_{i \in N} x_{ihk} = \sum_{i \in N} x_{hjk}, \forall h \in M \cup R, k \in K, \tag{3}$$

$$\omega_{jk} \geq \omega_{ik} + 1 + \infty \times (1 - x_{ijk}), \forall i, j \in N, k \in K, \tag{4}$$

$$x_{ijk} = 0, \forall i \in N, k \in K, j \in S, \tag{5}$$

$$x_{ijk} = 0, \forall j \in N, k \in K, i \in E, \tag{6}$$

$$\sum_{i \in S(k)} y_{ik} = 1; \sum_{i \in E(k)} y_{ik} = 1, \forall k \in K. \tag{7}$$

Time windows constraints:

$$o_{ik} + \Delta_{ik} + t_{ijk} - \infty \times (1 - x_{ijk}) \le o_{jk}, \forall i, j \in N, k \in K, \tag{8}$$

$$a_i \le o_{ik}, o_{ik} + \Delta_{ik} \le b_i, \forall k \in K, \forall i \in S \cup R \cup E, \tag{9}$$

$$a_i \le o_{ik}, o_{ik} + \Delta_{ik} \le b_i, \forall k \in K, \text{ for all reconnaissance mission } i, \tag{10}$$

$$o_{ik} \le a_i, b_i \le o_{ik} + \Delta_{ik}, \forall k \in K, \text{ for all surveillance mission } i. \tag{11}$$

Resource constraints:

$$\alpha_{jk} \le \beta_{ik} - c_{ijk} + \infty \times (1 - x_{ijk}), \forall i, j \in N, k \in K, \tag{12}$$

$$\beta_{ik} = A_{ik}, \forall i \in S \cup R, \qquad \text{(fuelled/refuelled)} \tag{13}$$

$$\beta_{ik} = \alpha_{ik}, \forall i \in M \cup E, \qquad \text{(not refuelled)} \tag{14}$$

$$\alpha \ge 0, \beta \ge 0. \tag{15}$$

Also we consider the following constraint:

$$\sum_{i \in \mu(u)} \sum_{k \in K} y_{ik} \le 1, \forall u \in F. \tag{16}$$

Our purpose is to maximize a global criterion which is a balance between the satisfaction of the FIR and the expense. Thus, we have the following optimization problem :

$$\left\{ \begin{array}{l} \max\limits_{x,y,u,o,\alpha,\beta} \left(\sum\limits_{u \in F} p_u \sum\limits_{i \in \mu(u)} \sum\limits_{k \in K} y_{ik} g_{ik} - \sum\limits_{i \in N} \sum\limits_{j \in N} \sum\limits_{k \in K} \widetilde{c}_{ijk} x_{ijk} \right) \\ \text{subject to: from (1) to (16).} \end{array} \right. \tag{17}$$

This is a linear mixed 0-1 programming. This problem is NP-complete, since it is a generalization of VRPTW. Therefore, our considered problem is very hard to solve, even for reasonably sized cases.

3   A Column Generation Approach

While several successful methods for solving several VRPTW variants have been proposed in the literature [1–3, 6, 7, 20, 21], one of the most famous approach is column generation. The embedding of column generation techniques within a linear-programming-based branch-and-bound framework, introduced by Desrosiers et al. [6] for solving the VRPTW, became classic. It contributed as the key step in the design of exact algorithms for a large class of integer programs [15]. Nowadays, column generation is a prominent method to cope with a huge number of variables, and numerous integer programming column generation applications have been developed (see e.g. [15] for an overview). As a generalization of the VRPTW, our sensor planning has some good properties (for instance, trajectory constraints and time windows constraints) for a column generation based approach. Therefore, we will investigate the column generation approach for solving the problem (17) in this section.

3.1   Column Generation

Applying the methodology described in [6], the column generation approach will be based on the notion of feasible routes for the sensors. A feasible route of a sensor $k \in K$ is a route starting from a compatible departure, going to a compatible endpoint, satisfying all constraints and visiting *at least one mission* $i \in M$. We denote by Ω_k the set of all feasible routes for sensor k, and $\Omega = \bigcup_{k \in K} \Omega_k$ the set of all feasible routes.

Let $r = (r_1, r_2, ..., r_m) \in \Omega_k \subset \Omega$ be a route, where $r_1, \cdots, r_m \in N$ are the states visited by the sensor k. The performance of this route, denoted by $f(r)$, is computed as follows:

$$f(r) = \underbrace{\sum_{u \in F : \mu(u) \cap \{r_2, ..., r_{m-1}\} = \{r_h\}} p_u g_{r_h, k}}_{g(r)} - \underbrace{\sum_{i=1}^{m-1} \widetilde{c}_{r_i, r_{i+1}, k}}_{c(r)}. \tag{18}$$

In this formula, $g(r)$ is the gain of route r, and $c(r)$ is the cost of route r.

Now we define the parameter $a_{ru}, u \in F$ by:

$$a_{ru} = \begin{cases} 1 & \text{if } \mu(u) \cap \{r_2, ..., r_{m-1}\} \neq \emptyset, \\ 0 & \text{otherwise.} \end{cases} \tag{19}$$

The sensor planning problem (17) is reformulated as:

$$\begin{cases} \max \sum_{k \in K} \sum_{r \in \Omega_k} f(r).\theta_r \\ \text{s.t. } \sum_{k \in K} \sum_{r \in \Omega_k} a_{ru}\theta_r \leq 1, \forall u \in F, \\ \quad\quad \sum_{r \in \Omega_k} \theta_r \leq 1, \forall k \in K, \\ \quad\quad \theta_r \in \{0, 1\}, \forall r \in \Omega. \end{cases} \tag{20}$$

The variable $\theta_r \in \{0, 1\}$ is a decision variable which describes whether a route r is chosen or not. The first constraint specifies that each requirement $u \in F$ is satisfied at most one time while the second constraint ensures that each sensor $k \in K$ does at most one feasible route.

Because of the first constraint, the condition $\theta_r \in \{0, 1\}, \forall r \in \Omega$ can be replaced by $\theta_r \in \mathbb{N}, \forall r \in \Omega$. The linear relaxation of problem (20), i.e., with $\theta_r \geq 0, \forall r \in \Omega$, is called Master Problem (MP), which is an instrument for evaluating the feasible route generated at each iteration. The methodology of column generation approach can be described as follows.

Let $\Omega_k^1 \subset \Omega_k, k \in K$, and $\Omega^1 = \bigcup_{k \in K} \Omega_k^1$. We consider the restriction of the Master Problem, denoted MP(Ω^1):

$$\begin{cases} \max \sum_{k \in K} \sum_{r \in \Omega_k^1} f(r).\theta_r \\ \text{s.t. } \sum_{k \in K} \sum_{r \in \Omega_k^1} a_{ru}\theta_r \leq 1, \forall u \in F, \\ \quad\quad \sum_{r \in \Omega_k^1} \theta_r \leq 1, \forall k \in K, \\ \quad\quad \theta_r \geq 0, \forall r \in \Omega^1. \end{cases} \tag{21}$$

The dual program of (21), denoted by D(Ω^1), is:

$$\begin{cases} \min \sum\limits_{u \in F} \lambda_u + \sum\limits_{k \in K} \mu_k \\ \text{s.t.} \sum\limits_{u \in F} a_{ru}\lambda_u + \mu_k \geq f(r), \forall r \in \Omega_k^1, k \in K, \\ \lambda_u \geq 0, \forall u \in F, \\ \mu_k \geq 0, \forall k \in K. \end{cases} \tag{22}$$

Now suppose that $(\bar{\lambda}, \bar{\mu}) = (\bar{\lambda}_1, ..., \bar{\lambda}_F, \bar{\mu}_1, ..., \bar{\mu}_K)$ is an optimal solution of the dual problem $D(\Omega^1)$. Then, we have:

$$\sum_{u \in F} a_{ru}\bar{\lambda}_u + \bar{\mu}_k \geq f(r), \forall r \in \Omega_k^1, k \in K.$$

It is clear that if this condition holds for all $r \in \Omega_k, k \in K$, then $(\bar{\lambda}, \bar{\mu})$ is also the optimal solution of the dual program of (MP). Otherwise, we will look for a route $r \in \Omega_k \backslash \Omega_k^1$, for a $k \in K$ such that:

$$\sum_{u \in F} a_{ru}\bar{\lambda}_u + \bar{\mu}_k < f(r). \tag{23}$$

This is called the *sub-problem*.

The column generation algorithm for solving the problem (20) can be described as follows:

Column generation algorithm for solving (20):
Step 1. Generate initial sets Ω_k^1 for $k \in K$,
Step 2. Solve the problem (21) in order to obtain the optimal solution and its dual solution $(\bar{\lambda}, \bar{\mu})$,
Step 3. For each $k \in K$, find a route $r \in \Omega_k \backslash \Omega_k^1$ satisfying the condition (23) and update $\Omega_k^1 := \Omega_k^1 \cup \{r\}$,
Step 4. Iterate step 2-3 until there is no route satisfying the condition (23),
Step 5. Solve (20) with $\Omega := \Omega^1$.

3.2   A Label Correcting Method for Solving the Sub-problem

In [17], we have proposed an approach using CPLEX for the MILP formulation of the sub-problem in Step 3. In this section, we investigate another method for solving the sub-problem: the label correcting method. This method is based on the ideas of Feillet et al. (2004) [8] developed for treating the Elementary Shortest Path Problem with Resource Constraints. The principle of this method is to use the dynamic programming.

For a sensor k fixed, we consider $F(k) = \{FIR_1, \ldots, FIR_m\}$ the set of associated requirements. For each requirement $FIR_u \in F(k)$, we denote $\mu_k(FIR_u)$ the set of missions which can be performed by the sensor k in order to satisfy this requirement. Additionally, we denote $R(k) = \{R_1, ..., R_n\}$ the set of compatible refuelling centres and $E(k)$ the set of compatible endpoints corresponding to this sensor k. Since the position of sensor k is known, we also use the notation k to represent its position, and call $V^k = F(k) \cup R(k) \cup \{k\}$ the set of nodes.

Definition 1. *Each path P_{kv} from the position of sensor k to a node $v \in V^k \backslash \{k\}$ associates a state $H_v = (T_v^1, T_v^2, X_v^1, ..., X_v^m, Y_v^1, ..., Y_v^n, Z_v)$ and a performance $f_v = f(P_{kv})$. Here, the two first parameters T_v^1, T_v^2 correspond to the quantity of time and fuel resources used by the path. The parameters $X_v^1, ..., X_v^m$ represent* the visitation of requirement ($X_v^u = i \neq 0$ if the path visits the requirement FIR_u by performing the mission $i \in \mu_k(FIR_u)$, 0 otherwise), and the parameters $Y_v^1, ..., Y_v^n$ represent the visitation of refuelling centre ($Y_v^i = 1$ if the path visits the refuelling centre R_i, 0 otherwise). The last parameter Z_v shows the ability to reach an endpoint, i.e., $Z_v = 1$, if after visiting node v, the sensor k can go to some endpoint, 0 otherwise. The couple $\lambda_v = (H_v, f_v)$ is said to be a label on the node v.

Definition 2. Let P_{kv} and $\bar{P}_{kv}$ be two paths from the position of sensor k to a node v with associated labels $(H_v, f_v), H_v = (T_v^1, T_v^2, X_v^1, ..., X_v^m, Y_v^1, ..., Y_v^n, Z_v)$ and $(\bar{H}_v, \bar{f}_v), \bar{H}_v = (\bar{T}_v^1, \bar{T}_v^2, \bar{X}_v^1, ..., \bar{X}_v^m, \bar{Y}_v^1, ..., \bar{Y}_v^n, \bar{Z}_v)$. We say that P_{kv} dominates $\bar{P}_{kv}$ if:

$$T_v^i \leq \bar{T}_v^i, \forall i = 1, 2, id(X_v^i) \geq id(\bar{X}_v^i), \forall i = 1, 2, ..., m,$$

$$Y_v^i \leq \bar{Y}_v^i, \forall i = 1, 2, ..., n, Z_v \geq \bar{Z}_v, f_v \geq \bar{f}_v.$$

Here, $id(x) = 1$, if $x \neq 0; id(x) = 0$ otherwise.

We use the following notations to describe the algorithm:

- Λ_v: List of labels on node v.
- $Succ(v)$: Set of successors of node v.
- L: List of nodes waiting to be treated.
- $Extend(\lambda_v, \tilde{v})$: Multivalued function that returns the labels resulting from the extension of label $\lambda_v = (H_v, f_v) \in \Lambda_v$ towards node $\tilde{v}$ (with respect to the missions at $\tilde{v}$) when the extension is possible, nothing otherwise. More precisely, suppose that $\lambda_v = (H_v, f_v) \in \Lambda_v$ is a label on v, with $H_v = (T_v^1, T_v^2, X_v^1, ..., X_v^m, Y_v^1, ..., Y_v^n, Z_v)$. We will distinguish two cases of $\tilde{v}$ as follows:
  - If $\tilde{v}$ is a FIR, and $\mu_k(\tilde{v}) = \{m_1, ..., m_j\}$ is the set of missions corresponding, then we extend this label with respect to each mission $m_i, i = 1, ..., j$ in order to obtain the new label $\lambda_{\tilde{v}} = (H_{\tilde{v}}, f_{\tilde{v}})$ as follows.
  + If m_i is a reconnaisance mission

$$T_{\tilde{v}}^1 = \begin{cases} T_v^1 + t_{v,m_i,k} + \Delta_{m_i,k} & \text{if } a_{m_i} < T_v^1 + t_{v,m_i,k} < T_v^1 + t_{v,m_i,k} + \Delta_{m_i,k} \leq b_{m_i} \\ a_{m_i} + \Delta_{m_i,k} & \text{if } T_v^1 + t_{v,m_i,k} \leq a_{m_i}, \end{cases}$$

$$(24)$$

  + If m_i is a surveillance mission

$$T_{\tilde{v}}^1 = T_v^1 + t_{v,m_i,k} + \Delta_{m_i,k} \text{ if } T_v^1 + t_{v,m_i,k} \leq a_{m_i},$$

$$b_{m_i} \leq T_v^1 + t_{v,m_i,k} + \Delta_{m_i,k}, \quad (25)$$

$$T_{\tilde{v}}^2 = T_v^2 + c_{v,m_i,k} \text{ if } T_v^2 + c_{v,m_i,k} \leq A_k, \quad (26)$$

$$X_{\tilde{v}}^{\tilde{v}} = m_i, \quad (27)$$

$Z_{\tilde{v}} = 1$ if the sensor can go to an endpoint after

performing the mission m_i, otherwise 0, $\quad (28)$

$$f_{\tilde{v}} = f_v + g_{m_i,k} - \epsilon_1(c_{v,m_i,k} + \epsilon_2 d_{v,m_i,k}). \quad (29)$$

Here, A_k is the capacity of sensor k. Of course, if the conditions in (24), or (25) or (26) are violated, there is no extension. Therefore, from a label λ_v, after extension procedure we get at most $|\mu(\tilde{v})|$ new labels on node $\tilde{v}$.
  - If $\tilde{v}$ is a refuelling centre, we extend the label $\lambda_v = (H_v, f_v)$ to obtain a new label $\lambda_{\tilde{v}} = (H_{\tilde{v}}, f_{\tilde{v}})$ on $\tilde{v}$ by updating the following parameters:

$$T_{\tilde{v}}^1 = \begin{cases} T_v^1 + t_{v,\tilde{v},k} + \Delta_{\tilde{v},k} & \text{if } a_{\tilde{v}} < T_v^1 + t_{v,\tilde{v},k} < T_v^1 + t_{v,\tilde{v},k} + \Delta_{\tilde{v},k} \leq b_{\tilde{v}} \\ a_{\tilde{v}} + \Delta_{\tilde{v},k} & \text{if } T_v^1 + t_{v,\tilde{v},k} \leq a_{\tilde{v}}, \end{cases}$$

$$\tag{30}$$

$$T_{\tilde{v}}^2 = 0 \text{ if } T_v^2 + c_{v,\tilde{v},k} \leq A_k, \tag{31}$$

$$Y_{\tilde{v}}^{\tilde{v}} = 1, \tag{32}$$

$Z_{\tilde{v}} = 1$ if the sensor can go to an endpoint after

refuelling at $\tilde{v}$, otherwise 0,   (33)

$$f_{\tilde{v}} = f_v - \epsilon_1(c_{v,\tilde{v},k} + \epsilon_2 d_{v,\tilde{v},k}). \tag{34}$$

If the conditions in (30) or (31) are violated, then there is no extension.
- $F_{v,\tilde{v}}$: Set of labels extended from node v to node $\tilde{v}$.
- $EFF(\Lambda)$: Procedure that keeps only non-dominated labels in the list of labels Λ.

The label correcting procedure for solving the subproblem can be described as follows.

LabelCorrecting(k):
    Set $\Lambda_k = (0, 0, ..., 0)$ and $\Lambda_v = \emptyset$ for all $v \in V^k \backslash \{k\}$
    Set $L = \{k\}$
    Repeat
        Choose $v \in L$
        for all $\tilde{v} \in Succ(v)$
            Set $F_{v,\tilde{v}} = \emptyset$
            for all $\lambda_v = (H_v, f_v) \in \Lambda_v$, with $H_v = (T_v^1, T_v^2, X_v^1, ..., X_v^m, Y_v^1, ..., Y_v^n, Z_v)$
                if $X_v^{\tilde{v}} = 0$ or $Y_v^{\tilde{v}} = 0$ **then**
                    $F_{v,\tilde{v}} := F_{v,\tilde{v}} \cup \{Extend(\lambda_v, \tilde{v})\}$
                endif
            endfor
            $\Lambda_{\tilde{v}} = EFF(\Lambda_{\tilde{v}} \cup F_{v,\tilde{v}})$
            if $\Lambda_{\tilde{v}}$ has changed **then**
                $L = L \cup \{\tilde{v}\}$
            endif
        endfor
        Set $L = L \backslash \{v\}$
    Until finding a label $\lambda_v = (H_v, f_v)$ satisfying the following condition:
                f_v satisfies (23) and $Z_v = 1$.

Remark 1. In practice, to prevent the explosion of number of labels, we should limit the number of labels on each node at each iteration. We denote l_v the maximum labels on node v. After the step "$\Lambda_{\tilde{v}} = EFF(\Lambda_{\tilde{v}} \cup F_{v,\tilde{v}})$", if $\text{card}(\Lambda_{\tilde{v}}) > l_{\tilde{v}}$ then we only remain $l_{\tilde{v}}$ labels which have more requirements visited.

4   Experiments and Numerical Results

Our algorithm is written in MATLAB 2010, and is tested on a PC 64 bits Windows 7, Intel(R) Xeon (R) CPU X5690 @ 3.47 GHz 3.47 GHz, 24G of RAM. CPLEX 12.4 is used for solving the linear program (21), and the problem (20) in Step 5. In order to evaluate the performance of this approach, we compare

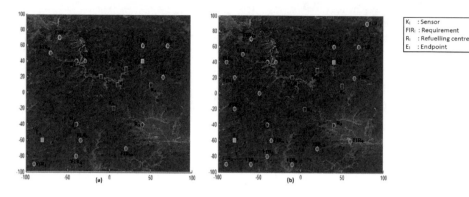

Fig. 1. Plans

the results obtained by our approach with a purely CPLEX-based approach (applying directly to the problem (17)).

In our scenarios, we assume that the sensors are starting from unique starting points, *i.e.* $\#S(k) = 1$, and that the sensors are endowed with the same autonomy level $A_{ik} = A$ after (re-)fuelling. The costs are also identically valued by $c_{ijk} = 20$. Therefore, if $A = 100$, then each sensor can visit less or equal to 5 states without refuelling. The priority of requirement is determined as follows: if the requirement u has the priority 1 (resp. 2), then $p_u = 100$ (resp. $p_u = 1$). We also define $\Delta_{ik} = 20$ (minutes), $t_{ijk} = 20$ (minutes) and $\epsilon_1 = \epsilon_2 = 10^{-4}$.

The set of initial routes for the column generation method is generated as follows: for each requirement $u \in F$, we choose a mission $i \in \mu(u)$ and a compatible sensor $k \in K$ that performs the maximum gain. Then, we choose an arrival point e which implies the smallest corrected cost, thus obtaining the route: "$s \to i \to e$".

4.1   The First Data

We have $|F| = 10$ requirements, $|M| = 15$ missions, $|R| = 3$ refuelling centres, $|E| = 2$ arrival points and $|K| = 5$ sensors (see Fig. 1 (a)). Tables 1-3 present the parameters of missions, refuelling centres and arrival points respectively. Table 4 presents the gains of missions performed by the sensors. In this case, we have MILPs with 3250 binary variables, 465 continuous variables and 10730 constraints. The maximum number of labels $l_v = 100, \forall v \in V^k, \forall k \in K$. The computational time of label correcting algorithm for each sensor is limited to 10 seconds.

Table 5 gives the comparative results between the column generation method and CPLEX for different values of parameter A. From Table 5, we see that the column generation method produced very good solutions. The relative error of objective value between the two methods varies from 0.00% to 0.09% (0.03% in average). Moreover, the column generation method is slightly faster than the

Table 1. Parameters of missions in Data 1

Requirement	Mission	Type of mission	Priority	Compatible type of sensor	Start	End
1	1	R	1	2, 4	110	360
	2	R	1	2, 4	110	360
2	3	R	1	2, 4	250	375
	4	R	1	2, 4	250	375
3	5	R	2	2, 4	425	540
	6	R	2	2, 4	425	540
4	7	R	1	4	320	370
	8	R	1	2, 4	320	370
5	9	R	1	2, 4	120	470
	10	R	1	2, 4	120	470
6	11	S	1	4	280	300
7	12	S	2	1, 4	320	340
8	13	S	1	4	360	380
9	14	S	1	4	400	420
10	15	S	2	4	445	465

Table 2. Parameters of refueling centers in Data 1

Refuelling centre	Compatible type of sensor	Start	End
1	2, 4	100	600
2	1, 2, 3, 4	100	600
3	1, 2, 3, 4	100	600

Table 3. Parameters of endpoints in Data 1

Endpoint	Compatible type of sensor	Start	End
1	4	100	600
2	1, 2, 3	100	600

Table 4. The gains of missions given by sensors in Data 1

Mission	K_1 (Type 1)	K_2 (Type 2)	K_3 (Type 3)	K_4 (Type 4)	K_5 (Type 1)
1	0	10	0	6	0
2	0	12	0	5	0
3	0	18	0	15	0
4	0	17	0	17	0
5	0	7	0	17	0
6	0	8	0	16	0
7	0	0	0	14	0
8	0	20	0	13	0
9	0	18	0	13	0
10	0	18	0	13	0
11	0	0	0	12	0
12	4	0	0	14	5
13	0	0	0	11	0
14	0	0	0	9	0
15	0	0	0	18	0

Table 5. Numerical results for Data 1

Test	A	CPLEX		Column Generation			
		Objective value	CPU time (s)	Objective value	CPU time (s)	Iteration	Column
1	120	10039.9760	41.19	10039.9740	52.94	65	87
2	100	10039.9720	38.42	10030.9720	31.48	46	80
3	80	10030.9700	39.61	10030.9700	27.88	53	85

pure CPLEX approach: the average of CPU time of column generation method is 37.43 seconds while this of CPLEX is 39.74 seconds.

4.2   The Second Data

In this section, we tested the performance of our approach on a large scale scenario. We have $|F| = 18$ requirements, $|M| = 34$ missions, $|R| = 3$ refuelling centres, $|E| = 2$ arrival points and $|K| = 6$ sensors (see Fig. 1 (b)). The refuelling centres and arrival points are the same as in the first data set. Tables 6 presents the parameters of missions. Table 7 presents the gains of missions performed by the sensors. In this case, we have MILPs with 12420 binary variables, 1032 continuous variables and 44106 constraints.

Here, the sensors K_1, K_2 are of the same type (Type 1) and located in the same position (depot), and so are the sensors K_3, K_4 (Type 2). As is done classically, same-type sensors have been solved by only one sub-problem. The maximum number of labels $l_v = 200, \forall v \in V^k, \forall k \in K$. The computational time of label correcting algorithm for each sensor is limited to 20 seconds. Also, we use a stopping criteria (gap $\leq 1\%$) when implementing the purely CPLEX-based approach.

From Table 8, we see that the column generation method once again produces quite good solutions in acceptable time. The relative error of objective value between the two methods varies from 0.61% to 1.69% (1.16% in average).

Table 6. Parameters of missions in Data 2

Requirement	Mission	Type of mission	Priority	Compatible type of sensor	Start	End
1	1	R	1	2, 4	110	360
	2	R	1	2, 4	110	360
2	3	R	1	2, 4	250	375
	4	R	1	2, 4	250	375
3	5	R	2	2, 4	425	540
	6	R	2	2, 4	425	540
4	7	R	1	4	320	370
	8	R	1	2, 4	320	370
5	9	R	1	2, 4	120	470
	10	R	1	2, 4	120	470
6	11	R	1	1, 2	110	360
	12	R	1	2, 3	110	360
7	13	R	1	2, 3	250	375
	14	R	1	1, 2	250	375
8	15	R	2	1, 2, 4	425	540
	16	R	2	2, 3, 4	425	540
9	17	R	1	4	320	370
	18	R	1	1, 2, 4	320	370
10	19	R	1	1, 3, 4	120	470
	20	R	1	1, 2, 4	120	470
11	21	R	1	2, 4	220	300
	22	S	1	1, 2, 4	220	240
	23	R	1	2, 3, 4	220	280
12	24	R	2	1, 2, 4	260	360
	25	R	2	1, 3, 4	260	340
	26	R	2	1, 3, 4	260	300
13	27	R	1	1, 2, 3, 4	280	360
	28	R	1	1, 4	280	390
	29	R	1	1, 4	280	350
14	30	S	1	1, 4	280	300
15	31	S	2	1, 4	320	340
16	32	S	1	2, 4	360	380
17	33	S	1	1, 4	400	420
18	34	S	2	2, 4	445	465

Table 7. The gains of missions given by sensors in Data 2

Mission	K_1 (Type 1)	K_2 (Type 1)	K_3 (Type 2)	K_4 (Type 2)	K_5 (Type 3)	K_6 (Type 4)
1	0	0	17	17	0	16
2	0	0	17	17	0	18
3	0	0	15	15	0	14
4	0	0	13	13	0	15
5	0	0	14	14	0	14
6	0	0	16	16	0	12
7	0	0	0	0	0	16
8	0	0	13	13	0	16
9	0	0	9	9	0	13
10	0	0	10	10	0	14
11	11	11	9	9	0	0
12	0	0	11	11	15	0
13	0	0	12	12	14	0
14	17	17	13	13	0	0
15	17	17	14	14	0	13
16	0	0	15	15	14	15
17	0	0	0	0	0	19
18	12	12	18	18	0	18
19	4	4	0	0	8	7
20	5	5	8	8	0	9
21	0	0	11	11	0	12
22	17	17	9	9	0	11
23	0	0	10	10	13	9
24	4	4	13	13	0	11
25	5	5	0	0	7	10
26	5	5	0	0	8	12
27	11	11	15	15	15	15
28	10	10	0	0	0	14
29	12	12	0	0	0	16
30	14	14	0	0	0	10
31	8	8	0	0	0	14
32	0	0	15	15	0	9
33	11	11	0	0	0	13
34	0	0	7	7	0	8

Table 8. Numerical results for Data 2

Test	A	CPLEX		Column Generation			
		Objective value	CPU time (s)	Objective value	CPU time (s)	Iteration	Column
1	120	19751.9480	5292.98	19632.9540	605.86	38	145
2	100	19761.9500	1496.26	19532.9540	377.62	42	141
3	80	19760.9500	1479.12	19432.9520	405.47	48	156

5   Conclusion

In this paper, we have proposed a column generation approach for solving the optimal sensors management in an information collection process, where a label correcting algorithm has been developped for treating the sub-problem. The comparative results with CPLEX have demonstrated the efficiency of our proposed approach. It found a near-optimal solution within acceptable time for even large-scale problems. In the future, we plan to study some dedicated heuristics and meta-heuristics for the search of column candidate. Also, we intend to parallelize the Step 3 so as to speed up our algorithm.

Acknowledgement. The authors would like to thank the DGA (Délégation Générale pour l'Armement) for its support to this research.

References

1. Bräysy, O., Gendreau, M.: Vehicle routing problem with time windows, Part I: Route construction and local search algorithms. Transport. Sci. 39(1), 104–118 (2005)
2. Bräysy, O., Gendreau, M.: Vehicle routing problem with time windows, Part II: Metaheuristics. Transport. Sci. 39(1), 119–139 (2005)

3. Campbell, A., Savelsbergh, M.: Efficient insertion heuristics for vehicle routing and scheduling problems. Transport. Sci. 38(3), 369–378 (2004)
4. Chakrabarty, K., Iyengar, S.S., Qi, H., Cho, E.: Grid coverage for surveillance and target location in distributed sensor networks. IEEE Transactions on Computers 51, 1448–1453 (2002)
5. Dambreville, F.: Cross-entropic learning of a machine for the decision in a partially observable universe. Journal of Global Optimization 37(4), 541–555 (2007)
6. Desrosiers, J., Soumis, F., Desrochers, M.: Routing With Time Windows by Column Generation. Networks 14, 545–565 (1984)
7. Desrosiers, J., Soumis, F., Sauvé, M.: Lagrangian Relaxation Methods for Solving the Minimum Fleet Size Multiple Traveling Salesman Problem With Time Windows. Mgmt. Sci. 34, 1005–1022 (1988)
8. Feillet, D., Dejax, P., Gendreau, M., Gueguen, C.: An exact algorithm for the elementary shortest path problemwith resource constraints: application to some vehicle routing problems. Networks 44(3), 216–229 (2004)
9. Frost, J.R.: Principle of search theory. Technical report, Soza & Company Ltd. (1999)
10. Haley, K.B., Stone, L.D.: Search Theory and Applications. Plenum Press, New York (1980)
11. Janez, F.: Optimization method for sensor planning. Aerospace Science and Technologie 11, 310–316 (2007)
12. Jayaweera, S.K.: Optimal node placement in decision fusion wireless sensor networks for distributed detection of a randomly-located target. In: IEEE Military Communications Conference, pp. 1–6 (2007)
13. Koopman, B.O.: The theory of search. iii. the optimum distribution of searching effort. Operations Research 5(5), 613–626 (1957)
14. Le Thi, H.A., Nguyen, D.M., Pham, D.T.: A DC programming approach for planning a multisensor multizones search for a target. Computers & Operations Research, Online first (July 2012)
15. Lübbecke, M., Desrosiers, J.: Selected Topics in Column Generation. Operations Research 53(6), 1007–1023 (2005)
16. Ng, G.W., Ng, K.H.: Sensor management – what, why and how. Information Fusion 1(2), 67–75 (2000)
17. Nguyen, D.M., Dambreville, F., Toumi, A., Cexus, J.C., Khenchaf, A.: A column generation method for solving the sensor management in an information collection process. Submitted to Optimization (October 2012)
18. Schaefer, C.G., Hintz, K.J.: Sensor management in a sensor rich environment. In: Proceedings of the SPIE International Symposium on Aerospace/Defense Sensing and Control, Orlando, FL, vol. 4052, pp. 48–57 (2000)
19. Simonin, C., Le Cadre, J.-P., Dambreville, F.: A hierarchical approach for planning a multisensor multizone search for a moving target. Computers and Operations Research 36(7), 2179–2192 (2009)
20. Solomon, M.: Algorithms for the Vehicle Routing and Scheduling Problem with Time Window Constraints. Operations Research 35, 254–365 (1987)
21. Toth, P., Vigo, D.: An exact algorithm for the vehicle routing problem with backhauls. Transport. Sci. 31, 372–385 (1997)
22. Tremois, O., Le Cadre, J.-P.: Optimal observer trajectory in bearings-only tracking for maneuvering sources. Sonar and Navigation 146(1), 1242–1257 (1997)
23. Washburn, A.R.: Search for a moving target: The FAB algorithm. Operations Research 31(4), 739–751 (1983)
24. Xiong, N., Svensson, P.: Multi-sensor management for information fusion: issues and approaches. Information Fusion 3(2), 163–186 (2002)

Planning Sensors with Cost-Restricted Subprocess Calls: A Rare-Event Simulation Approach

Frédéric Dambreville*

Lab-STICC UMR CNRS 6285, ENSTA Bretagne
2 rue François Verny, 29806 Brest Cedex 9, France
submit@fredericdambreville.com

Abstract. This paper deals with optimal sensor planning in the context of an observation mission. In order to accomplish this mission, the observer may request some intelligence teams for preliminary prior information. Since team requests are expensive and resources are bound, the entire process results in a two-level optimization, the first level being an experiment devoted to enhance the criterion modelling. The paper proposes a solve of this problem by rare-event simulation, and a mission scenario is addressed.

1 Introduction

The main background of this paper is the optimal planning of sensors in the context of an acquisition mission. Typically, the acquisition mission may result in the localisation of a target, with the final purpose of intercepting this target. In this work, we focus especially on dealing with the modelling errors of the sensor planning problem. Then, the question of interest is: *how to spend resources optimally in order to reduce the model errors, and how does that affect the sensor planning problem?*

Sensor planning, especially in order to localize a target, has been thoroughly studied in the literature. First works in this domain track back to the works of Koopman during World War II [1,2]. This seminal works has been extended in various manner, so as to take into account motion models [3,4], or reactive behaviours of the target [5,6]. Sensor planning now deals with the general domain of search and surveillance [7,8]. The combination of multiple sensors with their constraints is addressed by some works and in various application contexts: optimizing the performance of a sensor network [9,10]; optimizing the tasks-to-sensors affectation in the context of an intelligence collection process [11,12,13,14]. Another major issue in sensor planning is also to maximize the positive effect of subsequent data processing in regards to mission objectives. In [15], entropic-based criterion are used in order to take into account optimal post-processing of the collected information (typically data fusion). A more direct approach has also been addressed by means of Partially Observable Markov Decision Processes [16,17]. From this last point of view, sensor planning is clearly related to the domain of robotic.

---

* Corresponding author.

N.T. Nguyen, T. Van Do, and H.A. Le Thi (Eds.): *ICCSAMA 2013*, SCI 479, pp. 91–104.
DOI: 10.1007/978-3-319-00293-4_8   © Springer International Publishing Switzerland 2013

Thus, a variety of approach have been investigated for many contexts of the sensor planning. Nevertheless, there is not as much works dedicated to the question of modelling the sensor planning. In the inspiring work[18,19], Koopman addressed initially this formalisation, priorly to sensor planning problem. In [20], Le Cadre studied various practical case of use of the model of Koopman, and deduced related parametrization of the models. Whatever, it appears that a minimal effort is necessary for acquiring a good estimation of the parameters modelling our sensor planning. In the case of a reproducible scenario, it is possible to learn such parameters. However, there are cases where a prior learning of the parameters is clearly impossible. Such cases hold typically when the planning team has a limited control on the sensors, and relies on sub-processes or on sub-teams in order to implement the sensors or compute their performance parameters. Learning the parameters is generally not possible in such case, since any experiment on the sensors is a request to a sub-process, which is generally done at the expense of limited resources.

The main purpose of this paper is to handle the sensor planning as a bi-level optimization, involving:

– The improvement of the prior knowledge on the mission. This is done by planning probing experiments, which result in requests to sub-processes,
– The optimal sensor planning on the basis of the enhanced prior.

This problem is related to some issues in optimal experiment planning. Especially in [21,22], approaches (inspired from kriging) are proposed in order to plan experiments when the model of measure is known imperfectly. In such approach, the experiments are optimized in order to both enhance the measure model and the measure plan. The problem considered in this paper is somewhat different: the resources allocable for enhancing the models are distinct to the resources allocable for performing the mission.

In the first section 2 of this paper, we propose a general formalisation of the sensor planning with experiment sub-processes. In section 3, a rare-event simulation approach is proposed for solving this bi-level sensor planning. Section 4 presents a scenario and numerical results. Section 5 concludes.

2   Sensor Planning with Experiment Sub-processes

We are interested in the general problem of optimizing the planning of a sensors so as to accomplish an observation mission. A main and first issue in such optimization problem is the modelling of the formal optimization criterion and constraints. This is prerequisite to any practical sensor planning process, and it appears that the models are known with significant model noise. Two consequences are implied. First, it is not necessary to obtain an accurate optimum for a function when it is known to be noisy; smoothed criteria, derived from the expectation of the model, are much more relevant. Second, it is interesting to harvest additional information , so as to reduce this model noise. This is obtained by probing experiments, which are resources expensive. A balance has to be decided between the experiment expense and the final accuracy of the optimal sensor planning.

Our approach to this problem is formalized in section 2.2. In this paper, we model the mission criterion to be optimized by means of the function $(d, \epsilon) \mapsto f(d, \epsilon)$, which is dependent on both a decision parameter d to be optimized and on a noise parameter ϵ which encompass the uncertainty about the model. It is interesting to present as an introduction the well known *Efficient Global Optimization*, which is a reference method applicable to a sub-case of this problem.

2.1 Efficient Global Optimization

The EGO is a method for optimizing an unknown function by planning efficiently the point-evaluations of this function: a point evaluation is seen as an experiment which will enhance a modelling of the actual but unknown criterion function. EGO as introduced in [21], is based on a kriging interpolation model, with a spatial Gaussian noise, of the criterion function, which takes the form of the following functional prior:

$$f(\mathbf{d}) = \mathbf{p}(\mathbf{d})^T \mathbf{b} + Z(\mathbf{d}) \,,$$

where $\mathbf{b} = b_{1:N}$ is a model parameter (typically known with a flat prior), $\mathbf{p} = p_{1:N}$ is a predefined functional basis by which the function f is interpolated, and Z is a model spatial Gaussian law with zero-mean and a covariance $\mathrm{Cov}(Z(\mathbf{d}), Z(\mathbf{d}')) = K(\mathbf{d} - \mathbf{d}')$, which is typically dependant on a distance between the decision parameters. Being given this prior model, the estimate of the function (and of its minimizer) is computed with increasing accuracy by evaluating the real criterion function on a sequence of experimental decisions $\{\mathbf{d}_k\}$. Of course, each experiment implies a cost, and the sequence of experiment has to be optimized. Jones and al [21] proposed to optimize each step of experiment by maximizing a criterion based on the *estimated function* $\mathbf{d} \mapsto \widehat{f}(\mathbf{d})$ and the *variance of the prediction error*, $\mathbf{d} \mapsto \widehat{\sigma}(\mathbf{d})$ computed from the model and previous experimental measures. A common criterion for choosing a new experimental decision $\mathbf{d}_{k+1}$ is to maximize the *Expected Improvement* (EI), which is given by:

$$\mathbf{d}_{k+1} \in \arg \max_{\mathbf{d}} EI(\mathbf{d}) \,, \quad \text{with} \quad EI(\mathbf{d}) = \widehat{\sigma}(\mathbf{d}_{k+1})(u\Phi(u) + \phi(u)) \,,$$

where:

$$u = \frac{\min_{i=1:k} f(\mathbf{d}_i) - \widehat{f}(\mathbf{d})}{\widehat{\sigma}(\mathbf{d})} \,.$$

There has been many successful applications and extensions of the EGO algorithm during the last years [22].

In this paper, however, we will consider a different optimization scheme, in the sense that the experiment processes and the functional evaluation will not work on the same variables: we will not be able to probe the decision $\mathbf{d}_{k+1}$ directly; instead, we will request and experiment $\mathbf{r}_{k+1}$ which is not in the same space than $\mathbf{d}_{k+1}$. For solving this problem, a direct simulation-based approach will be proposed. Notice that it is probably possible to consider extensions of the EGO algorithm to the problem formalized subsequently, for example by handling variables $\mathbf{r}, \mathbf{d}$ as a joint variable, and defining a function prior on this joint variable. This approach is not considered in this paper.

2.2 Formalisation of a Direct Approach

From now on, we are studying a bi-level sensor planning, involving a first stage of model improvement by means of experiment request and a second stage of sensor planning on the basis of the corrected model. This problem is characterized by:

variables: process variables; noise variables; control variable, including decisions and experiment requests,
Known functions and parameters: noise-dependant objective function; cost function; cumulative cost bound,
Prior probabilistic laws: model noise; measure law.

Criterion and Constraints

Definition of the variables

 - $d \in D$ is a variable describing the decision of the sensor planner. This variable is intended to be optimized. The set D encompasses all the possible control decision of the planner,
 - $\epsilon \in E$ is a variable describing the error of the model. The value ϵ is obtained from a known random process, and the planner cannot control this value.
 - $r_{1:N} \in R$ are variables describing a sequence of N experiments requested by the planner. These variables are intended to be optimized, but N is assumed as a known parameter of the problem. The set R encompasses all possible experiments likely to be required by the planner. The planner does these requests before deciding for a control of the sensor. These experiments are intended to reduce the uncertainty about the noise ϵ,
 - $m_{1:N} \in M$ are variables describing a sequence of N measures resulting from the requested experiments $r_{1:N}$. The set M encompasses all possible measures.

Definition of the parameters and functions

 - $(d, \epsilon) \mapsto f(d, \epsilon)$ is the objective function to be maximized. It depends both on the decision variable and on the model error,
 - $r \mapsto \gamma(r)$ is a *positively* valued cost function. This function evaluates the cost of the experiments,
 - Γ is the cumulative cost bound. The sum of all experiment costs cannot exceed this value.

Definition of the laws

- $\epsilon \mapsto p(\epsilon)$ is the law of the error of model.
- $(m, r, \epsilon) \mapsto p(m|r, \epsilon)$ is the law of measure conditionally to the request and model error. It is assumed that the measures are obtained independently.

Criterion and Constraints

Criterion. The success of the mission is evaluated by means of the criterion function f. The purpose is to optimize the decision d so as to maximize the expected success; the expectation is computed according to the law of the model error, conditionally to the requested experiments and resulting measures:

$$\max_{d \in D} \int_{\epsilon \in E} p(\epsilon|m_{1:N}, r_{1:N}) f(d, \epsilon)\, d\epsilon \;.$$

The entire bi-level planning also involves the choice of a sequence of experiments, priorly to the mission:

$$\max_{r_{1:N} \in R} \int_{m_{1:N} \in M} p(m_{1:N}|r_{1:N}) \max_{d \in D} \int_{\epsilon \in E} p(\epsilon|m_{1:N}, r_{1:N}) f(d, \epsilon)\, d\epsilon\, dm_{1:N} \;.$$

Combining the model and measure law in a same joint law, the entire criterion is equivalently rewritten:

$$\max_{r_{1:N} \in R} \int_{m_{1:N} \in M} \max_{d \in D} \int_{\epsilon \in E} p(\epsilon, m_{1:N}|r_{1:N}) f(d, \epsilon)\, d\epsilon\, dm_{1:N} \;, \tag{1}$$

where:

$$p(\epsilon, m_{1:N}|r_{1:N}) = p(\epsilon) \prod_{n=1:N} p(m_n|r_n, \epsilon) \;. \tag{2}$$

The optimization (1) may as well be rewritten:

$$\max_{r_{1:N}} \max_{m_{1:N} \mapsto d} \int_{m_{1:N} \in M, \epsilon \in E} p(\epsilon, m_{1:N}|r_{1:N}) f(d(m_{1:N}), \epsilon)\, d\epsilon\, dm_{1:N} \;, \tag{3}$$

Constraints. The only constraint is resulting form the cumulative cost bound for the experiments:

$$\sum_{n=1:N} \gamma(r_n) \le \Gamma \;. \tag{4}$$

Bi-Level Optimization Problems. Summing up both (1) and (4), the optimization problem comes as follows:

$$\text{Solve } \arg\max_{r_{1:N}} \max_{m_{1:N} \mapsto d} \int_{m_{1:N} \in M, \epsilon \in E} p(\epsilon, m_{1:N}|r_{1:N}) f(d(m_{1:N}), \epsilon)\, d\epsilon\, dm_{1:N} \;, \tag{5}$$

$$\text{Under constraint } \sum_{n-1:N} \gamma(r_n) \le \Gamma \;. \tag{6}$$

Sometime, it is useful to reformulate this problem as an optimization on parametric laws:

$$\text{Solve} \ \underset{\pi \in \Pi}{\arg\max} \int_{m_{1:N} \in M, \epsilon \in E} p(\epsilon, m_{1:N} | r_{1:N}) \pi(d, r_{1:N} | m_{1:N}) f(d, \epsilon) \, d\epsilon \, dm_{1:N} \,, \tag{7}$$

where Π is a family of conditional laws $\pi(d, r_{1:N} | m_{1:N})$ which are compliant with constraint (4). As shown in [17], reformulations based on parametric laws are efficiently used for approximating such optimization problem. These questions are outside the scope of this paper however.

Sub-case of interest. In section 4, two scenarios are proposed where the measures are reduced to a *detection/non detection paradigm*. For convenience, it is also assumed that:

$$\Gamma = N \text{ and } \gamma = 1 \,, \tag{8}$$

so that:

$$\text{Constraint (6) is removed.} \tag{9}$$

All measures are assumed independent, so that multiple experiment will multiplicatively decrease the probability of non detection. There are two way to handle this, depending whether the measure processes are discrete or continuous.

Discrete case: In this case, each experiment $r \in R$ is related to a predicate $X_r(\epsilon)$ which may be true or false depending on the value of ϵ. Conditionally to the hypothesis that X_r is true, it is assumed that each request to experiment r will result in a positive confirmation (*i.e.* detection $\mathbf{d}$) with probability $\theta(r) \in [0, 1]$. Otherwise, the confirmation is negative (*i.e.* non detection $\mathbf{nd}$). The measure set and the measure probability are then defined as follows:

$$M = \{\mathbf{d}, \mathbf{nd}\} \text{ and } \begin{cases} p(\mathbf{d} | r, \epsilon) = \theta(r) \text{ if } X_r(\epsilon) = \textbf{true} \,. \\ p(\mathbf{d} | r, \epsilon) = 0 \text{ if } X_r(\epsilon) = \textbf{false} \,. \end{cases} \tag{10}$$

Continuous case: In this case, it is considered that the requests are implemented continuously, so that *a request takes the form of a ratio of time dedicated to an experiment*. It is defined the set K of experiments (in this case, the experiments $k \in K$ are distinguished from the requests). Instead of making N sequential requests, we will do a single, but *vectorial*, request. The (single) measure is a vector of confirmation for all possible experiment. As a consequence, N, M and R are defined as follows:

$$N = 1 \,, \tag{11}$$

$$M = \{\mathbf{d}, \mathbf{nd}\}^K \,, \tag{12}$$

$$R = \left\{ \rho_K \in \mathbb{R}^{+K} \Big/ \sum_{k \in K} c_k \rho_k = C \right\} \,, \tag{13}$$

where c_k is a cost rate for request k and C is a cumulative cost bound.

Now, an experiment $k \in K$ is related to a predicate $X_k(\epsilon)$ which may be true or false depending on the value of ϵ. Conditionally to the hypothesis that X_k is true, it is assumed that each request ρ_k to an experiment k will result in a positive confirmation (*i.e.* detection **d**) with the exponential probability $1 - \exp(-\omega_k \rho_k)$, where ω_k is a detection rate characterizing the infinitesimal probability of detection. Otherwise, the confirmation is negative (*i.e.* non detection **nd**). The measure probability is then defined as follows:

$$\text{For any } m_K \in M, \ p(m_K|\rho_K, \epsilon) = \prod_{k \in K} p_k(m_k|\rho_k, \epsilon) \qquad (14)$$

$$\text{where } \begin{cases} p_k(\mathbf{d}|\rho_k, \epsilon) = 1 - \exp(-\omega_k \rho_k) \text{ if } X_k(\epsilon) = \mathbf{true} \ . \\ p_k(\mathbf{d}|\rho_k, \epsilon) = 0 \text{ if } X_r(\epsilon) = \mathbf{false} \ . \end{cases} \ .$$

$$(15)$$

These cases of interest will be implemented in the scenarios of section 4.

3 A Rare-Event Simulation-Based Implementation

A mathematical approach for solving problem (5) and corollaries is not straightforward, and would need more refinement on the model. On the other hand, this problem is well suited to simulation approaches, especially as the optimization criterion is obtained by means of an expectation. Especially, we are interested in model-based simulation approaches, which encompass the cross-entropy method (CE) created by Rubinstein [23], or the model reference adaptive search method (MRAS) [24]. In the current stage of this work, the cross-entropy method (CE) is implemented. The MRAS method seems promising but is not considered for this paper.

3.1 The Cross-Entropy Method

It is assumed a $\mathbb{R}$-valued function $y \mapsto \varphi(y)$ to be optimized for $y \in Y$. The domain Y is probabilized. The purpose is to optimize y so as to maximize $\varphi(y)$:

$$\max_{y \in Y} \varphi(y) \ .$$

For solving this optimization, model-based simulation approaches have been proposed, based on the following general synopsis:

- Generate samples by means of a parametric distribution,
- Evaluate the quality of the samples in accordance with the criterion function,
- Update the parametrized distribution by learning from the samples graded with their quality.

Especially, the implementation of the *Cross-Entropy* method will involve the following elements:

- A sampling distributions family, $\pi(\cdot|\lambda)$ with $\lambda \in \Lambda$, which applies on variable y,
- An increasing selection function, $\sigma : \mathbb{R} \to [0,1]$,
- A smoothing prameter $\theta \in]0,1]$.

The CE algorithm for maximizing $\varphi(y)$ on the basis of π is derived as follows:

1. Initialize $\lambda \in \Lambda$,
2. Generate S samples $y_{1:S}$ by means of $\pi(\cdot|\lambda)$,
3. Compute the weighting parameters $\sigma_s = \sigma(\varphi(y_s))$ for all samples y_s,
4. Learn $\widetilde{\lambda}$, by minimizing the Kullback-Leibler divergence with the weighted samples:
$$\widetilde{\lambda} \in \arg\max_{\lambda \in \Lambda} \sum_{s=1:S} \sigma_s \ln \pi(y_s|\lambda) , \qquad (16)$$
5. Set $\lambda = \theta\lambda + (1-\theta)\widetilde{\lambda}$,
   (it is assumed that this operation makes sense in Λ)
6. Repeat from step 2 until convergence.

It is noticed that the selection function may evolve with the iteration step and the samples statistic. In the classical implementation of the CE for example, the sample selection is based on the quantiles: being given the selection rate $\rho \in]0,1[$, the $\lfloor \rho S \rfloor$ best samples are selected. In this case, the selection function is computed as follows:

- Build $\gamma \in \mathbb{R}$ and $\Sigma \subset [\![1,S]\!]$ such that:

  $\text{card}(\Sigma) = \lfloor \rho S \rfloor ,$ and $\varphi(y) \leq \gamma \leq \varphi(z)$ for any $y \in [\![1,S]\!] \setminus \Sigma$ et $z \in \Sigma$,

- Define $\sigma(\varphi) = I[\varphi \geq \gamma]$, where:

$$I[\mathbf{true}] = 1 \text{ and } I[\mathbf{false}] = 0 . \qquad (17)$$

This selection principle will be used in this work.

3.2   Implementation of the Sub-cases of Interest

The point here is to define the evaluation function, the sampling family and the learning step for the discrete case and the continuous case. The choice of the selection rate and of the smoothing parameter is not difficult in practice.

In the scenario of section 4, the decisions d are same-dimension real vectors.

Subcase of Interest: Discrete Case

Evaluation function. The evaluation function is defined by:

$$\varphi((m_{1:N} \mapsto d), r_{1:N}) = \int_{m_{1:N} \in M, \epsilon \in E} p(\epsilon, m_{1:N}|r_{1:N}) f(d(m_{1:N}), \epsilon) \, d\epsilon \, dm_{1:N} . \qquad (18)$$

By defining explicitly a function of measure $m = \mu(r, \epsilon, \nu)$, where ν is a noise of law $p(\nu)$, the function φ is equivalently rewritten:

$$\varphi((m_{1:N} \mapsto d), r_{1:N}) = \int_{\nu, \epsilon} p(\epsilon) p(\nu) f\Big(d\big(\mu(r_1, \epsilon, \nu), \cdots, \mu(r_N, \epsilon, \nu)\big), \epsilon\Big) \, d\epsilon \, d\nu \;, \tag{19}$$

which is computed by means of a Monte-Carlo simulation on the variables (ϵ, ν).

Sampling family. In our examples, the variable $((m_{1:N} \mapsto d), r_{1:N})$ is sampled by the means of the family:

$$(d, r; m) \mapsto N_d(d|\mu_m, \Sigma_m) \times \pi_r(r_{1:N}) \;, \tag{20}$$

where N_d is any multivariate Gaussian law on d (defined conditionally to m) and π_r is any discrete law defined on $r_{1:N}$. The family parameter is $\lambda = ((\mu_m, \Sigma_m)_m, \pi_r)$.

Distribution update. The optimisation (16) is easy and implies an empirical estimation of the law parameters:

$$\tilde{\pi}_r(r) = \sum_s \sigma_s I[r_s = r] \, \Big/ \, \sum_s \sigma_s \;, \tag{21}$$

$$\tilde{\mu}_m = \sum_s \sigma_s I[m_s = m] d_s \, \Big/ \, \sum_s \sigma_s I[m_s = m] \;, \tag{22}$$

$$\tilde{\Sigma}_m = \sum_s \sigma_s I[m_s = m](d_s - \tilde{\mu}_m)(d_s - \tilde{\mu}_m)^T \, \Big/ \, \sum_s \sigma_s I[m_s = m] \;. \tag{23}$$

The values (d_s, m_s, r_s) are issued from sample s.

Subcase of Interest: Continuous Case

Evaluation function. The evaluation function is defined by:

$$\varphi((m_K \mapsto d), \rho_K) = \int_{m_K \in M, \epsilon \in E} p(\epsilon, m_K | \rho_K) f\big(d(m_K), \epsilon\big) \, d\epsilon \, dm_K \;. \tag{24}$$

By defining explicitly a function of measure $m_k = \mu_k(\rho_k, \epsilon, \nu)$, where ν is a noise of law $p(\nu)$, the function φ is equivalently rewritten:

$$\varphi((m_K \mapsto d), \rho_K) = \int_{\nu, \epsilon} p(\epsilon) p(\nu) f\Big(d\big(\mu_k(\rho_k, \epsilon, \nu) | k \in K\big), \epsilon\Big) \, d\epsilon \, d\nu \;, \tag{25}$$

which is computed by means of a Monte-Carlo simulation on the variables (ϵ, ν).

Sampling family. In our examples, the variable ρ_K is derived by a bijective transform from a real vector ϱ_K of dimension card $(K)-1$. The variable $((m_K \mapsto d), \varrho_K)$ is sampled by the means of the family:

$$(d, r; m) \mapsto N_d(d|\mu_m, \Sigma_m) \times N_\varrho(\varrho|\mu_\varrho, \Sigma_\varrho) ,\qquad(26)$$

where N_d is any multivariate Gaussian law on d (defined conditionally to m) and N_ϱ is any multivariate Gaussian law on $r_{1:N}$. The family parameter is $\lambda = ((\mu_m, \Sigma_m)_m, \mu_\varrho, \Sigma_\varrho)$.

Distribution update. The optimisation (16) is easy and implies an empirical estimation of the law parameters:

$$\tilde\mu_\varrho = \sum_s \sigma_s \varrho_s \Big/ \sum_s \sigma_s ,\qquad(27)$$

$$\tilde\Sigma_\varrho = \sum_s \sigma_s(\varrho_s - \tilde\mu_\varrho)(\varrho_s - \tilde\mu_\varrho)^T \Big/ \sum_s \sigma_s ,\qquad(28)$$

$$\tilde\mu_m = \sum_s \sigma_s I[m_s = m]d_s \Big/ \sum_s \sigma_s I[m_s = m] ,\qquad(29)$$

$$\tilde\Sigma_m = \sum_s \sigma_s I[m_s = m](d_s - \tilde\mu_m)(d_s - \tilde\mu_m)^T \Big/ \sum_s \sigma_s I[m_s = m] .\qquad(30)$$

The values (d_s, m_s, ϱ_s) are issued from sample s.

4   Scenario and Numerical Results

4.1   Scenario

The mission is to intercept a target (symbolized by a smiley on picture 4), which is hidden within the theatre. In order intercept this target, the planer has to position a patrol as close as possible to the target. Then, this patrol will proceed to the search of the target and to its interception.

At the early beginning of the mission, the position of the target is known with uncertainty, and this uncertainty is characterized by means of a Gaussian distribution. In order to enhance this prior knowledge, the planner may request some teams, which will collect information in the neighbourhood about the target, and if it is in the neighbourhood, it will assert the presence of the target with a given probability.

The problem is then to:

1. Select the teams to request,
2. Plan the patrol in regards to the earned information.

Sensors are positioned regularly on a grid. The position, range and detection probabilities of the team are indicated in picture 1 and 2. These pictures respect the relative dimension of these parameters. These parameters, position, range (R), detection probability, are given subsequently in this order:

$(-1,1)$,	$R=1$,	0.3	$(0,1)$,	$R=1$,	0.5	$(1,1)$,	$R=1.5$,	0.25
$(-1,0)$,	$R=0.9$,	0.8	$(0,0)$,	$R=0.5$,	0.5	$(1,0)$,	$R=1$,	0.25
$(-1,-1)$,	$R=1.2$,	0.3	$(0,-1)$,	$R=1$,	0.8	$(1,-1)$,	$R=1$,	0.1

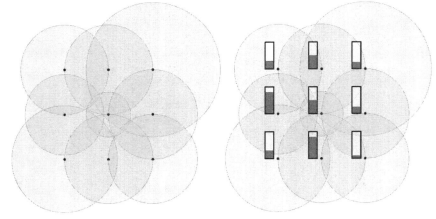

Sensors grid with range / detection probability

The target is known with a Gaussian uncertainty with mean μ_T and covariance Σ_T:

$$\mu_T = \begin{pmatrix} 0.5 \\ 0.25 \end{pmatrix} \quad \text{and} \quad \Sigma_T = \begin{pmatrix} 2 & 1 \\ 1 & 1 \end{pmatrix}$$

The target uncertainty is indicated in picture 3. The evaluation criterion of the mission is the *estimated distance* between the patrol and the target, as indicated in picture 4.

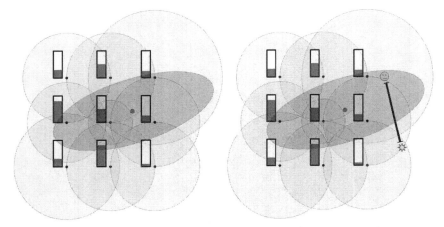

Target uncertainty / criterion: distance(target,sensor)

The parameters for the CE optimization are $\rho = \alpha = 0.15$ and the number of samples $S = 100$. The Monte-Carlo expectation is computed by means of 1000 particles. For the subsequent examples, the convergence is considered achieved after 100 to 200 iterations.

4.2   Results

Test 1 and test 2. Test 1 and test 2 are both about discrete requests of experiment. In test 1, however, only 1 request is done, while 8 are done in test 2.

Picture 5 indicate the result of the planning for test 1. Team 4 (in green) is requested and it is shown the decided patrol positioning: this position depends on detection (moon) or non-detection (sun). These results are compliant with the setting of the problem.

Picture 6 indicate the result of the planning for test 2. Teams 4 ($2\times$), 5 ($1\times$), 7 ($3\times$), 8 ($1\times$) and 9 ($1\times$) (in green) are requested. It is not possible to give here the patrol positioning, since there are actually 32 possible cases. Again, these results are compliant with the setting of the problem.

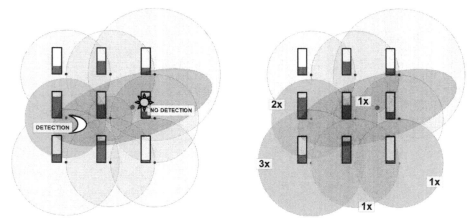

Planning with 1 request / Planning with 8 request

Test 3. This test implements the subcase with continuous requests of experiments. The scenario parameters are identical, with 9 possible requests, $K = \{1 : 9\}$. In addition, the cumulative cost bound is $C = 10$, and the cost rates c_K and detection rates ω_K are given by the subsequent table:

$c_1 = 5,\quad \omega_1 = 0.36$	$c_2 = 25,\quad \omega_2 = 0.69$	$c_3 = 10,\quad \omega_3 = 0.29$
$c_4 = 10,\quad \omega_4 = 1.61$	$c_5 = 30,\quad \omega_5 = 0.69$	$c_6 = 2,\quad \omega_6 = 0.29$
$c_7 = 10,\quad \omega_7 = 0.36$	$c_8 = 20,\quad \omega_8 = 1.61$	$c_9 = 5,\quad \omega_9 = 0.11$

As a result, the following table indicates the optimized efforts ρ_K :

$\rho_1 = 0.24$	$\rho_2 = 0.03$	$\rho_3 = 0.09$
$\rho_4 = 0.58$	$\rho_5 = 0.01$	$\rho_6 = 0.10$
$\rho_7 = 0.02$	$\rho_8 = 0.03$	$\rho_9 = 0.03$

Interpretation of these last results is not so easy, although it is noticed that the optimization does the balance between the cost and detection rates.

5 Conclusion

In this paper we considered a bi-level optimization problem consisting in a first experiment request stage and in a final mission optimization stage. The first stage is dedicated to the improvement of the prior model, which is known with parameter uncertainty and condition the main objective. This problem is related to the domain of experiment plan optimization. We propose an original formalization and optimization method for this problem. Our solving approach is based on simulation methods. Our algorithm has been tested on a target search and interception scenario. The result is promising.

References

1. Koopman, B.O.: The theory of search. iii. the optimum distribution of searching effort. Operations Research 5(5), 613–626 (1957)
2. de Guenin, J.: Optimum distribution of effort: An extension of the Koopman basic theory. Operations Research 9, 1–7 (1961)
3. Washburn, A.R.: Search for a moving target: The FAB algorithm. Operations Research 31(4), 739–751 (1983)
4. Brown, S.S.: Optimal search for a moving target in discrete time and space. Operations Research 28(6), 1275–1289 (1980)
5. Iida, K., Hohzaki, R., Furui, S.: A search game for a mobile target with the conditionally deterministic motion defined by paths. Journal of the Operations Research of Japan 39(4), 501–511 (1996)
6. Dambreville, F., Le Cadre, J.-P.: Search game for a moving target with dynamically generated informations. In: Int. Conf. on Information Fusion (Fusion 2002), Annapolis, Maryland, pp. 243–250 (July 2002)
7. Frost, J.R.: Principle of search theory. Technical report, Soza & Company Ltd. (1999)
8. Haley, K.B., Stone, L.D.: Search Theory and Applications. Plenum Press, New York (1980)
9. Chakrabarty, K., Iyengar, S.S., Qi, H., Cho, E.: Grid coverage for surveillance and target location in distributed sensor networks. IEEE Transactions on Computers 51, 1448–1453 (2002)
10. Jayaweera, S.K.: Optimal node placement in decision fusion wireless sensor networks for distributed detection of a randomly-located target. In: IEEE Military Communications Conference, pp. 1–6 (2007)
11. Le Thi, H.A., Nguyen, D.M., Pham, D.T.: A DC programming approach for planning a multisensor multizones search for a target. Computers & Operations Research, Online first (July 2012)
12. Simonin, C., Le Cadre, J.-P., Dambreville, F.: A hierarchical approach for planning a multisensor multizone search for a moving target. Computers and Operations Research 36(7), 2179–2192 (2009)
13. Janez, F.: Optimization method for sensor planning. Aerospace Science and Technologie 11, 310–316 (2007)

14. Nguyen, D.M., Dambreville, F., Toumi, A., Cexus, J.C., Khenchaf, A.: A column generation method for solving the sensor management in an information collection process. Submitted to Optimization (October 2012)
15. Céleste, F., Dambreville, F., Le Cadre, J.-P.: Optimized trajectories for mobile robot with map uncertainty. In: IFAC Symp. on System Identification (SYSID 2009), Saint-Malo, France, pp. 1475–1480 (July 2009)
16. Tremois, O., Le Cadre, J.-P.: Optimal observer trajectory in bearings-only tracking for maneuvering sources. Sonar and Navigation 146(1), 1242–1257 (1997)
17. Dambreville, F.: Cross-entropic learning of a machine for the decision in a partially observable universe. Journal of Global Optimization 37, 541–555 (2007)
18. Koopman, B.O.: The theory of search, part i. kinematic bases. Operations Research 4(5), 324–346 (1956)
19. Koopman, B.O.: The theory of search, part ii. target detection. Operations Research 4(5), 503–531 (1956)
20. Le Cadre, J.-P.: Approximations de la probabilité de détection d'une cible mobile. In: Actes du Colloque GRETSI, Toulouse (September 2001)
21. Jones, D.R., Schonlau, M.J., Welch, W.J.: Efficient global optimization of expensive black-box function. J. Glob. Optim. 13(4), 455–492 (1998)
22. Marzat, J., Walter, E., Piet-Lahanier, H.: Worst-case global optimization of black-box functions through Kriging and relaxation. J. Glob. Optim. (2012)
23. De Boer, P.T., Kroese, D.P., Mannor, S., Rubinstein, R.Y.: A tutorial on the cross-entropy method. Annals of Operations Research 134 (2002)
24. Hu, J., Fu, M.C., Marjus, S.I.: A model Reference Adaptive Search Method for Global Optimization. Oper. Res. 55, 549–568 (2007, 2008)

Large Scale Image Classification
with Many Classes, Multi-features
and Very High-Dimensional Signatures

Thanh-Nghi Doan[1], Thanh-Nghi Do[3], and François Poulet[1,2]

[1] IRISA,[2] Université de Rennes I, Campus de Beaulieu, 35042 Rennes Cedex, France
{thanh-nghi.doan,francois.poulet}@irisa.fr
[3] Institut Telecom; Telecom Bretagne UMR CNRS 3192 Lab-STICC,
Université européenne de Bretagne, France, Can Tho University, Vietnam
tn.do@telecom-bretagne.eu

Abstract. The usual frameworks for image classification involve three steps: extracting features, building codebook and encoding features, and training the classifiers with a standard classification algorithm. However, the task complexity becomes very large when performing on a large dataset ImageNet [1] containing more than 14M images and 21K classes. The complexity is about the time needed to perform each task and the memory. In this paper, we propose an efficient framework for large scale image classification. We extend LIBLINEAR developed by Rong-En Fan [2] in two ways: (1) The first one is to build the balanced bagging classifiers with under-sampling strategy. Our algorithm avoids training on full data, and the training process rapidly converges to the solution, (2) The second one is to parallelize the training process of all classifiers with a multi-core computer. The evaluation on the 100 largest classes of ImageNet shows that our approach is 10 times faster than the original LIBLINEAR, 157 times faster than our parallel version of LIBSVM and 690 times faster than OCAS [3]. Furthermore, a lot of information is lost in quantization step and the obtained bag-of-words is not enough discriminative power for classification. Therefore, we propose a novel approach using several local descriptors simultaneously.

Keywords: Large Scale Visual Classification, High Performance Computing, Balanced Bagging, Parallel Support Vector Machines.

1 Introduction

Image classification is one of the important research topics in the areas of computer vision and machine learning. Local image features and bag-of-words model (BoW) is the core of state-of-the-art image classification systems. The usual image classification frameworks involve tree steps: 1) extracting features, 2) building codebook and encoding features, and 3) training classifiers. Step 1 is to extract local image features: the recent popular choices are SIFT [4], SURF [5],

N.T. Nguyen, T. Van Do, and H.A. Le Thi (Eds.): *ICCSAMA 2013*, SCI 479, pp. 105–116.
DOI: 10.1007/978-3-319-00293-4_9   © Springer International Publishing Switzerland 2013

and dense SIFT (DSIFT) [6]. Step 2 is to build codebook and encode features: k-means clustering algorithm is the popular choice to build codebook, BoW model is used to encode features. Step 3 is to train classifiers: many systems choose either non-linear or linear kernel SVMs. All these frameworks are evaluated on small datasets, e.g. Caltech 101 [7], Caltech 256 [8], and PASCAL VOC [9] that can fit into desktop memory. However, the emergence of ImageNet makes the complexity of image classification become very large. This challenge motivates us to study an efficient framework in both computation time and classification accuracy. In this paper, we show how to address this challenge and achieve good results over the usual frameworks. Our key contributions include:

1. Develop a balanced bagging algorithm for training the binary classifiers of LIBLINEAR. Our algorithm avoids training on full dataset, and the training process rapidly converges to the optimal solution.

2. Parallelize the training process of the binary classifiers of LINLINEAR based on high performance computing models. In the training step, we apply our balanced bagging algorithm to get the best performance.

3. Propose a novel approach using different feature types and show how to combine them by using multi-feature and multi-codebook approach.

2   Related Work

Large Scale Image Classification: Many previous works on image classification have relied on BoW model [10], local feature quantization, and support vector machines. This model can be enhanced by multi-scale spatial pyramids (SPM) [11] on BoW. Fergus *et al.* [12] study semi-supervised learning on 126 hand labeled Tiny Images categories, Wang *et al.* [13] show classification on a maximum of 315 categories. Li *et al.* [14] do research with landmark classification on a collection of 500 landmarks and 2 million images. On a small subset of 10 classes, they could improve BoW classification by increasing the visual vocabulary up to 80K visual words. To make large scale learning more practical, many researchers are beginning to study strategies where the data is first transformed by a nonlinear mapping induced by a particular kernel and then efficient linear classifiers are trained in the resulting space [15]. However, the state-of-the-art techniques [16], [9] need a lot of time and thus it is difficult to scale-up to large datasets. The difference between our work and previous studies is related to take into account parallel algorithms to speedup two processes: extracting features and training classifiers.

Image Signature: Computing the signature of an image based on BoW model includes three steps: 1) feature detection, 2) feature description, and 3) codebook generation. Recent works have studied these steps and achieved impressive improvements. However, in each processing step there exist a significant amount of lost information, and the resulting visual-words are not enough discriminative for image classification. Many approaches have been proposed to solve this challenge. In the feature detection step, multiple local features are grouped to obtain a more global and discriminative feature. In the feature description step, high-dimensional descriptors enhanced by other information have been studied to get more image information [17]. In the codebook generation step, many approaches have proposed efficient quantizers or codebooks that reduce quantization errors and preserve more information of feature descriptors [18]. We have a more general view for all these steps and propose a novel approach that combines both multi-feature and multi-codebook approach to construct the final image signature. Our approach aims to increase the discriminative power of image signatures by embedding more useful information from the original image features. In multi-feature and multi-codebook approach, first BoW of images for each feature channel is constructed based on their corresponding codebook. The result is a bag-of-BoW for all feature types and we call it a bag-of-visual packets or a bag-of-packets. Finally, all BoW in the bag-of-packets are concatenated to form the image signature, as shown in Fig. 1. Our approach is novel in the way the image signature is constructed, that improves the discriminative power of image signature. This is the major difference between our approach and previous studies.

3   Classifiers

The first step of our framework is to extract SIFT, SURF and DSIFT. These features have been proven to be successful in vision tasks. Before training classifiers, we use multi-feature and multi-codebook approach to construct image signatures.

3.1   Multi-feature and Multi-codebook

The high intraclass variability of images in the same class of ImageNet is a real challenge for image classification systems. Many previous works try to design a robust image feature which is invariant to image transformation, illumination and scale changes. There are some improvements when using robust features, but it is clear that none of feature descriptors have the same discriminative power for all classes. For instance, the features based on shape information might be useful when classifying photos with the same geometric direction. However, it will not be sufficient when the images are rotated or the objects are taken a shot in different camera angles. In this case, the appropriate choice should be the features based on interesting keypoints (e.g. SIFT). Obviously, instead of using a single feature type for all classes we can combine many different feature types to get higher classification accuracy. In this section, we present a novel multi-feature and multi-codebook approach and show how to combine these features.

Let a set of all different feature descriptor types extracted from an image i be $F = \{f_i^j\}$, where f_i^j are the descriptors of feature type j extracted from image i, M is the number of feature types, and $j = 1, .., M$. In our approach BoW histograms of each feature type are constructed based on their corresponding codebook, as shown in Fig. 1. Instead of using a single codebook for constructing the final image signature, we use multiple codebooks $\{C^1, C^2, ..., C^M\}$ that are

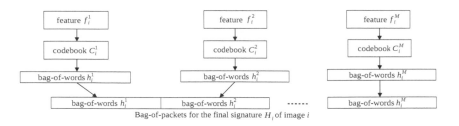

Fig. 1. Construct bag-of-packets based on multi-feature and multi-codebook approach

built from different feature types. More specifically, the codebook C^j is used
to construct BoW histogram h_i^j for feature descriptors $f_i^j \in F$. Then all BoW
histograms h_i^j are concatenated to form the final image signature H_i. As a result,
for each image i, we obtain H_i with M elements $H_i = \{h_i^1, h_i^2, ..., h_i^M\}$. For
simplicity, we call H_i a "bag-of-packets" (BoP) that is the final image signature
constructed based on different codebooks of the original image i. A BoP is more
discriminative than an usual BoW because two BoP H_i and H_j are considered
identical if and only if their corresponding BoW are identical. Formally, it takes
the intersection of the BoW elements from multiple features:

$$(H_i = H_j) \equiv (h_i^1 = h_j^1) \wedge (h_i^2 = h_j^2) \wedge ... \wedge (h_i^M = h_j^M) \qquad (1)$$

3.2 LIBLINEAR Support Vector Machines

Let us consider a binary linear classification task with m datapoints in a n-dimensional input space $x_1, x_2, ..., x_m$ having corresponding labels $y_i = \pm 1$.
SVM classification algorithm [19] aims to find the best separating surface as
being furthest from both classes. It can simultaneously maximize the margin
between the support planes for each class and minimize the error. This can be
performed through the quadratic program (2).

$$min_\alpha (1/2) \sum_{i=1}^{m} \sum_{j=1}^{m} y_i y_j \alpha_i \alpha_j \langle x_i \cdot x_j \rangle - \sum_{i=1}^{m} \alpha_i$$

$$s.t. \begin{cases} \sum_{i=1}^{m} y_i \alpha_i = 0 \\ 0 \le \alpha_i \le C \quad \forall i = 1, 2, ..., m \end{cases} \qquad (2)$$

where C is a positive constant used to tune the margin and the error.

The support vectors (for which $\alpha_i > 0$) are given by the solution of the
quadratic program (2), and then, the separating surface and the scalar b are
determined by the support vectors. The classification of a new data point x is
based on:

$$sign(\sum_{i=1}^{\#SV} y_i \alpha_i \langle x \cdot x_i \rangle - b) \qquad (3)$$

Variations on SVM algorithms use different classification functions. No algorithmic changes are required from the usual linear inner product $\langle x \cdot x_i \rangle$ to kernel function $K\langle x \cdot x_i \rangle$ other than the modification of the kernel evaluation, including a polynomial function of degree d, a RBF (Radial Basis Function) or a sigmoid function. We can get different support vector classification models.

LIBLINEAR proposed by [2] uses a dual coordinate descent method for dealing with linear SVM using L1- and L2-loss functions. And then, LIBLINEAR is simple and reaches an ϵ-accurate solution in $O(\log(1/\epsilon))$ iterations. The algorithm is much faster than state of the art solvers such as LibSVM [20] or SVMperf [21].

3.3   Improving LIBLINEAR for Large Number of Classes

Most SVM algorithms are only able to deal with a two-class problem. There are several extensions of binary classification SVM solver to multi-class (k classes, $k \geq 3$) classification tasks. The state-of-the-art multi-class SVMs are categorized into two types of approaches. The first one is considering the multi-class case in one optimization problem [22], [23]. The second one is decomposing multi-class into a series of binary SVMs, including one-versus-all [19], one-versus-one [24] and Decision Directed Acyclic Graph [25]. Recently, hierarchical methods for multi-class SVM [26], [27] start from the whole data set, hierarchically divide the data into two subsets until every subset consists of only one class.

In practice, one-versus-all, one-versus-one are the most popular methods due to their simplicity. Let us consider k classes ($k > 2$). The one-versus-all strategy builds k different classifiers where the ith classifier separates the ith class from the rest. The one-versus-one strategy constructs $k(k-1)/2$ classifiers, using all the binary pairwise combinations of the k classes. The class is then predicted with a majority vote.

When dealing with very large number of classes, e.g. hundreds of classes, the one-versus-one strategy is too expensive because it needs to train many thousands of classifiers. Therefore, the one-versus-all strategy becomes popular in this case. LIBLINEAR SVM algorithm also uses the one-versus-all approach to train independently k binary classifiers. However, the current LIBLINEAR SVM needs very long time to classify very large number of classes.

Due to this problem, we propose two ways for speed-up learning tasks of LIBLINEAR SVM. The first one is to build the balanced bagging classifiers with sampling strategy. The second one is to parallelize the training task of all classifiers with multi-core computers.

Balanced Bagging LIBLINEAR. In the one-versus-all approach, the learning task of LIBLINEAR SVM is try to separate the ith class (positive class) from the $k - 1$ other classes (negative class). For very large number of classes, e.g. 100 classes, this leads to the extreme unbalance between the positive class and the negative class. The problem is well-known as the class imbalance. As summarized by the review papers [28], [29], [30] and the very comprehensive papers [31], [32], solutions to the class imbalance problems were proposed both at the data and algorithmic level. At the data level, these algorithms change the class distribution, including over-sampling the minority class or under-sampling the majority class. At the algorithmic level, the solution is to re-balance the error rate by weighting each type of error with the corresponding cost. Our balanced bagging LIBLINEAR SVM belongs to the first approach (forms of re-sampling). Furthermore, the class prior probabilities in this context are highly unequal (e.g. the distribution of the positive class is 1% in the 100 classes classification problem), and over-sampling the minority class is very expensive. We propose the balanced bagging LIBLINEAR SVM using under-sampling the majority class (negative class).

For separating the ith class (positive class) from the rest (negative class), the balanced bagging LIBLINEAR SVM trains T models as shown in algorithm 1.

Algorithm 1. Balanced bagging LIBLINEAR SVM

    input :
               D_p the training data of the positive class
               D_n the training data of the negative class
               T the number of base learners
    output:
               LIBLINEAR SVM model

    Learn:
    for $k \leftarrow 1$ **to** T **do**
        1. The subset D'_n is created by sampling without replacement $|D'_n|$ negative
           datapoints from D_n (with $|D'_n| = |D_p|$)
        2. Build a LIBLINEAR SVM model using the training set (including D_p
        and D'_n)
    end
    combine T models into the aggregated **LIBLINEAR SVM model**

We remark that the margin can be seen as the minimum distance between two convex hulls, H_p of the positive class and H_n of the negative class (the farthest distance between the two classes). Under-sampling the negative class (D'_n) done by balanced bagging can increase the minimum distance between H_p and H'_n (the reduced convex hull of H_n). It can be easier to achieve the largest margin than learning on the full dataset. Therefore, the training task of LIBLINEAR SVM is fast to converge to the solution. According to our experiments, by setting $T = \sqrt{\frac{|D_n|}{|D_p|}}$, the balanced bagging LIBLINEAR SVM achieves good results in very fast training speed.

Parallel LIBLINEAR Training. Although LIBLINEAR SVM and balanced bagging LIBLINEAR SVM deal with very large dataset with high speed, they do not take into account the benefits of high performance computing, e.g. multi-core computers. Furthermore, both LIBLINEAR SVM and balanced bagging

LIBLINEAR SVM train independently k binary classifiers for k classes problems. This is a nice property for parallel learning. Our investigation aims to speed-up training tasks of multi-class LIBLINEAR SVM, balanced bagging LIBLINEAR SVM with multi-processor computers. The idea is to learn k binary classifiers in parallel.

Algorithm 2. OpenMP parallel LIBLINEAR SVM

---

input :
> D the training dataset with k classes

output:
> **LIBLINEAR SVM model**

Learn:

#pragma omp parallel for
for $i \leftarrow 1$ **to** k **do**
> Build a binary LIBLINEAR SVM model using the training set D to separate the positive class i^{th} from the rest.

end

The parallel programming is currently based on two major models, Message Passing Interface (MPI) [33] and Open Multiprocessing (OpenMP) [34]. MPI is a standardized and portable message-passing mechanism for distributed memory systems. MPI remains the dominant model (high performance, scalability, and portability) used in high-performance computing today. However, a MPI process loads the whole dataset ($\sim$ 20GB) into memory during learning tasks, making it intractable. The simplest development of parallel LIBLINEAR SVM algorithms is based on the shared memory multiprocessing programming model OpenMP. The parallel LIBLINEAR SVM algorithm is described in algorithm 2.

4   Experiments and Results

Our approach is evaluated on the 100 largest classes of ImageNet. We sample 50% images for training and 50% images for testing.

4.1   Parallel Extracting Feature and Constructing Bag-of-packets

We perform experiments on an Intel(R) Xeon(R), 2.67GHz computer. Depending on parameters setting, extracting time of features (e.g. SIFT) of an image ranges from 0.46 to 1 second. To process the 100 largest classes, it would take from 1 to 2 days. Therefore, it is difficult to scale-up to the full ImageNet because if it takes 1 second per image for extracting features then we need 14M x 1 second $\simeq$ 162 days. To reduce extracting time, we apply parallel algorithms.

SIFT/DSIFT. VLFeat, a free version for extracting SIFTs, can be downloaded from the author's homepage (www.vlfeat.org). We use 8 CPU cores to extract

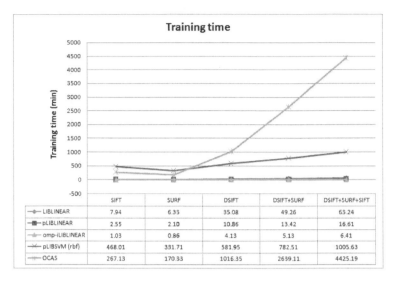

Fig. 2. Training time of SVM classifiers

features in a parallel way. We need 3 hours 30 minutes to extract more than 3 billions DSIFTs from the training dataset. That means it takes $\simeq$ 0.14 second to extract features from an image. Therefore, with the full ImageNet, it would take 0.14s x 14M $\simeq$ 22 days.

Parallel SURF. Parallel SURF is a fast parallel version of SURF maintained by David Gossow [35]. We also use 8 CPU cores to extract features. We need 3 hours 18 minutes to extract more than 72 millions SURFs from the training dataset. That means it takes $\simeq$ 0.13 second to extract features from an image. Therefore, with the full ImageNet, it would take 0.13s x 14M $\simeq$ 21 days.

Parallel Constructing Bag-of-packets. In BoW model, one of the steps that takes a long time is to build codebook. With a large dataset we need to get a large amount of datapoints to build a discriminative codebook, so this task becomes very large in time complexity. One of the popular choices is k-means algorithm. The original implementation of k-means takes many days to convergence, so we use parallel k-means of Wei Dong (http://www.cs.princeton.edu/~wdong/-kmeans). When we use n codebooks for constructing BoP of images, it means we need n more times to finish. To reduce the computation time, we perform this process in a parallel way. Consequently, the total time is the same as the largest individual standard approach.

4.2   Training Time and Classification Accuracy

The linear kernel on the classical histogram based feature gives very poor accuracy on image classification. Therefore, once BoW histograms are constructed, some recent image classification systems use feature map to convert BoW his-

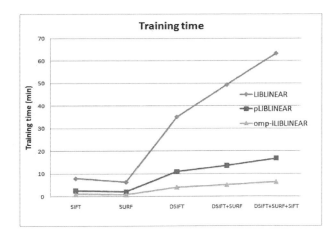

Fig. 3. Training time of linear SVM classifiers

tograms from initial space to higher-dimensional space. This step is useful when one want to stick to the efficient linear classifiers [36]. The result is the image signatures in high-dimension space that ensure non-linear separability of the classes. Notice that before training classifiers, we should normalize BoW histograms, so that the image size does not influence histogram counts. In the experiments we use L1-Norm to normalize BoW histograms and then convert them to higher-dimensional space by using homogeneous kernel map from [16]. In this section we want to compare the performance of two parallel versions of LIBLINEAR with the original one, LIBSVM and OCAS in terms of training time and classification accuracy. The training time of LIBSVM is too high, so we develop a parallel version of LIBSVM (pLIBSVM) by parallelizing the task of computing kernel values in the matrices of various formulations. This allow us train pLIBSVM on a large dataset in reasonable time. In the experiments, we use RBF kernel to train pLIBSVM classifiers and use 8 CPU cores on the same computer as in section 4.1.

OpenMP LIBLINEAR. To evaluate the performance of OpenMP version of LIBLINEAR (omp-LIBLINEAR), we compare it with LIBLINEAR, OCAS, and pLIBSVM. In terms of training time, omp-LIBLINEAR achieves a significant speedup in training process with 8 OpenMP threads. As shown in Fig. 2, in the case of combination of 3 feature types, our implementation is 3.8 times faster than the original LIBLINEAR, 61 times faster than pLIBSVM and 266 times faster than OCAS. Furthermore, LIBLINEAR is comparable with pLIBSVM and OCAS in terms of classification accuracy, as shown in Fig. 4.

OpenMP Balanced Bagging LIBLINEAR. By applying the balanced bagging algorithm to OpenMP version of LIBLINEAR (omp-iLIBLINEAR), we also significantly speedup the training process. As shown in Fig. 2 and 3, in the case of combination of 3 feature types and with 8 OpenMP threads, omp-iLIBLINEAR

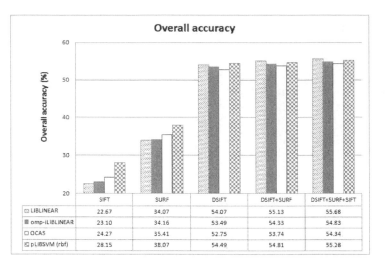

Fig. 4. Overall accuracy of SVM classifiers. The final image signature is converted to high-dimensional space.

is 10 times faster than LIBLINEAR, 157 times faster than pLIBSVM and 690 times faster than OCAS. Although omp-iLIBLINEAR run much faster than the original one and other SVMs, it does not (or very few) compromise classification accuracy, as shown in Fig. 4. This result confirms that our approach has a great ability to scaleup to full ImageNet dataset.

Multi-feature and Multi-codebook. To evaluate the performance of multi-feature and multi-codebook approach, we conduct the experiments for each single feature SIFT, SURF and DSIFT. Then we perform classification by using simultaneously different feature types DSIFT+SURF and DSIFT+SURF+SIFT. As shown in Fig. 4, in the case of training LIBLINEAR on the combination of 3 feature types, we significantly improve the performance of overall classification accuracy to +33.01%, compared to a single feature type SIFT (this is a relative improvement of more than 145%).

5   Conclusion and Future Work

We have proposed an efficient framework for large scale image classification. In this framework, we have developed two parallel versions of LIBLINEAR. The first one is to build the balanced bagging classifiers with under-sampling strategy. The second one is to parallelize the training process of all classifiers on a multi-core computer. In parallel versions we have applied our balanced bagging algorithm to obtain the best performance. We have also presented a novel approach using several different local features simultaneously to improve the classification accuracy.

Our approach has been evaluated on the 100 largest classes of ImageNet. By setting the number of threads to 8 on our computer, we achieves a significant speedup in training time without (or very few) compromise the classification accuracy. Our implementation is 10 times faster than the original LIBLINEAR, 157 times faster than pLIBSVM and 690 times faster than OCAS. It is a roadmap towards very large scale visual classification. In the future, we plan to develop a hybrid MPI/OpenMP for LIBLINEAR and study how to combine global features with local features to get more discriminative power of image signatures. That will be a promising research for large scale image classification.

References

1. Deng, J., Dong, W., Socher, R., Li, L.J., Li, K., Li, F.F.: Imagenet: A large-scale hierarchical image database. In: IEEE Computer Society Conference on Computer Vision and Pattern Recognition, pp. 248–255 (2009)
2. Fan, R.E., Chang, K.W., Hsieh, C.J., Wang, X.R., Lin, C.J.: Liblinear: A library for large linear classification. Journal of Machine Learning Research 9, 1871–1874 (2008)
3. Franc, V., Sonnenburg, S.: Optimized cutting plane algorithm for support vector machines. In: International Conference on Machine Learning, pp. 320–327 (2008)
4. Lowe, D.G.: Distinctive image features from scale-invariant keypoints. International Journal of Computer Vision 60(2), 91–110 (2004)
5. Bay, H., Ess, A., Tuytelaars, T., Gool, L.J.V.: Speeded-up robust features (surf). Computer Vision and Image Understanding 110(3), 346–359 (2008)
6. Bosch, A., Zisserman, A., Muñoz, X.: Image classification using random forests and ferns. In: International Conference on Computer Vision, pp. 1–8 (2007)
7. Li, F.F., Fergus, R., Perona, P.: Learning generative visual models from few training examples: An incremental bayesian approach tested on 101 object categories. Computer Vision and Image Understanding 106(1), 59–70 (2007)
8. Griffin, G., Holub, A., Perona, P.: Caltech-256 Object Category Dataset. Technical Report CNS-TR 2007-001, California Institute of Technology (2007)
9. Everingham, M., Van Gool, L., Williams, C.K.I., Winn, J., Zisserman, A.: The pascal visual object classes (voc) challenge. International Journal of Computer Vision 88(2), 303–338 (2010)
10. Csurka, G., Dance, C.R., Fan, L., Willamowski, J., Bray, C.: Visual categorization with bags of keypoints. In: Workshop on Statistical Learning in Computer Vision, ECCV, pp. 1–22 (2004)
11. Lazebnik, S., Schmid, C., Ponce, J.: Beyond bags of features: Spatial pyramid matching for recognizing natural scene categories. In: IEEE Computer Society Conference on Computer Vision and Pattern Recognition, pp. 2169–2178 (2006)
12. Fergus, R., Weiss, Y., Torralba, A.: Semi-supervised learning in gigantic image collections. In: Advances in Neural Information Processing Systems, pp. 522–530 (2009)
13. Wang, C., Yan, S., Zhang, H.J.: Large scale natural image classification by sparsity exploration. In: Proceedings of the IEEE International Conference on Acoustics, Speech, and Signal Processing, pp. 3709–3712. IEEE (2009)
14. Li, Y., Crandall, D.J., Huttenlocher, D.P.: Landmark classification in large-scale image collections. In: IEEE 12th International Conference on Computer Vision, pp. 1957–1964. IEEE (2009)
15. Deng, J., Berg, A.C., Li, K., Fei-Fei, L.: What does classifying more than 10,000 image categories tell us? In: Daniilidis, K., Maragos, P., Paragios, N. (eds.) ECCV 2010, Part V. LNCS, vol. 6315, pp. 71–84. Springer, Heidelberg (2010)

16. Vedaldi, A., Gulshan, V., Varma, M., Zisserman, A.: Multiple kernels for object detection. In: IEEE 12th International Conference on Computer Vision, pp. 606–613. IEEE (2009)
17. Winder, S.A.J., Brown, M.: Learning local image descriptors. In: CVPR (2007)
18. Philbin, J., Chum, O., Isard, M., Sivic, J., Zisserman, A.: Lost in quantization: Improving particular object retrieval in large scale image databases. In: IEEE Conference on Computer Vision and Pattern Recognition (2008)
19. Vapnik, V.: The Nature of Statistical Learning Theory. Springer (1995)
20. Chang, C.C., Lin, C.J.: LIBSVM – a library for support vector machines (2001), http://www.csie.ntu.edu.tw/$\sim$cjlin/libsvm
21. Joachims, T.: Training linear svms in linear time. In: Proc. of the ACM SIGKDD Intl. Conf. on KDD, pp. 217–226. ACM (2006)
22. Weston, J., Watkins, C.: Support vector machines for multi-class pattern recognition. In: Proceedings of the Seventh European Symposium on Artificial Neural Networks, pp. 219–224 (1999)
23. Guermeur, Y.: Svm multiclasses, théorie et applications (2007)
24. Krebel, U.: Pairwise classification and support vector machines. In: Advances in Kernel Methods: Support Vector Learning, pp. 255–268 (1999)
25. Platt, J., Cristianini, N., Shawe-Taylor, J.: Large margin dags for multiclass classification. In: Advances in Neural Information Processing Systems, vol. 12, pp. 547–553 (2000)
26. Vural, V., Dy, J.: A hierarchical method for multi-class support vector machines. In: Proceedings of the Twenty-First International Conference on Machine Learning, pp. 831–838 (2004)
27. Benabdeslem, K., Bennani, Y.: Dendogram-based svm for multi-class classification. Journal of Computing and Information Technology 14(4), 283–289 (2006)
28. Japkowicz, N. (ed.): AAAI'Workshop on Learning from Imbalanced Data Sets. Number WS-00-05 in AAAI Tech Report (2000)
29. Weiss, G.M., Provost, F.: Learning when training data are costly: The effect of class distribution on tree induction. Journal of Artificial Intelligence Research 19, 315–354 (2003)
30. Visa, S., Ralescu, A.: Issues in mining imbalanced data sets - A review paper. In: Midwest Artificial Intelligence and Cognitive Science Conf., Dayton, USA, pp. 67–73 (2005)
31. Lenca, P., Lallich, S., Do, T.-N., Pham, N.-K.: A Comparison of Different Off-Centered Entropies to Deal with Class Imbalance for Decision Trees. In: Washio, T., Suzuki, E., Ting, K.M., Inokuchi, A. (eds.) PAKDD 2008. LNCS (LNAI), vol. 5012, pp. 634–643. Springer, Heidelberg (2008)
32. Pham, N.K., Do, T.N., Lenca, P., Lallich, S.: Using local node information in decision trees: coupling a local decision rule with an off-centered entropy. In: International Conference on Data Mining, pp. 117–123. CSREA Press, Las Vegas (2008)
33. MPI-Forum.: Mpi: A message-passing interface standard
34. OpenMP Architecture Review Board: OpenMP application program interface version 3.0 (2008)
35. Gossow, D., Decker, P., Paulus, D.: An Evaluation of Open Source SURF Implementations. In: Ruiz-del-Solar, J. (ed.) RoboCup 2010. LNCS, vol. 6556, pp. 169–179. Springer, Heidelberg (2010)
36. Chatfield, K., Lempitsky, V., Vedaldi, A., Zisserman, A.: The devil is in the details: an evaluation of recent feature encoding methods. In: British Machine Vision Conference, pp. 76.1–76.12 (2011)

Partitioning Methods to Parallelize Constraint Programming Solver Using the Parallel Framework Bobpp*

Tarek Menouer, Bertrand Le Cun, and Pascal Vander-Swalmen

University of Versailles Saint-Quentin-en-Yvelines
tarek.menouer@prism.uvsq.fr

Abstract. This paper presents a parallelization of a Constraint Programming solver, OR-Tools[1], using the parallel framework Bobpp [2].

An argument in support of this approach is that the parallelization of algorithms searching for solutions in the research area is extensively studied over the world.

The novelty presented here is the study of a parallelization for which the control of the OR-Tools sequential search is limited. Using OR-Tools, it is possible to record the path from the tree's root to a node so as to stop the search at a precise node. However, to start the search on a sub-tree, the entire path from the root of the main tree to the root of the sub-tree has to be replayed. This suggests that this leads to additional costs during the search.

To thwart this problem, different strategies of load balancing are tried to reduce the extra costs due to the redundant branches.

Keywords: Parallelism, Dynamic load balancing, Combinatorial Optimization.

1 Introduction

The innovations in hardware architectures as the multi-core parallel machines or MIC (*Many Integrated Cores*) and the progress in the field of parallelism (architecture, systems, languages, execution environments and algorithms) have created new challenges to design parallel tools such as a *Constraint Programming* solver.

Constraint Programming is a method used to solve Constraint Satisfaction Problems (CSPs), defined as mathematical processes used to solve artificial intelligence problems or Operational Research problems.

CSPs are represented by both a set of variables and a set of constraints. Each variable may receive a value defined in a domain. The aim is to affect values to variables to satisfy all the constraints. As example, OR-Tools is a programming

---

* This work is funded by "PAJERO" OSEO-ISI project.
[1] Developed by Google.

N.T. Nguyen, T. Van Do, and H.A. Le Thi (Eds.): *ICCSAMA 2013*, SCI 479, pp. 117–127.
DOI: 10.1007/978-3-319-00293-4_10   © Springer International Publishing Switzerland 2013

constraint library developed by Google. It is a library that includes a set of tools for Operational Research developed in C++.

OR-Tools [16] is an open source solver (Apache license 2.0) providing the possibility to control the search from an external program. On one hand, the algorithms used by this tool are executed sequentially, building a search-space to find solutions. On the other hand, there are several frameworks which link solvers and parallel machines in order to parallelize the search-space, as example: BCP [21], PEBBL [6], PICO [10], ALPS [28,18,24], PUBB [23,22], PPBB [25], Bobpp [9,14,4]. Numerous studies of parallelization of such algorithms were performed as the studies of Divide and Conquer [11,7,8], Branch and Bound and all its variations [3,12,5,4,13,1,19,24,9]. For CSP, several researches are available [17,20,15,26].

The team who develops OR-Tools is very dynamic, the library is regularly extended with new algorithms to improve the portion of *Constraint Programming*. OR-Tools has been used in parallel, however, the parallelization is based on the principle of portfolio.

The portfolio principle [27] is used to solve a problem by performing several models of research (according to different strategies), the first model finding the solution stops the search for other models. But the disadvantage is that each model performs the search in sequential.

The present study is about using the parallel framework Bobpp to parallelize the OR-Tools library without changing the library itself. Indeed, the parallelization in this case is not planned from the beginning because of its difficulty without a fundamental change in the source code. The notion of node is critical to parallelize a tree-search based algorithm. To migrate a search from a core to another one, we need to represent it in an object; for example, all data about the history of the path and the changing data during the exploration of the search-tree. The initial design of OR-Tools makes the library sequential, so this notion of node could be clearly identified. However, the OR-Tools library provides mechanisms called monitor used to control research. Indeed, it is possible to save the search branches and to stop or replay a branch.

The aim is to parallelize a search using multiple instances of the solver running on different computing cores. The Bobpp framework is used as the runtime support. It proposes a set of search algorithms like Branch and Bound, Divide and Conquer and A* algorithms using different methods of parallelism. The purpose of this framework is to provide a unique environment for most classes of Combinatorial Optimization problems which can be solved using different programming environments like POSIX threads as well as MPI, or more specialized libraries such as Athapascan/Kaapi [9].

Section 2 is about different strategies used to partition the search-tree. Section 3 presents some experiments with Bobpp for several types of problems modeled by OR-Tools on two types of parallel machines. Finally, a conclusion and some perspectives are presented in section 4.

2 Parallelization of the Search-Space

To parallelize any constraint programming solver we must take in consideration:

- The number of cores used,
- The type of communication between cores,
- If the constraints are shared between cores.

The majority of algorithms used to solve constraint problems creates a search-space under the form of a search-tree, it explores the search-tree in order to retrieve the first or all the possible solutions. The search-tree is made of three types of nodes: the root, the internal nodes and the leaves. The root represents the original problem with all constraints, however, the internal node represents a partial solution of the original problem and satisfies some constraints, finally, the leaves are either solutions or failures, failure means that from this node it is impossible to find a solution (at least one constraint will never be satisfied).

So to parallelize the search-tree two methods are commonly used:

- Static Partitioning: partitioning of the search-space before the execution then assignment of each sub-tree on one core,
- Dynamic Partitioning: choose the partitioning of the search-space and the assignments during the execution of the algorithm.

2.1 Static Partitioning

This method consists in exploring the first levels of the search-tree providing a sufficient number of nodes (each node represents a root of a sub-tree to explore). Then a second phase follows to assign these nodes on different cores to perform the search. Since this method is simple to implement and there are no many communications between the cores of the machine, a Bobpp prototype allowing to parallelize the OR-Tools search-space has been developed in this study.

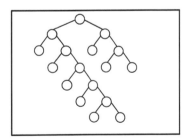

Fig. 1. Imbalanced search-tree

Generally, the search-tree is imbalanced (see Figure 1). This is very difficult, even impossible, to estimate the size of a sub-tree accessible from one node. This strategy implies that, sometimes, only one computing core performs almost all the search, the other computing cores wait this one ends.

To avoid this, a dynamic partitioning/assignment has been proposed.

2.2 Dynamic Partitioning (Work Stealing)

The principle of this method is that the different cores of a machine share the work via a global priority queue. The search-tree is partitioned and allocated to the cores on demand. The threads perform the search locally and sequentially using the OR-Tools solver. When a thread finishes the search on its sub-tree, it gets function from the global priority queue. If the global priority queue is empty, the thread will declare itself as a pending thread. The other threads, performing a search on their sub-trees, test if pending threads exist. In that case, the search on the left branch is stopped, then creates a *BOB-node*, inserts the *BOB-node* in the global priority queue and resumes the search on the right node. The pending threads take effect by the insertion of a new *BOB-node* in the priority queue.

In the progress of this algorithm presented in the figures 2, 3 and 4, a *BOB-node* is actually an object that stores the path from the root node of the initial search-tree to the considered node.

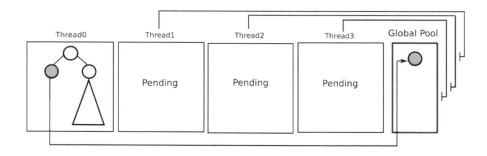

Fig. 2. Dynamic Partitioning: Step 1

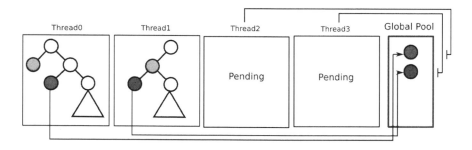

Fig. 3. Dynamic Partitioning: Step 2

In the figure 2, the thread 0 is loaded and all the other threads are pending, thread 0 stops the search on the left node, creates a *BOB-node*, inserts it in the global priority queue, then continues the search on the right node.

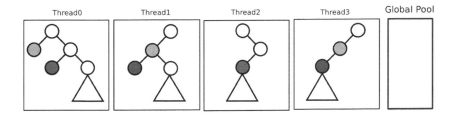

Fig. 4. Dynamic Partitioning: Step 3

In figure 3 the thread 0 detects the threads 2 and 3 are pending, the search is stopped on the left node, a *BOB-node* is created and inserted in the global priority queue. Then, the thread 0 continues the search on the right node. The thread 1 had the same behavior and shared a node too.

In figure 4 all the threads are in function which explains a good load balancing.

The OR-Tools library provides a mechanism called monitor allowing some control on the search. Indeed, it is possible to save all the history of the search, but it is also possible to stop a search on specific node and also replay the search on a path. Then in our solution, each thread uses one OR-Tools solver that uses a specific OR-Tools Monitor to store the path from the root node to a specific node.

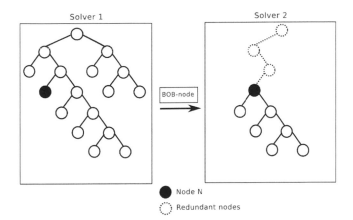

Fig. 5. Migration of node

For technical reasons, in the OR-Tools library, when node n is migrated from the solver 1 to the solver 2 as presented in figure 5, the solver 1 stops the search on the node n. The path from the root node to the node n is saved in what we call a *BOB-node*. The *BOB-node* is communicated to the solver 2. Then the solver 2 replays the path from the root node to the node n in order to set all internal data, then the search can continues on the sub-tree rooted on the node n.

Algorithm 1. Migration test

Require: S, the dynamic partitioning threshold
  Let P, the depth of the current node
  if ∃ at least one pending thread AND $P < S$ **then**
    Stop the search on the left branch
    Create a *BOB-node*
    Insert the *BOB-node* in the global queue
    Restart the search on the right sub-tree
  else
    Continue the sequential exploration of the search-space
  end if

At each time a node is migrated from one solver to another, some redundant nodes are performed, this implies an overhead for each load balancing operation.

Some tests showed more than 300% redundant nodes relative to the search-tree. The following array represents experiences on a search-tree of 4,037,843 nodes. The first line represents the number of redundant nodes and the second line represents the percentage of redundant nodes in accordance to the number of nodes in the main search-tree.

Table 1. Number of redundant nodes according to the number of nodes explored by each thread

Number of cores	4	8	12	16	24
Redundant	21,260,207	38,966,876	52,904,931	54,074,345	72,592,214
Percentage	99	182	234	266	300

To reduce the number of redundant nodes, the length of the path which must be replayed for each created *BOB-node* has to be minimized. To limit the maximum depth of *BOB-nodes* a threshold is used. The algorithm 1 shows this.

The choice of the value of the threshold is a difficult problem. A small one makes the algorithm close to the static partitioning method whereas a high threshold makes the algorithm like the dynamical version without threshold. The current algorithm uses a threshold statically determined at the beginning of the search.

3 Experimentations

To validate the approach used in this study, some experiments have been performed using two different computers, the first one is a bi-processor Intel Xeon X5650 (2.67 GHz) with the Hyper-Threading technology (12 physical cores) with 48 GB of RAM and the second one is a quad-processor AMD Opteron 6176 (2.3 GHz) (48 physical cores) with 256 GB of RAM. Two different problems were addressed:

- N-Queens problem: this problem consists in placing N queens on N distinct squares on an N×N chess board, (the number of feasible solutions is counted). This is a Constraint Satisfaction Problem, performed on sizes 15 and 16.
- Golomb Ruler problem: find N points on a graduated ruler such that the distances between each pair of points are different. This is a Constraint Optimization Problem, performed on sizes 12 and 13.

Computation times, given in seconds, are an average of several runs.

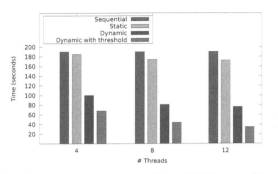

Fig. 6. Computation time for solving the problem of N-Queens (size 15) according the partitioning methods

The figure 6 shows a comparison between three partitioning methods: static, dynamic and dynamic partitioning with a threshold, solving the N-Queens problems (size 15) using 4, 8 and 12 cores.

The performance with a static partitioning is limited since it has a low speed-up compared with the dynamic partitioning using a migration threshold which performs the best load-balancing.

However, it is interesting to notice that the threshold determination remains a problem. The fact of the matter is that, choosing a very small threshold minimizes redundant nodes, but tends towards the static partitioning leading to limit the number of sub-trees to distribute. Conversely, choosing a too high threshold may potentially generates a lot of tasks, which facilitates the load balancing, but increases the redundant nodes.

Figure 7 is a confirmation of which precedes. Studies has shown that the threshold value varied according to the number of cores and especially depending on the problem (type and size).

The inescapable conclusion which emerges from that is the existence of a threshold value which leads to minimize the computation time and represents the best compromise to achieve a satisfactory load balancing and to limit redundant nodes.

In the following experiments, a dynamic method with the best threshold value (determined experimentally) was used. As example for the N-Queens problem with size 16 on a parallel machine with 12 cores, the best threshold is 29.

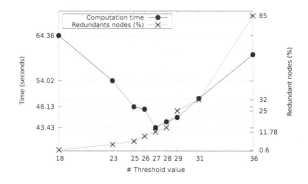

Fig. 7. Computation time for solving the problem of N-Queens (size 15) on 8 cores using different thresholds of migration

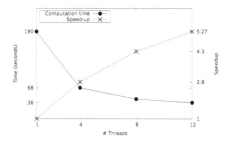

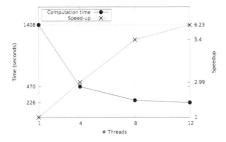

Fig. 8. Computation time and speed-up for solving the problem of N-Queens (size 15) using parallel machine with 12 cores

Fig. 9. Computation time and speed-up for solving the problem of N-Queens (size 16) using parallel machine with 12 cores

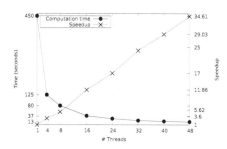

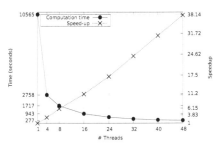

Fig. 10. Computation time and speed-up for solving the problem of Golomb Ruler (size 12) using parallel machine with 48 cores

Fig. 11. Computation time and speed-up for solving the problem of Golomb Ruler (size 13) using parallel machine with 48 cores

As shown in figures 8, 9, 10 and 11, the dynamic partitioning with migration threshold can improve performance and is able to parallelize the solver OR-Tools for a Constraint Satisfaction Problem and a Constraint Optimization Problem. Whenever the number of core increases, performance is gained. Speed-up of 38.14 is reached with the resolution of the Golomb Ruler problem using 48 cores and 6.23 for solving the N-Queens problem using 12 cores.

We also note in figures 12 and 13 that all threads worked and waited during an equivalent time and have almost the same number of nodes visited by OR-Tools. For both problems, a specific thread has no latency and explore more nodes. One might think that these threads have explored large sub-trees beyond the threshold in sequential.

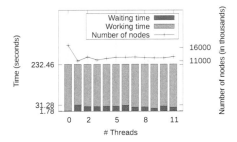

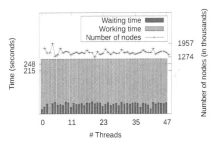

Fig. 12. Load balancing for solving the problem of N-Queens (size 16) using 12 cores

Fig. 13. Load balancing for solving the problem of Golomb Ruler (size 13) using 48 cores

4   Conclusion and Perspectives

This paper presents a parallelization of a Constraint Programming solver, called OR-Tools, with the parallel framework Bobpp.

The solution presents a dynamic partitioning method which performs a dynamic assignments of different parts of the search space during the execution of the algorithm. This solution obtain good performance despite the additional cost required when a node is migrated from one thread to another using a migration threshold.

It has been demonstrated that an optimal value for the threshold is determined so as to minimizing the computation time.

As a first perspective, it should be interesting to introduce a threshold which could automatically determine its value instead of fixing for each run.

Several possibilities can be imagined for the threshold calculation. First, the initial value of the threshold may be choosen in order to generate enough sub-trees for each threads. Then this threshold should be increased when some threads are pending or when the numbers of explored nodes by each thread are imbalanced. It would also be interesting to modify the dynamic partitioning algorithm by adding a structure containing candidate nodes that eventually be migrated.

Bobpp achieves good speed-up on shared memory architectures. These results are obtained on several types of combinatorial problems using different computers. Bobpp is also able to run on distributed memory machine mixing MPI and Pthreads. The parallelization of the dynamic partitioning with threshold presented here on the distributed memory machines is a second perspective.

References

1. Anstreicher, K., Brixius, N., Goux, J.-P., Linderoth, J.: Solving large quadratic assignment problems on computational grids (2000)
2. Bertrand Le Cun, P.V.-S., Menouer, T.: Bobpp, http://forge.prism.uvsq.fr/projects/bobpp
3. Gendron, B., Crainic, T.G.: Parallel branch-and-bound algorithms: Survey and synthesis. Operational Research 42(06), 1042–1066 (1994)
4. Cung, V.-D., Dowaji, S., Le, C.B., Mautor, T., Roucairol, C.: Concurrent data structures and load balancing strategies for parallel branch-and-bound / a* algorithms. In: III Annual Implementation Challenge Workshop, DIMACS, New Brunswick, USA (October 1994)
5. Miller, D.L., Pekny, J.F.: Results from a parallel branch and bound algorithm for the asymmetric traveling salesman problem. Technical report, Carnegie Mellon University, PITTSBURGH, PA 15212 (1989)
6. Eckstein, J., Phillips, C.A., Hart, W.E.: PEBBL 1.0 User Guide. RRR 19-2006, RUTCOR (August 2006)
7. Feldmann, R.: Game Tree Search on Massively Parallel Systems. PhD thesis, Department of mathematics and computer science, University of Paderborn, Germany (August 1993)
8. Finkel, R., Manber, U.: Dib- a distributed implementation of backtracking. ACM, Transaction on Programming Languages and Systems 9(02), 235–256 (1987)
9. Galea, F., Le Cun, B.: Bob++: a framework for exact combinatorial optimization methods on parallel machines. In: International Conference High Performance Computing & Simulation 2007 (HPCS 2007) and in conjunction with The 21st European Conference on Modeling and Simulation (ECMS 2007), pp. 779–785 (June 2007)
10. Eckstein, J., Phillips, C.A., Hart, W.E.: PICO: An object-oriented framework for parallel branch and bound. In: Scientific, E. (ed.) Proceedings of the Workshop on Inherently Parallel Algorithms in Optimization and Feasibility and their Applications. Studies in Computational Mathematics, pp. 219–265 (2001)
11. Kumar, V., Ramesh, K., Rao, V.N.: Parallel best-first search of state-space graphs : A summary of results. In: The AAAI Conference, pp. 122–127 (1987)
12. Lai, T., Sprague, A.: Performance of parallel branch-and-bound algorithms. IEEE Transactions On Computers C-34(10), 962–964 (1985)
13. Le Cun, B., Roucairol, C.: Concurrent data structures for tree search algorithms. In: Ferreira, K.A.A., Rolim, J. (eds.) IFIP WG 10.3, IRREGULAR 1994: Parallel Algorithms for Irregular Structured Problems, pp. 135–155 (September 1994)
14. Le Cun, B., Roucairol, C., the PNN team. Bob : a unified platform for implementing branch-and-bound like algorithms. RR 95/16, Laboratoire PRiSM, Université de Versailles - Saint Quentin en Yvelines (September 1995)
15. Michel, L., See, A., Van Hentenryck, P.: Transparent parallelization of constraint programming. Informs Journal on Computing 21(3), 363–382 (2009)

16. van Omme, V.F.N., Perron, L.: Or-tools. Technical report, Google (2012)
17. Gent, I.P., Jefferson, C., Miguel, I., Moore, N.C., Nightingale, P., Prosser, P., Unsworth, C.: A preliminary review of literature on parallel constraint solving. In: Proceedings PMCS 2011 Workshop on Parallel Methods for Constraint Solving (2011)
18. Ralphs, T., Ladányi, L., Saltzman, M.: A Library Hierarchy for Implementing Scalable Parallel Search algorithms. The Journal of Supercomputing 28(2), 215–234 (2004)
19. Ladányi, L., Ralphs, T.K., Trotter Jr., L.E.: Branch, cut, and price: Sequential and parallel. In: Jünger, M., Naddef, D. (eds.) Computat. Comb. Optimization. LNCS, vol. 2241, pp. 223–260. Springer, Heidelberg (2001)
20. Rolf, C.C.: Parallelism in Constraint Programming. PhD thesis, Department of Computer Science, Lund University (October 2011)
21. Saltzman, M.J.: COIN-OR: An Open Source Library for optimization. In: Advances in Computational Economics. Kluwer, Boston (2002)
22. Shinano, Y., Higaki, M., Hirabayashi, R.: An Interface Design for General Parallel Branch-and-Bound Algorithms. In: Workshop on Parallel Algorithms for Irregularly Structured Problems, pp. 277–284 (1996)
23. Shinano, Y., Higari, M., Hirabayashi, R.: Generalized utility for parallel branch-and-bound algorithms. In: Proceedings of the 1995 Seventh Symposium on Parallel and Distributed Processing, pp. 392–401. IEEE Computer Society Press, Los Alamitos (1995)
24. Ralphs, T.K., Ladányi, L., Saltzman, M.: Parallel Branch, Cut, and Price for Large-scale Discrete Optimization. Mathematical Programming 98(253) (2003)
25. Tschoke, S., Polzer, T.: Portable parallel branch-and-bound library user manual, library version 2.0. Technical report, Department of Computer Sciences, University of Paderborn (1996)
26. Vander-Swalmen, P., Dequen, G., Krajecki, M.: Designing a parallel collaborative sat solver. In: 17th International Conference on Parallel and Distributed Processing Techniques and Applications, USA, CSREA Press (2011)
27. Wetzel, G., Zabatta, F.: A constraint programming approach to portfolio selection. In: Proceeding of the 13th Biennial European Conference on Artificial Intelligence, pp. 263–264 (1998)
28. Xu, Y., Ralphs, T., Ladányi, L., Saltzman, M.: ALPS: A Framework for Implementing Parallel Search Algorithms. In: Proceedings of the Ninth INFORMS Computing Society Conference (2005)

Part II
Queueing Theory and Applications

Spectral Expansion Solution Methodology for QBD-M Processes and Applications in Future Internet Engineering

Tien Van Do[1], Ram Chakka[2], and János Sztrik[3]

[1] Department of Networked Systems and Services
Budapest University of Technology and Economics
Budapest, Hungary
[2] Meerut Institute of Engineering and Technology (MIET),
Meerut 250005, India
[3] Faculty of Informatics, University of Debrecen,
Egyetem tér 1, Po.Box 12, 4010 Debrecen, Hungary

Abstract. Quasi Simultaneous-Multiple Births and Deaths (QBD-M) Processes are used to model many of the traffic, service and related problems in modern communication systems. Their importance is on the increase due to the great strides that are taking place in telecommunication systems and networks. This paper presents the overview of the Spectral Expansion (SE) for the steady state solution of QBD-M processes and applications in future Internet engineering.

Keywords: QBD-M, Compound Poisson Process, Spectral Expansion.

1 Introduction

The concept of Quasi Birth-Death (QBD) processes, as a generalization of the classical birth and death M/M/1 queues was first introduced by [1] and [2] in the late sixties. The states of a QBD process are described by two dimensional random variables called a phase and a level [3–5] and transitions in a QBD process are only possible between adjacent levels. It is observed that QBD processes create a useful framework for the performability analysis of many problems in telecommunications and computer networks [6–11].

In the QBD process, if the nonzero jumps in levels are not accompanied with changes in a phase, then these processes are known as Markov-modulated Birth and Death processes . The infinite number of states involved makes the solution of these models nontrivial. There are several methods of solving these models, either the whole class of models or any of the subclasses.

Seelen has analysed a Ph/Ph/c queue in this frame work [12]. Seelen's method is an approximate one, the Markov chain is first truncated to a finite state which is an approximation of the original process. The resulting finite state Markov chain is then analysed, by exploiting the structure in devising an efficient iterative solution algorithm. The second method is to reduce the infinite state problem to a linear equation involving vector generating function and some unknown

N.T. Nguyen, T. Van Do, and H.A. Le Thi (Eds.): *ICCSAMA 2013*, SCI 479, pp. 131–142.
DOI: 10.1007/978-3-319-00293-4_11   © Springer International Publishing Switzerland 2013

probabilities. The latter are then determined with the aid of the singularities of the coefficient matrix. A comprehensive treatment of that approach, in the context of a discrete time process with a general M/G/1 type structure, is presented in [13]. The third way of solving these models is the well known matrix-geometric method, first proposed by Evans [2, 3]. In this method a nonlinear matrix equation is first formed from the system parameters and the minimal nonnegative solution R of this equation is computed by an iterative method. The invariant vector is then expressed in terms of the powers of R. Neuts claims this method has probablistic interpretation for the steps in computation. That is certainly an advantage. Yet, this method suffers from the fact that there is no way of knowing how many iterations are needed to compute R to a given accuracy. It can also be shown that for certain parameter values the computation requirements are uncertain and formidably large. The fourth method is known as the spectral expansion method. It is based on expressing the invariant vector of the process in terms of eigenvalues and left eigenvectors of a certain matrix polynomial. The generating function and the spectral expansion methods are closely related. However, the latter produces steady state probabilities directly using an algebraic expansion while the former provides them through a transform.

It is confirmed by a number of works that the spectral expansion method is better than the matrix geometric one from some aspects [4, 14, 15]. This paper gives the overview of the SE methodology and explains how the SE methodology is used towards the analysis of QBD-M processes and the performance evaluation of ICT systems and future Internet.

The rest of the paper is organized as follows. In Section 2, the terminology and definitions are presented. The spectral expansion methodology is provided in Section 3. Examples are given in Section 4. The paper is concluded in Section 5.

2   Definitions

Consider a two-dimensional continuous time, irreducible Markov chain $X=\{(I(t), J(t)), t \geq 0\}$ on a lattice strip.

- $I(t)$ is called the phase (e.g., the state of the environment) of the system at time t. Random variable $I(t)$ takes values from the set $\{0, 1, 2, \ldots, N\}$, where N is the maximum value of the phase variable.
- Random variable $J(t)$ is often called the level of the system at time t and takes a set of values $\{0, 1, \ldots, L\}$, where L can be finite or infinite.

The state space of the Markov chain X is $\{(i, j) : 0 \leq i \leq N, 0 \leq j \leq L\}$. Let $p_{i,j}$ denote the steady state probability of the state (i, j) as

$$p_{i,j} = \lim_{t \to \infty} \Pr(I(t) = i, J(t) = j); \quad (i = 0, \ldots, N; \quad j = 0, 1, \ldots, L).$$

Vector $\mathbf{v}_j$ is defined as

$$\mathbf{v}_j = (p_{0,j}, \ldots, p_{N,j}) \quad (j = 0, 1, \ldots, L).$$

Since the sum of all the probabilities $p_{i,j}$ is 1.0, we have the normalization equation as

$$\sum_{j=0}^{L} \mathbf{v}_j \mathbf{e}_{N+1} = 1 \,, \tag{1}$$

where $\mathbf{e}_{N+1}$ is a column vector of size $N + 1$ with all ones.

2.1   Continuous Time QBD Processes

Definition 1. *A continuous time Quasi-Birth-and-Death (QBD) process is formed when one-step transitions of the Markov chain X are allowed to states in the same level or in the two adjacent levels. That is, the dynamics of the process are driven by*

(a) *purely phase transitions. $A_j(i, k)$ denotes the transition rate from state (i, j) to state (k, j) $(0 \leq i, k \leq N; i \neq k; j = 0, 1, \ldots, L)$;*
(b) *one−step upward transitions. $B_j(i, k)$ is the transition rate from state (i, j) to state $(k, j + 1)$ $(0 \leq i, k \leq N; j = 0, 1, \ldots, L)$;*
(c) *one−step downward transitions. $C_j(i, k)$ is the transition rate from state (i, j) to state $(k, j - 1)$ $(0 \leq i, k \leq N; j = 0, 1, \ldots, L)$.*

Let A_j, B_j and C_j denote $(N + 1) \times (N + 1)$ matrices with elements $A_j(i, k)$, $B_j(i, k)$ and $C_j(i, k)$, respectively. Note that the diagonal elements of matrix A are zero. Let D^{A_j}, D^{B_j} and D^{C_j} be the diagonal matrices of size $(N+1) \times (N+1)$, defined by the ith $(i = 0, \ldots, N)$ diagonal element as follows

$$D^{A_j}(i, i) = \sum_{k=0}^{N} A_j(i, k); \ D^{B_j}(i, i) = \sum_{k=0}^{N} B_j(i, k); \ D^{C_j}(i, i) = \sum_{k=0}^{N} C_j(i, k).$$

For the convenience of the presentation we define matrices $B_{-1} = 0$, $B_L = 0$ and $C_0 = 0$.

The steady state balance equations satisfied by the vectors $\mathbf{v}_j$ are

$$\mathbf{v}_j[D^{A_j} + D^{B_j} + D^{C_j}] = \mathbf{v}_{j-1}B_{j-1} + \mathbf{v}_j A_j + \mathbf{v}_{j+1}C_{j+1} \ \forall j \,. \tag{2}$$

Assume that there exist thresholds T_1^* and T_2^* such that

$$A_j = A \ (T_2^* \geq j \geq T_1^*),$$
$$B_j = B \ (T_2^* \geq j \geq T_1^* - 1),$$
$$C_j = C \ (T_2^* + 1 \geq j \geq T_1^*).$$

D^A, D^B and D^C are the corresponding diagonal matrices with the diagonal elements as

$$D^A(i, i) = \sum_{k=1}^{N} A(i, k), \ D^B(i, i) = \sum_{k=1}^{N} B(i, k), \ D^C(i, i) = \sum_{k=1}^{N} C(i, k).$$

The generator matrix of the QBD process is written as

$$
\begin{bmatrix}
A_0^{(1)} & B_0 & 0 & 0 & \cdots & \cdots & \cdots & \cdots & \cdots \\
C_1 & A_1^{(1)} & B_1 & 0 & \cdots & \cdots & \cdots & \cdots & \cdots \\
0 & C_2 & A_2^{(1)} & B_2 & \cdots & \cdots & \cdots & \cdots & \cdots \\
\vdots & \vdots & \vdots & \ddots & \cdots & \cdots & \cdots & \cdots \\
0 & 0 & \cdots & C_{T_1^*-1} & A_{T_1^*-1}^{(1)} & B_{T_1^*-1} & 0 & 0 & \cdots \\
0 & 0 & \cdots & 0 & C_{T_1^*} & A_{T_1^*}^{(1)} & B_{T_1^*} & 0 & \cdots \\
0 & 0 & \cdots & 0 & 0 & C_{T_1^*+1} & A_{T_1^*+1}^{(1)} & B_{T_1^*+1} & \cdots \\
\vdots & \vdots & \vdots & \vdots & \cdots & \cdots & \ddots & \cdots & \cdots
\end{bmatrix}
$$

$$
=
\begin{bmatrix}
A_0^{(1)} & B_0 & 0 & 0 & \cdots & \cdots \cdots \cdots \cdots \cdots \\
C_1 & A_1^{(1)} & B_1 & 0 & \cdots & \cdots \cdots \cdots \cdots \cdots \\
0 & C_2 & A_2^{(1)} & B_2 & \cdots & \cdots \cdots \cdots \cdots \cdots \\
\vdots & \vdots & \vdots & \ddots & \cdots & \cdots \cdots \cdots \cdots \\
0 & 0 & \cdots & C_{T_1^*-1} & A_{T_1^*-1}^{(1)} & Q_0\ 0\ 0\ \cdots \cdots \cdots \\
0 & 0 & \cdots & 0 & C_{T_1} & Q_1\ Q_0\ 0\ \cdots \cdots \cdots \\
0 & 0 & \cdots & 0 & 0 & Q_2\ Q_1\ Q_0\ \cdots \cdots \cdots \\
0 & 0 & \cdots & 0 & 0 & 0\ Q_2\ Q_1\ Q_0\ \cdots \\
\vdots & \vdots & \vdots & \vdots & \cdots & \cdots\ \ddots \cdots \cdots \cdots
\end{bmatrix},
$$

where $A_j^{(1)} = A_j - D^{A_j} - D^{B_j} - D^{C_j}$.

The j-independent balance equations can be rewritten as follows

$$
\mathbf{v}_{j-1}Q_0 + \mathbf{v}_j Q_1 + \mathbf{v}_{j+1}Q_2 = 0 \quad (T_1^* \le j \le T_2^*), \tag{3}
$$

where $Q_0 = B$, $Q_1 = A - D^A - D^B - D^C$, $Q_2 = C$.

2.2   Continuous Time QBD-M Processes

Definition 2. *The Markov chain X is called a continuous time quasi simultaneous-bounded-multiple births and simultaneous-bounded-multiple deaths (QBD-M) process if the balance equation for level j can be written as*

$$
\sum_{i=0}^{y} \mathbf{v}_{j-y_1+i}Q_i = 0 \quad (T_1 \le j \le T_2), \tag{4}
$$

where y, y_1, T_1 and T_2 are integer constants for a specific system, while Q_i are j-independent matrices of size $(N+1) \times (N+1)$.

2.3   Generalized Exponential Distribution

Definition 3. *The versatile Generalized Exponential (GE) distribution is given in the following form:*

$$
F(t) = P(W \le t) = 1 - (1 - \phi)e^{-\mu t} \quad (t \ge 0), \tag{5}
$$

where W is the GE random variable with parameters μ, ϕ.

Thus, the GE parameter estimation can be by obtained by $1/\nu$, the mean, and C^2_{coeff}, the squared coefficient of variation of the inter-event time of the sample as

$$1 - \phi = 2/(C^2_{coeff} + 1) \; ; \; \mu = \nu(1 - \phi) . \tag{6}$$

Remarks. For $C^2_{coeff} > 1$, the GE model is a mixed-type probability distribution having the same mean and coefficient of variation, and with one of the two phases having zero service time, or a bulk type distribution with an underlying counting process equivalent to a Batch (or Bulk) Poisson Process (BPP) with batch-arrival rate μ and geometrically distributed batch size with mean $1/(1-\phi)$ and SCV $(C^2_{coeff} - 1)/(1 + C^2_{coeff})$ (see [16]). It can be observed that there is an infinite family of BPP's with the same GE-type inter-event time distribution. It is shown that, among them, the BPP with geometrically distributed bulk sizes (referred as the CPP) is the only one that constitutes a renewal process (the zero inter-event times within a bulk/batch are *independent* if the bulk size distribution is geometric [17]). The GE distribution is versatile, possessing pseudo-memoryless properties which make the solution of many GE-type queuing systems analytically tractable [17]. The choice of the GE distribution is often motivated by the fact that measurements of actual inter-arrival or service times may be generally limited and so only a few parameters (for example the mean and variance) can be computed reliably. Typically, when only the mean and variance can be relied upon, a choice of a distribution which implies least bias is that of GE-type distribution [16, 17].

Definition 4 (CPP). *The inter-arrival time distribution of customers of the Compound Poisson Process (CPP) is GE with parameters (σ, θ). That is, the inter-arrival time probability distribution function is $1 - (1 - \theta)e^{-\sigma t}$.* □

Thus, the arrival *point*-process has batches arriving at each point having independent and geometric batch-size distribution. Specifically the probability that a batch is of size s is $(1 - \theta)\theta^{s-1}$.

3   The Spectral Expansion Method for QBD-M Processes

Let $Q(\lambda)$ denote the characteristic matrix polynomial associated with the balance equation (4) as

$$Q(\lambda) = \sum_{i=0}^{y} Q_i \lambda^i. \tag{7}$$

If $(\lambda, \boldsymbol{\psi})$ is the eigenvalue and left-eigenvector pair of the characteristic matrix-polynomial, the following equation holds

$$\boldsymbol{\psi} Q(\lambda) = 0; \; det[Q(\lambda)] = 0. \tag{8}$$

Assume that $Q(\lambda)$ has d pairs of eigenvalue-eigenvectors. For the kth ($k = 1, \ldots, d$) non-zero eigenvalue-eigenvector pair, $(\lambda_k, \boldsymbol{\psi}_k)$, by substituting $\mathbf{v}_j = \boldsymbol{\psi}_k \lambda_k^j$ ($T_1 - y_1 \leq j \leq T_2 - y_1 + y$) in the equations (4), it can be seen that this set of equations is satisfied. Hence, that is a particular solution. The equations can even be satisfied with $\psi_k \lambda_k^{j+l_k}$ for any real l_k. It is easy to prove that the general solution for $\mathbf{v}_j$ is the linear sum of all the factors $(\psi_k \lambda_k^{j-T_1+y_1})$ as

$$\mathbf{v}_j = \sum_{l=1}^{d} a_l \psi_l \lambda_l^{j-T_1+y_1} \quad (j = T_1 - y_1, T_1 - y_1 + 1, \ldots, T_2 - y_1 + y), \quad (9)$$

where a_l $(l = 1, \ldots, d)$ are constants.

Therefore, the steady state probability can be written as follows

$$p_{i,j} = \sum_{l=1}^{d} a_l \psi_l(i) \lambda_l^{j-T_1+y_1} \quad (j = T_1 - y_1, T_1 - y_1 + 1, \ldots, T_2 - y_1 + y). \quad (10)$$

An interesting property can be observed concerning the eigenvalues of $Q(\lambda)$ for QBD-M process X as follows. If (λ_k, ψ_k) is the left-eigenvalue and eigenvector pair of $Q(\lambda)$, then $(1/\lambda_k, \psi_k)$ is the left-eigenvalue and eigenvector pair of $\overline{Q}(\lambda) = \sum_{i=0}^{y} Q_{y-i} \lambda^i$, the characteristic matrix polynomial of the dual process of X (see [14]).

3.1   Infinite QBD-M Processes

When L and T_2 are infinite (unbounded), consider the probability sum

$$\sum_{j=T_1-y_1}^{\infty} p_{i,j} = \sum_{j=T_1-y_1}^{\infty} \sum_{l=1}^{d} a_l \psi_l(i) \lambda_l^{j-T_1+y_1}. \quad (11)$$

In order to ensure that this sum is less or equal to 1.0, the necessary condition is

$$a_k = 0, \; if \; |\lambda_k| \geq 1.$$

Thus, by renumbering the eigenvalues inside the unit circle, the general solution is obtained as

$$\mathbf{v}_j = \sum_{l=1}^{\chi} a_l \psi_l \lambda_l^{j-T_1+y_1} \quad (j = T_1 - y_1, T_1 - y_1 + 1, \ldots), \quad (12)$$

$$p_{i,j} = \sum_{l=1}^{\chi} a_l \psi_l(i) \lambda_l^{j-T_1+y_1} \quad (j = T_1 - y_1, T_1 - y_1 + 1, \ldots). \quad (13)$$

where χ is the number of eigenvalues that are present strictly within the unit circle. These eigenvalues appear some as real and others as complex-conjugate pairs, and as do the corresponding eigenvectors.

In order to determine the steady state probabilities, the unknown constants a_l are to be determined. Their number is χ. We still have other unknowns

$\mathbf{v}_0, \mathbf{v}_1, \ldots, \mathbf{v}_{T_1-y_1-1}$. These unknowns are determined with the aid of the state dependent balance equations (their number is $T_1(N+1)$) and the normalization equation (1), out of which $T_1(N+1)$ are linearly independent. These equations can have a unique solution if and only if $(T_1 - y_1)(N+1) + \chi = T_1(N+1)$, or equivalently

$$\chi = y_1(N+1) \tag{14}$$

holds.

3.2   Finite QBD-M Processes

In order to compute the steady state probabilities, the unknown constants a_l are to be determined. Their number is d. We still have other unknowns $\mathbf{v}_0, \mathbf{v}_1, \ldots, \mathbf{v}_{T_1-y_1-1}, \mathbf{v}_{T_2-y_1+y+1}, \mathbf{v}_{T_2-y_1+y+2}, \ldots, \mathbf{v}_L$. Therefore, the number of unknowns is

$$d + (T_1 - y_1)(N+1) + (L - T_2 + y_1 - y)(N+1).$$

These unknowns are determined with the aid of the state dependent balance equations (their number is $T_1(N+1) + (L-T_2)(N+1)$) and the normalization equation, out of which $T_1(N+1) + (L-T_2)(N+1)$ are linearly independent. These equations can have a unique solution if and only if

$$d + (T_1 - y_1)(N+1) + (L - T_2 + y_1 - y)(N+1) = T_1(N+1) + (L-T_2)(N+1),$$

equivalently

$$d = y(N+1) \tag{15}$$

holds.

4   Examples and Applications

Example 1 (M/M/c/L queue with breakdowns and repairs). The queue with an infinite buffer is described by the Markov chain $\{I(t), J(t)\}$, where $I(t)$ -the operative state of the system- represents the number of operative servers at time t and $J(t)$ is the number of jobs in the system at time t, including those being served. The maximum number of operative servers is c. The Markov chain is irreducible with state space $\{0, 1, \ldots, c\} \times \{0, 1, \ldots, L\}$. Note that in this example the phase is numbered from 0 and the transition rate matrices are of size $(c+1) \times (c+1)$. The number of phases is $N+1 = c+1$. Jobs arrive according to an independent Poisson process with rate σ. The service rate of an operative server is denoted by μ. Processors break down independently at rate ξ and are repaired at rate η. When a new job arrives or when a completed job departs from the system, the operative state does not change.

The matrices A_j and A are given by

$$A = A_j = \begin{bmatrix} 0 & c\eta & & & \\ \xi & 0 & (c-1)\eta & & \\ 2\xi & 0 & \ddots & & \\ & & \ddots & \ddots & \eta \\ & & & c\xi & 0 \end{bmatrix} \quad (j = 0, 1, \ldots). \tag{16}$$

The one-step upward transitions are created by the arrivals of single jobs. Therefore, B and B_j the one-step upward transition rate matrices are

$$B = B_j = diag[\sigma, \sigma, \ldots, \sigma](j = 0, 1, \ldots). \tag{17}$$

The one-step downward transitions take place by the departures of single jobs, after their service completion. The departure rate $(C_j(i,i))$ of jobs at time t depends on $I(t) = i$ and $J(t) = j$. If $i > j$, then a server is assigned to every job and not all operative servers are occupied, hence the departure rate $C_j(i,i) = j\mu$. If $i \le j$, then all the operative processors are occupied by jobs, hence the departure rate $C_j(i,i) = i\mu$. Note that $C_j(i,i)$ does not depend on j if $j \ge i$. Therefor, C_j does not depend on j if $j \ge c$.

$$C_j = diag[0, min(j,1)\mu, min(j,2)\mu, \ldots, min(j,c)\mu] \quad (0 < j < c),$$
$$C = diag[0, \mu, 2\mu, \ldots, c\mu] \quad (j \ge c),$$
$$C_0 = 0. \tag{18}$$

The M/M/c/L queue with breakdowns and repairs is an example of the QBD process, where the coefficient matrices of the characteristic matrix polynomial are $Q_0 = B = B = diag[\sigma, \sigma, \ldots, \sigma]$, $Q_1 = A - D^A - D^B - D^C$, $Q_2 = C$.

Example 2 (Retrial queues to model DHCP [18]). The size of the pool (i.e.: the number of allocatable IP addresses) is c. The fix lease time value sent by the DHCP server is denoted by T_l. The inter-arrival times of DHCP requests are exponentially distributed with a mean inter-arrival time $1/\lambda$.

Assume that the holding times (i.e.: how long does a client need an IP address) of clients are represented by random variable H with a cumulative distribution function $\Pr(H < x) = F(x)$. Upon the expiration of the lease time, the previously allocated address at the DHCP server becomes free and can be allocated to another client unless the client extends the use of a specific IP address before the expiration of the lease time. Let a denote the probability that DHCP clients leave (i.e.: switch off the computer) the system or do not renew the allocated IP address after the expiration of its lease time. We can write

$$a = \Pr(H < T_l) = F(T_l).$$

Let $I(t)$ denote the number of allocated IP addresses at time t. Note that $0 \le I(t) \le c$ holds. A client who does not receive the allocation of an IP address because the shortage (when $I(t) = c$) of IP addresses sets a timer to wait for a limited time and will retry the request for an IP address upon the expiration of backoff time. We model this phenomenon as the client joins the "virtual orbit". $J(t)$ represents the number of DHCP clients in the "orbit" at time t and takes values from 0 to ∞.

Lease times are exponentially distributed with a mean lease time $1/\mu = T_l$. Clients waiting in the orbit repeat the request for the DHCP server with rate ν (i.e.: the inter-repetition times are exponentially distributed with parameter ν), which is independent of the number of waiting clients in the orbit.

The evolution of the system is driven by the following transitions.

(a) $A_j(i, k)$ denotes a transition rate from state (i, j) to state (k, j) $(0 \leq i, k \leq c; j = 0, 1, \ldots)$, which is caused by either the arrival of DHCPDISCOVERY requests or by the expiration of the lease time without the renewal of an allocated IP address. Matrix A_j is defined as the matrix with elements $A_j(i, k)$. Since A_j is j-independent, it can be written as

$$A_j = A = \begin{bmatrix} 0 & \lambda & 0 & \ldots & 0 & & 0 & 0 \\ a\mu & 0 & \lambda & \ldots & 0 & & 0 & 0 \\ \vdots & \vdots & \vdots & \vdots & \vdots & & \vdots & \vdots \\ 0 & 0 & & \ldots & a(c-1)\mu & 0 & \lambda \\ 0 & 0 & & \ldots & 0 & & ac\mu & 0 \end{bmatrix} \quad \forall j \geq 0;$$

(b) $B_j(i, k)$ represents one step upward transition from state (i, j) to state $(k, j+1)$ $(0 \leq i, k \leq c; j = 0, 1, \ldots)$, which is due to the arrival of DHCPDIS-COVERY requests when no free IP address is available in the IP address pool. In the similar way, matrix B_j (B) with elements $B_j(i, k)$ is defined as

$$B_j = B = \begin{bmatrix} 0 & 0 & 0 & \ldots & 0 & 0 & 0 \\ 0 & 0 & 0 & \ldots & 0 & 0 & 0 \\ \vdots & \vdots & \vdots & \vdots & & \vdots & \vdots & \vdots \\ 0 & 0 & & \ldots & 0 & 0 & 0 \\ 0 & 0 & & \ldots & 0 & 0 & \lambda \end{bmatrix} \quad \forall j \geq 0;$$

(c) $C_j(i, k)$ is the transition rate from state (i, j) to state $(k, j-1)$ $(0 \leq i, k \leq c; j = 1, \ldots)$, which is due to the successful retrial of a request from the orbit. Matrix C_j $(\forall j \geq 1)$ with elements $C_j(i, k)$ is written as

$$C_j = C = \begin{bmatrix} 0 & \nu & 0 & \ldots & 0 & 0 & 0 \\ 0 & 0 & \nu & \ldots & 0 & 0 & 0 \\ \vdots & \vdots & \vdots & \vdots & & \vdots & \vdots & \vdots \\ 0 & 0 & & \ldots & 0 & 0 & \nu \\ 0 & 0 & & \ldots & 0 & 0 & 0 \end{bmatrix} \quad \forall j \geq 1.$$

The infinitesimal generator matrix of Y can be written as follows

$$
\begin{bmatrix}
A_{00} & B & 0 & \cdots & \cdots & \cdots & \cdots \\
C & Q_1 & B & 0 & \cdots & \cdots & \cdots \\
0 & C & Q_1 & B & 0 & \cdots & \cdots \\
0 & 0 & C & Q_1 & B & 0 & \cdots \\
\vdots & \vdots & \vdots & \vdots & \vdots & \vdots & \vdots \\
\cdots & \cdots & \cdots & \cdots & \cdots & \cdots & \cdots
\end{bmatrix},
\tag{19}
$$

where D^A and D^C are diagonal matrices whose diagonal elements are the sum of the elements in the corresponding row of A and C, respectively. Note that $A_{00} = A - D^A - B$, $Q_1 = A - D^A - B - D^C$.

5   Conclusions

We have presented an overview for the spectral expansion method to solve QBD-M processes which can be applied to evaluate the performance of various systems, services in information and communication technology (ICT)systems and future Internet. The spectral expansion method is proved to be a mature technique for the performance analysis of various problems [4, 6, 7, 14, 19–38]. The examples include the performance evaluation of Optical Burst/Packet (OBS) Switching networks [24, 39], MPLS networks [23, 30], the Apache web server [7], and wireless networks [6, 24, 28, 40].

Acknowledgement. The publication was supported by the TÁMOP-4.2.2.C-11/1/KONV-2012-0001 project. The project has been supported by the European Union, co-financed by the European Social Fund.

References

1. Wallace, V.L.: The Solution of Quasi Birth and Death Processes Arising from multiple Access Computer Systems. PhD thesis, University of Michigan (1969)
2. Evans, R.V.: Geometric Distribution in some Two-dimensional Queueing Systems. Operations Research 15, 830–846 (1967)
3. Neuts, M.F.: Matrix Geometric Soluctions in Stochastic Model. Johns Hopkins University Press, Baltimore (1981)
4. Mitrani, I., Chakka, R.: Spectral expansion solution for a class of Markov models: Application and comparison with the matrix-geometric method. Performance Evaluation 23, 241–260 (1995)
5. Latouche, G., Ramaswami, V.: Introduction to Matrix Analytic Methods in Stochastic Modeling. ASA-SIAM Series on Statistics and Applied Probability (1999)
6. Chakka, R., Do, T.V.: The MM $\sum_{k=1}^{K} CPP_k/GE/c/L$ G-Queue with Heterogeneous Servers: Steady state solution and an application to performance evaluation. Performance Evaluation 64, 191–209 (2007)

7. Do, T.V., Krieger, U.R., Chakka, R.: Performance modeling of an apache web server with a dynamic pool of service processes. Telecommunication Systems 39(2), 117–129 (2008)
8. Krieger, U.R., Naoumov, V., Wagner, D.: Analysis of a Finite FIFO Buffer in an Advanced Packet-Switched Network. IEICE Trans. Commun. E81-B, 937–947 (1998)
9. Naoumov, V., Krieger, U.R., Warner, D.: Analysis of a Multi-Server Delay-Loss System With a General Markovian Arrival Process. In: Chakravarthy, S.R., Alfa, A.S. (eds.) Matrix-analytic Methods in Stochastic Models. Lecture Notes in Pure and Applied Mathematics, vol. 183, pp. 43–66. Marcel Dekker (1997)
10. Rosti, E., Smirni, E., Sevcik, K.C.: On processor saving scheduling policies for multiprocessor systems. IEEE Trans. Comp. 47, 47–2 (1998)
11. Wierman, A., Osogami, T., Harchol-Balter, M., Scheller-Wolf, A.: How many servers are best in a dual-priority M/PH/k system? Perform. Eval. 63(12), 1253–1272 (2006)
12. Seelen, L.P.: An Algorithm for Ph/Ph/c queues. European Journal of Operational Research 23, 118–127 (1986)
13. Gail, H.R., Hantler, S.L., Taylor, B.A.: Spectral analysis of M/G/1 type Markov chains. Technical Report RC17765, IBM Research Division (1992)
14. Chakka, R.: Performance and Reliability Modelling of Computing Systems Using Spectral Expansion. PhD thesis, University of Newcastle upon Tyne (Newcastle upon Tyne) (1995)
15. Grassmann, W.K., Drekic, S.: An analytical solution for a tandem queue with blocking. Queueing System (1-3), 221–235 (2000)
16. Skianis, C., Kouvatsos, D.: An Information Theoretic Approach for the Performance Evaluation of Multihop Wireless Ad Hoc Networks. In: Kouvatsos, D.D. (ed.) Proceedings of the Second International Working Conference on Performance Modelling and Evaluation of Heterogeneous Networks (HET-NETs 2004), Ilkley, UK, P81/1–13 (July 2004)
17. Kouvatsos, D.D.: A maximum entropy analysis of the G/G/1 Queue at Equilibrium. Journal of Operations Research Society 39, 183–200 (1998)
18. Do, T.V.: An Efficient Solution to a Retrial Queue for the Performability Evaluation of DHCP. Computers & OR 37(7), 1191–1198 (2010)
19. Chakka, R.: Spectral Expansion Solution for some Finite Capacity Queues. Annals of Operations Research 79, 27–44 (1998)
20. Chakka, R., Harrison, P.G.: Analysis of MMPP/M/c/L queues. In: Proceedings of the Twelfth UK Computer and Telecommunications Performance Engineering Workshop, Edinburgh, pp. 117–128 (1996)
21. Chakka, R., Harrison, P.G.: A Markov modulated multi-server queue with negative customers - the MM CPP/GE/c/L G-queue. Acta Informatica 37, 881–919 (2001)
22. Chakka, R., Harrison, P.G.: The MMCPP/GE/c queue. Queueing Systems: Theory and Applications 38, 307–326 (2001)
23. Chakka, R., Do, T.V.: The $MM \sum_{k=1}^{K} CPP_k/GE/c/L$ G-Queue and Its Application to the Analysis of the Load Balancing in MPLS Networks. In: Proceedings of the 27th Annual IEEE Conference on Local Computer Networks (LCN 2002), Tampa, FL, USA, November 6-8, pp. 735–736 (2002)
24. Chakka, R., Do, T.V., Pandi, Z.: A Generalized Markovian Queue and Its Applications to Performance Analysis in Telecommunications Networks. In: Kouvatsos, D. (ed.) Performance Modelling and Analysis of Heterogeneous Networks, pp. 371–387. River Publisher (2009)

25. Chakka, R., Ever, E., Gemikonakli, O.: Joint-state modeling for open queuing networks with breakdowns, repairs and finite buffers. In: 15th International Symposium on Modeling, Analysis, and Simulation of Computer and Telecommunication Systems (MASCOTS), pp. 260–266. IEEE Computer Society (2007)
26. Chakka, R., Mitrani, I.: Multiprocessor systems with general breakdowns and repairs. In: SIGMETRICS, pp. 245–246 (1992)
27. Chakka, R., Mitrani, I.: Heterogeneous multiprocessor systems with breakdowns: Performance and optimal repair strategies. Theor. Comput. Sci. 125(1), 91–109 (1994)
28. Do, T.V., Chakka, R., Harrison, P.G.: An integrated analytical model for computation and comparison of the throughputs of the UMTS/HSDPA user equipment categories. In: Proceedings of the 10th ACM Symposium on Modeling, Analysis, and Simulation of Wireless and Mobile Systems, MSWiM 2007, pp. 45–51. ACM, New York (2007)
29. Van Do, T., Do, N.H., Chakka, R.: Performance evaluation of the high speed downlink packet access in communications networks based on high altitude platforms. In: Al-Begain, K., Heindl, A., Telek, M. (eds.) ASMTA 2008. LNCS, vol. 5055, pp. 310–322. Springer, Heidelberg (2008)
30. Do, T.V., Papp, D., Chakka, R., Truong, M.X.T.: A Performance Model of MPLS Multipath Routing with Failures and Repairs of the LSPs. In: Kouvatsos, D. (ed.) Performance Modelling and Analysis of Heterogeneous Networks, pp. 27–43. River Publisher (2009)
31. Drekic, S., Grassmann, W.K.: An eigenvalue approach to analyzing a finite source priority queueing model. Annals OR 112(1-4), 139–152 (2002)
32. Ever, E., Gemikonakli, O., Chakka, R.: A mathematical model for performability of beowulf clusters. In: Annual Simulation Symposium, pp. 118–126. IEEE Computer Society (2006)
33. Ever, E., Gemikonakli, O., Chakka, R.: Analytical modelling and simulation of small scale, typical and highly available beowulf clusters with breakdowns and repairs. Simulation Modelling Practice and Theory 17(2), 327–347 (2009)
34. Grassmann, W.K.: The use of eigenvalues for finding equilibrium probabilities of certain markovian two-dimensional queueing problems. INFORMS Journal on Computing 15(4), 412–421 (2003)
35. Grassmann, W.K., Drekic, S.: An analytical solution for a tandem queue with blocking. Queueing Syst. 36(1-3), 221–235 (2000)
36. Mitrani, I.: Approximate solutions for heavily loaded markov-modulated queues. Perform. Eval. 62(1-4), 117–131 (2005)
37. Tran, H.T., Do, T.V.: Computational Aspects for Steady State Analysis of QBD Processes. Periodica Polytechnica, Ser. El. Eng., 179–200 (2000)
38. Zhao, Y., Grassmann, W.K.: A numerically stable algorithm for two server queue models. Queueing Syst. 8(1), 59–79 (1991)
39. Do, T.V., Chakka, R.: A New Performability Model for Queueing and FDL-related Burst Loss in Optical Switching Nodes. Computer Communications 33(S), 146–151 (2010)
40. Do, T.V., Chakka, R., Do, N., Pap, L.: A Markovian queue with varying number of servers and applications to the performance comparison of HSDPA user equipment. Acta Informatica 48, 243–269 (2011)

On the Properties of Generalised Markovian Queues with Heterogeneous Servers

Tien Van Do[1] and Ram Chakka[2]

[1] Department of Networked Systems and Services
Budapest University of Technology and Economics
Budapest, Hungary
[2] Meerut Institute of Engineering and Technology (MIET),
Meerut 250005, India

Abstract. This paper provides the proof for some fundamental properties of the generalised Markovian queue - HetSigma, which has been proposed in order to model nodes in modern telecommunication networks. The fundamental properties serve as the background for the efficient computational algorithm to calculate the steady state probabilities of the HetSigma queue.

Keywords: QBD, HetSigma, Compound Poisson Process, Spectral Expansion.

1 Introduction

We have proposed a new generalised multi-server queue, referred here as the HetSigma queue [1], in the Markovian framework. The queue has many of the necessary properties such as, joint (or, individual) Markov modulation of the arrival and service processes, superposition of K CPP (compound Poisson process) streams of (positive) customer arrivals, and a CPP of negative customer arrival stream in each of the modulating phases, a multi-server with c non-identical (can also be identical) servers, GE (generalised exponential) service times in each of the modulating phases and a buffer of finite or infinite capacity. Thus, the model can accommodate correlations of the inter-arrival times of batches, geometric as well as non-geometric batch size distributions of customers in both arrivals and services. The use of negative customers can facilitate modelling server failures, packet losses, load balancing, channel impairment in wireless networks, and in many other applications. An exact and computationally efficient solution of this new queue for the steady state probabilities and performance measures is developed. A closed form for the coefficient matrices of the characteristic matrix polinomial is derived. The fundamental properties which serve as the background for the efficient computational algorithm to obtain the steady state probabilities of the HetSigma queue are presented and proved.

The proposed model does provide a useful tool for the performance analysis of many problems of the emerging telecommunication systems and networks. The queuing model and its variants were successfully used to model Optical Burst/Packet (OBS) Switching networks [2–4], MPLS networks [5, 6] and another variant to successfully compute the performance of the Apache web server [7]. The *HetSigma* model has been applied to model wireless networks [1, 2, 8, 9].

N.T. Nguyen, T. Van Do, and H.A. Le Thi (Eds.): *ICCSAMA 2013*, SCI 479, pp. 143–155.
DOI: 10.1007/978-3-319-00293-4_12　　ⓒ Springer International Publishing Switzerland 2013

This paper revisits the HetSigma queue and provides the formal proof of some properties which form the foundation of the efficient solution presented in [1]. The rest of this paper is organized as follows. In Section 2, we provide the short overview of the HetSigma queue (note that the detailed description of the HetSigma queue can be found in [1]). In Section 3 we present the *rigorous proof* of the properties of the HetSigma queue.

2   The HetSigma Queue

The *HetSigma* queue is defined as the MM $\sum_{k=1}^{K} CPP_k$/GE/c/L G-queue with heterogenous servers. The arrival process is the superposition of the MM $\sum_{k=1}^{K} CPP_k$ and an independent CPP of negative customers (denoted by G). The MM$\sum_{k=1}^{K} CPP_k$ is obtained by Markov modulation of the parameters of the superposition of K independent CPP streams. That is, the K independent CPP's are jointly Markov modulated by a single modulating Markov process. L is the capacity of the system.

We consider a case where the arrival (of positive and negative customers) and service processes are modulated by the same continuous time, irreducible Markov phase process. The system is a multi-server queue with c heterogeneous servers.

The arrival process is modulated by a continuous time, irreducible Markov process with $N + 1$ states (or phases of modulation). Let Q be the generator matrix of this process, given by

$$Q = \begin{bmatrix} -q_0 & q_{0,2} & \cdots & q_{0,N} \\ q_{1,0} & -q_1 & \cdots & q_{1,N} \\ \vdots & \vdots & \ddots & \vdots \\ q_{N,0} & q_{N,1} & \cdots & -q_N \end{bmatrix},$$

where $q_{i,k} (i \neq k)$ is the instantaneous transition rate from phase i to phase k, and the diagonal elements, $-q_i = -\sum_{j=0}^{N} q_{i,j} (i = 0, \ldots, N)$, where $q_{i,i} = 0 \ \forall i$.

The arrival process is the superposition of K independent CPP [10] arrival streams of (positive) customers. The positive customers of the different arrival streams are not distinguishable. In the modulating phase i, the parameters of the GE inter-arrival time distribution of the kth $(k = 1, 2, \ldots, K)$ positive customer arrival stream are $(\sigma_{i,k}, \theta_{i,k})$. That is, the inter- arrival time probability distribution function is $1 - (1 - \theta_{i,k})e^{-\sigma_{i,k}t}$, in phase i, for the kth stream of positive customers. Thus, all the K arrival *point*-processes can be seen as batch-Poisson, with batches arriving at each point having geometric size distribution. Specifically, the probability that a batch is of size s is $(1 - \theta_{i,k})\theta_{i,k}^{s-1}$, in phase i, for the kth stream of positive customers.

The arrival process is the superposition of K independent CPP [10] arrival streams of (positive) customers and an independent CPP of negative customers. In the modulating phase i, the parameters of the GE inter-arrival time distribution of the negative customer arrival process are (ρ_i, δ_i). That is, the inter- arrival time probability distribution function is $1 - (1 - \delta_i)e^{-\rho_i t}$ for the negative customers in phase i.

The service facility has c *heterogeneous* servers in parallel. A number of scheduling policies can be thought of. Though, in principle, a number of scheduling policies can indeed be modelled by following our methodology, the one that we have adopted in this Chapter, for illustration and detailed study, is as follows. A set of service priorities is chosen by giving each server a unique service priority: 1 is the highest and c is the lowest. This set can be chosen arbitrarily from the $c!$ different possible ways.

Each server is then numbered, without loss of generality, by its own service priority. The GE-distributed service time parameters of the nth server ($n = 1, 2 \ldots, c$), in phase i, are denoted by $(\mu_{i,n}, \phi_{i,n})$.

The service discipline is FCFS (First Come First Scheduled for service) and each server serves at most one positive customer at any given time. Customers, on their completion of service, leave the system. When the number of customers in the system, j, (including those in service if any) is $\geq c$, then only c customers are served with the rest $(j - c)$ waiting for service. When $j < c$, only the first j servers, (*i.e.*, servers numbered $1, 2, \ldots, j$), are occupied and the rest are idle. This is made possible by what is known as customer switching. Thus, when server n becomes idle, an awaiting customer would be taken up for service. If there is no awaiting customer, then a customer that is being served by the lowest possible priority server (*i.e.*, among servers $(c, c - 1, \ldots, n + 1)$) switches to server n. In such a switching, the (batch) service time is governed by either *resume or repeat with resampling*, thus preserving the Markov property. The switching is instantaneous and the switching time is treated negligible. Negative customers neither wait in the queue, nor are served.

2.1   The Steady State Balance Equations

The state of the system at any time t can be specified completely by two integer-valued random variables, $I(t)$ and $J(t)$. $I(t)$ varies from 0 to N (known as operative states), representing the phase of the modulating Markov chain, and $0 \leq J(t) < L + 1$ represents the number of positive customers in the system at time t, including those in service. The system is now modelled by a continuous time discrete state Markov process, $\overline{Y}$ (Y if L is infinite), on a rectangular lattice strip. Let $I(t)$, the operative state, vary in the horizontal direction and $J(t)$, the queue length or the *level*, in the vertical direction.

We denote the steady state probabilities by $\{p_{i,j}\}$, where $p_{i,j} = \lim_{t \to \infty} \Pr(I(t) = i, J(t) = j)$, and let $\mathbf{v}_j = (p_{0,j}, \ldots, p_{N,j})$.

The process $\overline{Y}$ evolves due to the following instantaneous transition rates:

(a) $q_{i,k}$ – purely lateral transition rate – from state (i, j) to state (k, j), for all $j \geq 0$ and $0 \leq i, k \leq N$ ($i \neq k$), caused by a phase transition in the Markov chain governing the arrival phase process;

(b) $B_{i,j,j+s}$ – s-step upward transition rate – from state (i, j) to state $(i, j + s)$, for all phases i, caused by a new batch arrival of size s of positive customers. For a given j, s can be seen as bounded when L is finite and unbounded when L is infinite;

(c) $C_{i,j,j-s}$ – s-step downward transition rate – from state (i, j) to state $(i, j - s)$, $(j - s \geq c + 1)$ for all phases i, caused by either a batch service completion of size s or a batch arrival of negative customers of size s;

(d) $C_{i,c+s,c}$ – s-step downward transition rate – from state $(i, c + s)$ to state (i, c), for all phases i, caused by a batch arrival of negative customers of size $\geq s$ or a batch service completion of size s $(1 \leq s \leq L - c)$;

(e) $C_{i,c-1+s,c-1}$ – s-step downward transition rate, from state $(i, c - 1 + s)$ to state $(i, c-1)$, for all phases i, caused by a batch departure of size s $(1 \leq s \leq L-c+1)$;

(f) $C_{i,j+1,j}$ – 1-step downward transition rate, from state $(i, j + 1)$ to state (i, j), $(c \geq 2 ; 0 \leq j \leq c - 2)$, for all phases i, caused by a single departure.

The transition matrices can be obtained as follows

$$B_{i,j-s,j} = \sum_{k=1}^{K} (1 - \theta_{i,k})\theta_{i,k}^{s-1}\sigma_{i,k} \quad (\forall i ; 0 \leq j - s \leq L - 2 ; j - s < j < L) ;$$

$$B_{i,j,L} = \sum_{k=1}^{K} \sum_{s=L-j}^{\infty} (1 - \theta_{i,k})\theta_{i,k}^{s-1}\sigma_{i,k} = \sum_{k=1}^{K} \theta_{i,k}^{L-j-1}\sigma_{i,k} \quad (\forall i ; j \leq L - 1) ;$$

$$C_{i,j+s,j} = \sum_{n=1}^{c} \mu_{i,n}(1 - \phi_{i,n})\phi_{i,n}^{s-1} + (1 - \delta_i)\delta_i^{s-1}\rho_i$$

$$(\forall i ; c + 1 \leq j \leq L - 1 ; 1 \leq s \leq L - j)$$

$$= \sum_{n=1}^{c} \mu_{i,n}(1 - \phi_{i,n})\phi_{i,n}^{s-1} + \delta_i^{s-1}\rho_i \quad (\forall i ; j = c ; 1 \leq s \leq L - c)$$

$$= \sum_{n=1}^{c} \phi_{i,n}^{s-1}\mu_{i,n} \quad (\forall i ; j = c - 1 ; 1 \leq s \leq L - c + 1)$$

$$= 0 \quad (\forall i ; c \geq 2 ; 0 \leq j \leq c - 2 ; s \geq 2)$$

$$= \sum_{n=1}^{j+1} \mu_{i,n} \quad (\forall i ; c \geq 2 ; 0 \leq j \leq c - 2 ; s = 1) .$$

Define

$$B_{j-s,j} = \text{Diag}\,[B_{0,j-s,j}, B_{1,j-s,j}, \ldots, B_{N,j-s,j}] \quad (j - s < j \leq L) ;$$

$$B_s = B_{j-s,j} \quad (j < L)$$

$$= \text{Diag}\,\left[\sum_{k=1}^{K} \sigma_{0,k}(1 - \theta_{0,k})\theta_{0,k}^{s-1}, \ldots, \sum_{k=1}^{K} \sigma_{N,k}(1 - \theta_{N,k})\theta_{N,k}^{s-1}\right] ;$$

$$\Sigma_k = \text{Diag}\,[\sigma_{0,k}, \sigma_{1,k}, \ldots, \sigma_{N,k}] \quad (k = 1, 2, \ldots, K) ;$$

$$\Theta_k = \text{Diag}\,[\theta_{0,k}, \theta_{1,k}, \ldots, \theta_{N,k}] \quad (k = 1, 2, \ldots, K) ;$$

$$\Sigma = \sum_{k=1}^{K} \Sigma_k ;$$

$$R = \text{Diag}\,[\rho_0, \rho_1, \ldots, \rho_N] ; \quad \Delta = \text{Diag}\,[\delta_0, \delta_1, \ldots, \delta_N] ;$$

$$M_n = \text{Diag}\left[\mu_{0,n}, \mu_{1,n}, \ldots, \mu_{N,n}\right] \qquad (n = 1, 2, \ldots, c) \ ;$$

$$\Phi_n = \text{Diag}\left[\phi_{0,n}, \phi_{1,n}, \ldots, \phi_{N,n}\right] \qquad (n = 1, 2, \ldots, c) \ ;$$

$$C_j = \sum_{n=1}^{j} M_n \qquad (1 \leq j \leq c) \ ;$$

$$= \sum_{n=1}^{c} M_n = C \qquad (j \geq c) \ ;$$

$$C_{j+s,j} = \text{Diag}\left[C_{0,j+s,j}, C_{1,j+s,j}, \ldots, C_{N,j+s,j}\right] \ ;$$
$$E = \text{Diag}(e_N') \ .$$

Then, we get

$$B_s = \sum_{k=1}^{K} \Theta_k^{s-1}(E - \Theta_k)\Sigma_k \ ; \quad B_1 = B = \sum_{k=1}^{K}(E - \Theta_k)\Sigma_k \ ;$$

$$B_{L-s,L} = \sum_{k=1}^{K} \Theta_k^{s-1}\Sigma_k \ ;$$

$$C_{j+s,j} = \sum_{n=1}^{c} M_n(E - \Phi_n)\Phi_n^{s-1} + R(E - \Delta)\Delta^{s-1}$$

$$(c + 1 \leq j \leq L - 1 \ ; \ s = 1, 2, \ldots, L - j) \ ;$$

$$= \sum_{n=1}^{c} M_n(E - \Phi_n)\Phi_n^{s-1} + R\Delta^{s-1} \qquad (j = c \ ; \ s = 1, 2, \ldots, L - c) \ ;$$

$$= \sum_{n=1}^{c} M_n\Phi_n^{s-1} \qquad (j = c - 1 \ ; \ s = 1, 2, \ldots, L - c + 1) \ ;$$

$$= 0 \qquad (c \geq 2 \ ; \ 0 \leq j \leq c - 2 \ ; \ s \geq 2) \ ;$$

$$= C_{j+1} \qquad (c \geq 2 \ ; \ 0 \leq j \leq c - 2 \ ; \ s = 1) \ .$$

The steady state balance equations are:

(1) for the Lth row or level:

$$\sum_{s=1}^{L} \mathbf{v}_{L-s} B_{L-s,L} + \mathbf{v}_L \left[Q - C - R \right] = 0 ; \tag{1}$$

(2) for the jth row or level:

$$\sum_{s=1}^{j} \mathbf{v}_{j-s} B_s + \mathbf{v}_j \left[Q - \Sigma - C_j - R I_{j>c} \right] + \sum_{s=1}^{L-j} \mathbf{v}_{j+s} C_{j+s,j} = 0$$
$$(0 \le j \le L - 1) ; \tag{2}$$

(3) normalization

$$\sum_{j=0}^{L} \mathbf{v}_j \mathbf{e}_{N+1} = 1 . \tag{3}$$

Note that $I_{j>c} = 1$ if $j > c$ else 0, and $\mathbf{e}_{N+1}$ is a column vector of size $N+1$ with all ones.

One can observe that there are infinite number of equations in infinite number of unknowns, viz. $v_0, v_1, \ldots$, when $L = \infty$. Also, each of the balance equation is infinitely long containing all the infinite number of unknowns, viz. $v_0, v_1, \ldots$. The coefficient matrices of the unknown vectors are j-dependent. This is a very complex system of equations for which there is neither an existing solution (exact or approximate) nor a solution methodology. Hence, in the next section we *transform* this system of equations to a *computable* form.

2.2   Transforming the Balance Equations

Define the functions, $F_{K,l}$ $(l = 1, 2, \ldots, K)$ and $H_{c,n}$ $(n = 1, 2, \ldots, c)$ as

$$F_{K,l} = \sum_{1 \le k_1 < k_2 < \ldots < k_l \le K} \Theta_{k_1} \Theta_{k_2} \ldots \Theta_{k_l} \quad (l = 1, 2, \ldots, K)$$
$$= E \quad \text{if } l = 0$$
$$= 0 \quad \text{if } l \le -1 \text{ or } l > K, \tag{4}$$

$$H_{c,n} = \sum_{1 \le k_1 < k_2 < \ldots < k_n \le c} \Phi_{k_1} \Phi_{k_2} \ldots \Phi_{k_n} \quad (n = 1, 2, \ldots, c)$$
$$= E \quad \text{if } n = 0$$
$$= 0 \quad \text{if } n \le -1 \text{ or } n > c . \tag{5}$$

These functions have the following alternate definitions, properties and recursion by which they can be conceived and computed quite easily.

$$F_{k,0} = E , \quad F_{k,k} = \prod_{i=1}^{k} \Theta_i \ (k = 1, 2, \ldots, K);$$
$$F_{k,l} = 0 \ (k = 1, 2, \ldots, K; l < 0) ;$$
$$F_{k,l} = 0 \ (k = 1, 2, \ldots, K; l > k) \tag{6}$$

$$H_{m,0} = E \ , \quad H_{m,m} = \prod_{i=1}^{m} \Phi_i \ (m = 1, 2, \ldots, c);$$
$$H_{m,n} = 0 \ (m = 1, 2, \ldots, c; \ n < 0) \ ;$$
$$H_{m,n} = 0 \ (m = 1, 2, \ldots, c; \ n > m). \tag{7}$$

The recursion, then, is

$$F_{1,0} = E \ ; \quad F_{1,1} = \Theta_1 \ ;$$
$$F_{k,l} = F_{k-1,l} + \Theta_k F_{k-1,l-1} \ (2 \le k \le K \ , \ 1 \le l \le k - 1) \ ; \tag{8}$$

$$H_{1,0} = E \ ; \quad H_{1,1} = \Phi_1 \ ;$$
$$H_{m,n} = H_{m-1,n} + \Phi_m H_{m-1,n-1} \ (2 \le m \le c \ , \ 1 \le n \le m - 1) \ . \tag{9}$$

Transformation 1. *Modify simultaneously the balance equations for levels j ($L - 2 - c \ge j \ge c + K + 1$), by the transformation:*

$$<\mathbf{j}>^{(1)} \ \longleftarrow \ <\mathbf{j}> + \sum_{l=1}^{K} (-1)^l <\mathbf{j} - \mathbf{1}> F_{K,l} \quad (c + K + 1 \le j \le L - 2 - c);$$
$$<\mathbf{j}>^{(1)} \ \longleftarrow \ <\mathbf{j}> \qquad\qquad\qquad (j > L - 2 - c \ or \ j < c + K + 1).$$

The balance equation for level j after Transformation 1 is $<\mathbf{j}>^{(1)}$.

Transformation 2. *Modify simultaneously the balance equations for levels j ($L - 2 - c \ge j \ge c + K + 1$), by the transformation:*

$$<\mathbf{j}>^{(2)} \ \longleftarrow \ <\mathbf{j}>^{(1)} + \sum_{n=1}^{c} (-1)^n <\mathbf{j} + \mathbf{n}>^{(1)} H_{c,n};$$
$$(c + K + 1 \le j \le L - 2 - c)$$
$$<\mathbf{j}>^{(2)} \ \longleftarrow \ <\mathbf{j}>^{(1)} \qquad (j > L - 2 - c \ or \ j < c + K + 1).$$

The balance equation for level j after Transformation 2 is denoted by $<\mathbf{j}>^{(2)}$.

Transformation 3. *Modify simultaneously the balance equations for levels j ($L - 2 - c \ge j \ge c + K + 1$), by the transformation:*

$$<\mathbf{j}>^{(3)} \ \longleftarrow \ <\mathbf{j}>^{(2)} - <\mathbf{j} + \mathbf{1}>^{(2)} \Delta \quad (c + K + 1 \le j \le L - 2 - c),$$
$$<\mathbf{j}>^{(3)} \ \longleftarrow \ <\mathbf{j}>^{(2)} \qquad\qquad (j > L - 2 - c \ or \ j < c + K + 1) \ .$$

The balance equation for level j after Transformation 3 is denoted by $<\mathbf{j}>^{(3)}$.

2.3   The j-Independent Balance Equations

Theorem 1. *With these above three transformations, the transformed balance equation, $<\mathbf{j}>^{(3)}$'s, for the rows ($c + K + 1 \le j \le L - 2 - c$), will be of the form:*

$$\mathbf{v}_{j-K} Q_0 + \mathbf{v}_{j-K+1} Q_1 + \ldots + \mathbf{v}_{j+c+1} Q_{K+c+1} = 0$$
$$(j = L - 2 - c, L - 1 - c, \ldots, c + K + 1), \tag{10}$$

where

$$G_{K,c,m} = \sum_{\substack{l-n=m \\ l=-1,\dots,K \\ n=0,\dots,c}} (-1)^{l+n} [F_{K,l}H_{c,n} + F_{K,l+1}H_{c,n}\Delta]$$

$$= \sum_{n=0}^{c} (-1)^{m+2n} [F_{K,m+n} + F_{K,m+n+1}\Delta] H_{c,n}$$

$$= (-1)^m \sum_{n=0}^{c} [F_{K,m+n} + F_{K,m+n+1}\Delta] H_{c,n} \quad (m = -1 - c, \dots, K),$$

(11)

$$Q_{K-m} = \sum_{l=-1-c}^{m-1} \left[\sum_{n=1}^{K} \Theta_n^{m-l-1}(E - \Theta_n)\Sigma_n \right] G_{K,c,l} +$$

$$[Q - \Sigma - C_{j-m} - R] G_{K,c,m} + \sum_{l=m+1}^{K} [C_{j-m,j-l}] G_{K,c,l}$$

$$(m = j - L, \dots, -2, -1, 0, \dots, K, \dots, j), \quad (12)$$

$$Q_{K-m} = \sum_{l=-1-c}^{m-1} \left[\sum_{n=1}^{K} \Theta_n^{m-l-1}(E - \Theta_n)\Sigma_n \right] G_{K,c,l} + [Q - \Sigma - C - R] G_{K,c,m}$$

$$+ \sum_{l=m+1}^{K} \left[\sum_{n=1}^{c} M_n(E - \Phi_n)\Phi_n^{l-m-1} + R(E - \Delta)\Delta^{l-m-1} \right] G_{K,c,l}$$

$$(m = -1 - c, \dots, 0, \dots, K). \quad (13)$$

The proof of Theorem 1 is presented in [1].

3 The Important Properties of the HetSigma Queue

After obtaining $F_{K,l}$'s and $H_{c,n}$'s thus, $G_{K,c,k}$, $(k = -1 - c, \dots, K)$ can be computed from (11). Then, using them directly in (13), the required Q_l $(l = 0, 1, \dots, K + c + 1)$ can be computed.

An alternative way of computing the $G_{K,c,l}$'s is by the following properties and recursion which are obtained from (8), (9) and (11):

$$G_{k,n,l} = G_{k,n-1,l} - \Phi_n G_{k,n-1,l-1}$$
$$(2 \le k \le K, \ -1 \le l + c \le k + n \le k + c),$$
$$G_{k,c,l} = G_{k-1,c,l} - \Theta_k G_{k-1,c,l-1} \ (2 \le k \le K, \ -1 \le l \le k + c). \quad (14)$$

From (11) and (14), we have

$$G_{k+1,c,l} = G_{k,c,l} - \Theta_{k+1} G_{k,c,l-1}, \quad (15)$$
$$G_{k+1,c,l} = 0, \text{ if } l \le -2 - c \text{ or } l \ge k + 2.$$

Since, K itself is arbitrary in this section, let the Q_l's be designated differently to take that into account. Let $Q_{k-m}^{(k,h)}$ $(m = j - L, j - L + 1, \ldots, j)$ be the Q_l's of $<\mathbf{j}>^{(3)}$ when only the first k customer arrival streams and the first h servers are present and others are absent.

Theorem 2. *Referring to equation (12) for the row j $(c + K + 1 \leq j \leq L - 2 - c)$, for all K, $Q_{K-m}^{(K,c)} = 0$ $(j - L \leq m \leq -2 - c)$.*

Proof. Assume the proposition is *true* for any K=k. Hence, from (12), for the range $j - L \leq m \leq -2 - c$, we have

$$
\begin{aligned}
Q_{k-m}^{(k,c)} &= \sum_{l=m+1}^{k} C_{j-m,j-l} G_{k,c,l} = \sum_{l=-1-c}^{k} C_{j-m,j-l} G_{k,c,l} \\
&\qquad \text{(since } G_{k,c,l} = 0 \text{ if } l \leq -2 - c\text{)} \\
&= 0 \quad (j - L \leq m \leq -2 - c) \,.
\end{aligned}
\tag{16}
$$

For $K = k + 1$, from (12), we get

$$
Q_{k+1-m}^{(k+1,c)} = \sum_{l=m+1}^{k+1} C_{j-m,j-l} G_{k+c+1,l}
$$

$$
= \sum_{l=m+1}^{k+1} C_{j-m,j-l} G_{k,c,l} - \Theta_{k+1} \sum_{l=m+1}^{k+1} C_{j-m,j-l} G_{k,c,l-1} \quad \text{(substituting (15))}
$$

$$
= 0 - \Theta_{k+1} \sum_{l=m+1}^{k+1} C_{j-m,j-l} G_{k,c,l-1} \quad \text{(since } G_{k,c,k+1} = 0 \text{ \& using (16))}
$$

$$
= -\Theta_{k+1} \sum_{l-1=m}^{k} C_{j-(m-1)-1,j-(l-1)-1} G_{k,c,(l-1)} \quad \text{(rearranging)}
$$

$$
= -\Theta_{k+1} \sum_{l-1=(m-1)+1}^{k} C_{j-(m-1),j-(l-1)} G_{k,c,(l-1)}
$$

$$
\text{(since } G_{k,c,m} = 0 \text{ \& } C_{j-(m-1)-1,j-(l-1)-1} = C_{j-(m-1),j-(l-1)}\text{)}
$$

$$
= 0 \quad \text{(comparing with (16))} \,.
$$

Hence the proposition is true for $K = k + 1$. Also, the proposition can be proved for $K = 2$. Hence, the theorem is true for all values of $K \geq 2$.

Theorem 3. *Referring to equation (12) for the row j $(K + 1 \leq j \leq L - 2 - c)$, for all K, $Q_{K-m}^{(K,c)} = 0$ $(K + 1 \leq m \leq j)$.*

Proof. Assume the proposition is *true* for any $K = k$. Hence, from (12), we have

$$Q_{k-m}^{(k,c)} = \sum_{l=-1-c}^{m-1} \left[\sum_{n=1}^{k} \Theta_n^{m-l-1}(E - \Theta_n)\Sigma_n \right] G_{k,c,l} \quad (k+1 \le m \le j)$$

$$= \sum_{l=-1-c}^{k} \left[\sum_{n=1}^{k} \Theta_n^{m-l-1}(E - \Theta_n)\Sigma_n \right] G_{k,c,l} = 0$$

$$\text{(since } G_{k,c,l} = 0 \text{ for } l > k) . \tag{17}$$

Then, for $K = k + 1$, writing down the expression for $Q_{k+1-m}^{(k+1)}$ from (12), substituting (15) as before and expanding the terms, we get

$$Q_{k+1-m}^{(k+1,c)} = \sum_{l=-1-c}^{k+1} \left[\sum_{n=1}^{k} \Theta_n^{m-l-1}(E - \Theta_n)\Sigma_n + \Theta_{k+1}^{m-l-1}(E - \Theta_{k+1})\Sigma_{k+1} \right] G_{k,c,l}$$

$$-\Theta_{k+1} \sum_{l=-1-c}^{k+1} \left[\sum_{n=1}^{k} \Theta_n^{m-l-1}(E - \Theta_n)\Sigma_n + \Theta_{k+1}^{m-l-1}(E - \Theta_{k+1})\Sigma_{k+1} \right] G_{k,c,l-1}.$$

Using (17) and $G_{k,c,k+1} = 0$ in the above, it simplifies to

$$Q_{k+1-m}^{(k+1,c)} = \sum_{l=-1-c}^{k+1} \Theta_{k+1}^{m-l-1}(E - \Theta_{k+1})\Sigma_{k+1} G_{k,c,l}$$

$$-\Theta_{k+1} \sum_{l-1=-2-c}^{k} \left[\sum_{n=1}^{k} \Theta_n^{(m-1)-(l-1)-1}(E - \Theta_n)\Sigma_n \right] G_{k,c,l-1}$$

$$-\Theta_{k+1} \sum_{l-1=-2-c}^{k} \Theta_{k+1}^{m-l-1}(E - \Theta_{k+1})\Sigma_{k+1} G_{k,c,l-1} .$$

However the middle term of the R.H.S. above would be 0 for $k + 1 \le m - 1 \le j$ by comparing with (17) and by using $G_{k,c,-2-c} = 0$. Hence, we obtain, for $k + 1 \le m - 1 \le j$ and hence for $k + 2 \le m \le j$,

$$Q_{k+1-m}^{(k+1,c)} = \sum_{l=-1-c}^{k+1} \Theta_{k+1}^{m-l-1}(E - \Theta_{k+1})\Sigma_{k+1} G_{k,c,l}$$

$$- \sum_{l-1=-2-c}^{k} \Theta_{k+1}^{m-(l-1)-1}(E - \Theta_{k+1})\Sigma_{k+1} G_{k,c,l-1}$$

$$= 0 \quad (m = k+2, k+3, \ldots, j) \text{ (since } G_{k,c,k+1} = 0) . \tag{18}$$

Hence, the proposition is true for $K = k+1$. The proposition can be proved for $K = 1$. Hence, the theorem is true.

Theorem 4. *Referring to equation (12) for the row j ($K + 1 \le j \le L - 2 - c$), for all K, $Q_{K-m}^{(K,c)}$ ($m = -1 - c, 0, \ldots, K$) are j-independent.*

Proof. $Q_{K-m}^{(K,c)}$ for $m = -1 - c, 0, \ldots, K$ are separately derived in (13). From the R.H.S. of (13), it is clear that $Q_{K-m}^{(K)}$ $(m = -1 - c, \ldots, K)$ are j- independent.

Theorem 5. *Referring to equation (12) for the row j $(K + 1 \le j \le L - 2 - c)$, for all*

K, $\displaystyle\sum_{m=-1-c}^{K} Q_{K-m}^{(K,c)}$ *is singular.*

Proof. Assume the theorem is true for some $K = k$. The expressions for $Q_{k-m}^{(k,c)}$ and $Q_{k+1-m}^{(k+1,c)}$ are

$$
Q_{k-m}^{(k,c)} = \sum_{l=-1-c}^{m-1} \left[\sum_{n=1}^{k} \Theta_n^{m-l-1}(E - \Theta_n)\Sigma_n \right] G_{k,c,l}
$$

$$
+ \left[Q - \sum_{n=1}^{k} \Sigma_n - C - R \right] G_{k,c,m}
$$

$$
+ \sum_{l=m+1}^{k} \left[C(E - \Phi)\Phi^{l-m-1} + R(E - \Delta)\Delta^{l-m-1} \right] G_{k,c,l}
$$

$$
(m = -1 - c, 0, \ldots, k),
$$

$$
Q_{k+1-m}^{(k+1)} = \sum_{l=-1-c}^{m-1} \left[\sum_{n=1}^{k+1} \Theta_n^{m-l-1}(E - \Theta_n)\Sigma_n \right] G_{k+1,c,l}
$$

$$
+ \left[Q - \sum_{n=1}^{k+1} \Sigma_n - C - R \right] G_{k+1,c,m}
$$

$$
+ \sum_{l=m+1}^{k+1} \left[C(E - \Phi)\Phi^{l-m-1} + R(E - \Delta)\Delta^{l-m-1} \right] G_{k+1,c,l}
$$

$$
(m = -1 - c, 0, \ldots, k + 1).
$$

Substituting $G_{k+1,c,l} = G_{k,c,l} - \Theta_{k+1}G_{k,c,l-1}$ in the latter, we get

$$
Q_{k+1-m}^{(k+1,c)} = \sum_{l=-1-c}^{m-1} \left[\sum_{n=1}^{k+1} \Theta_n^{m-l-1}(E - \Theta_n)\Sigma_n \right] (G_{k,c,l} - \Theta_{k+1}G_{k,c,l-1})
$$

$$
+ \left[Q - \sum_{n=1}^{k+1} \Sigma_n - C - R \right] (G_{k,c,m} - \Theta_{k+1}G_{k,c,m-1})
$$

$$
+ \sum_{l=m+1}^{k+1} \left[C(E - \Phi)\Phi^{l-m-1} + R(E - \Delta)\Delta^{l-m-1} \right] (G_{k,c,l} - \Theta_{k+1}G_{k,c,l-1}).
$$

After expanding, rearranging and regrouping the terms, we get

$$Q_{k+1-m}^{(k+1,c)} = Q_{k-m}^{(k)} - Q_{k-(m-1)}^{(k)}\Theta_{k+1}$$
$$+ \sum_{l=-1-c}^{m-1} (E - \Theta_{k+1})\Sigma_{k+1}\left[\Theta_{k+1}^{m-l-1}\right]G_{k+1,c,l} - \Sigma_{k+1}G_{k+1,c,m}$$
$$(m = -1 - c, \ldots, k+1)$$
$$= Q_{k-m}^{(k)} - Q_{k-(m-1)}^{(k)}\Theta_{k+1}$$
$$+ (E - \Theta_{k+1})\Sigma_{k+1}\sum_{l=-1-c}^{m-1}\left[\Theta_{k+1}^{m-l-1}G_{k+1,c,l}\right] - \Sigma_{k+1}G_{k+1,c,m}$$
$$(m = -1 - c, \ldots, k+1).$$

Then, summing up the terms from $m = -1 - c, \ldots, k+1$, we get

$$\sum_{m=-1-c}^{k+1} Q_{k+1-m}^{(k+1,c)} = \left[\sum_{m=-1-c}^{k} Q_{k-m}^{(k)}\right][E - \Theta_{k+1}]$$
$$+ \Sigma_{k+1}\left[\sum_{l=-1-c}^{m-1}\Theta_{k+1}^{m-l-1}G_{k+1,c,l} - \sum_{l=-2-c}^{m}\Theta_{k+1}^{m-l}G_{k+1,c,l}\right].$$

By substituting $G_{k+1,c,l} = G_{k,c,l} - \Theta_{k+1}G_{k,c,l-1}$ in the r.h.s. of the above equation and expanding, the terms other than the first term cancel off, leaving,

$$\sum_{m=-1-c}^{k+1} Q_{k+1-m}^{(k+1,c)} = \left[\sum_{m=-1-c}^{k} Q_{k-m}^{(k)}\right][E - \Theta_{k+1}]. \tag{19}$$

The above r.h.s. expression is clearly a singular matrix if the theorem is true for $K = k$, that is, $\left[\sum_{m=-1-c}^{k} Q_{k-m}^{(k)}\right]$ is singular, since $[E - \Theta_{k+1}]$ is a diagonal matrix. The theorem can be easily proved for $K = 1$ and for $K = 2$. Hence the theorem is true for any K.

4   Conclusions

The HetSigma queue in the Markovian framework has been proposed in order to model nodes in modern telecommunication networks [1]. The HetSigma queue and its variants have been successfully applied to carry out the performance analysis of various problems in communication networks. We have provided the rigorous proofs for the fundamental properties of the HetSigma queue, which serves as the foundation for the efficient computational approach for the Hetsigma queue.

References

1. Chakka, R., Do. The, T.V.: MM $\sum_{k=1}^{K} CPP_k/GE/c/L$ G-Queue with Heterogeneous Servers: Steady state solution and an application to performance evaluation. Performance Evaluation 64, 191–209 (2007)
2. Chakka, R., Do, T.V., Pandi, Z.: A Generalized Markovian Queue and Its Applications to Performance Analysis in Telecommunications Networks. In: Kouvatsos, D. (ed.) Performance Modelling and Analysis of Heterogeneous Networks, pp. 371–387. River Publisher (2009)
3. Chakka, R., Do, T.V., Pandi, Z.: Exact Solution for the MM $\sum_{k=1}^{K} CPP_k/GE/c/L$ G-Queue and its Application to the Performance Analysis of an Optical Packet Switching Multiplexor. In: Proceedings of the 10th International Conference on Analytical and Stochastic Modelling Techniques and Applications, Nottingham, United Kingdom (June 2003)
4. Do, T.V., Chakka, R.: A New Performability Model for Queueing and FDL-related Burst Loss in Optical Switching Nodes. Computer Communications 33(S), 146–151 (2010)
5. Chakka, R., Do, T.V.: The $MM \sum_{k=1}^{K} CPP_k/GE/c/L$ G-Queue and Its Application to the Analysis of the Load Balancing in MPLS Networks. In: Proceedings of the 27th Annual IEEE Conference on Local Computer Networks (LCN 2002),, Tampa, FL, USA, November 6-8, pp. 735–736 (2002)
6. Do, T.V., Papp, D., Chakka, R., Truong, M.X.T.: A Performance Model of MPLS Multipath Routing with Failures and Repairs of the LSPs. In: Kouvatsos, D. (ed.) Performance Modelling and Analysis of Heterogeneous Networks, pp. 27–43. River Publisher (2009)
7. Do, T.V., Krieger, U.R., Chakka, R.: Performance modeling of an apache web server with a dynamic pool of service processes. Telecommunication Systems 39(2), 117–129 (2008)
8. Do, T.V., Chakka, R., Harrison, P.G.: An integrated analytical model for computation and comparison of the throughputs of the UMTS/HSDPA user equipment categories. In: Proceedings of the 10th ACM Symposium on Modeling, Analysis, and Simulation of Wireless and Mobile Systems, MSWiM 2007, pp. 45–51. ACM, New York (2007)
9. Do, T.V., Chakka, R., Do, N., Pap, L.: A markovian queue with varying number of servers and applications to the performance comparison of hsdpa user equipment. Acta Informatica 48, 243–269 (2011)
10. Kouvatsos, D.: Entropy Maximisation and Queueing Network Models. Annals of Operations Research 48, 63–126 (1994)

Performance Analysis of Edge Nodes with Multiple Path Routing

Hung Tuan Tran[1], Tien Van Do[1], and Yeon-Mo Yang[2]

[1] Department of Networked Systems and Services
Budapest University of Technology and Economics
Budapest, Hungary
do@hit.bme.hu
[2] Kumoh National Institute of Technology, South-Korea

Abstract. Multiprotocol label switching (MPLS) can flexibly establish one or several paths for traffic demands in the form of label switched paths (LSP). In this paper we propose a scheme for the multiple LSPs operation of edge nodes in MPLS networks. The proposal comprises the mechanisms of load-dependent path-decision, intervention and path selection policy to facilitate efficient LSP routing. We develop a performance model to evaluate the performance of the proposed scheme.

1 Introduction

There have been tremendous research efforts on traffic engineering to enhance the performance of IP networks [1]. These activities are partly motivated by the fact that traditional IP routing protocols calculate the shortest path based on link weights to set up routing tables to forward packets to their destination. Multiprotocol Label Switching (MPLS) [2] introduced by the IETF provides a flexible way to establish one or several paths for each traffic demand from a source to a destination in the form of label switched paths (LSP). Therefore, balanced traffic distribution can be achieved in networks [2]. Although the recommendations of MPLS include options for multipath routing, the search for mechanisms to establish paths taking into account traffic load remains an open research issue [3–5].

In this paper, we propose a scheme for the multipath (LSP- Label Switched Path) operation of edge nodes in MPLS networks. The proposal includes the mechanism of load-dependent path-decision, intervention, and path selection policy to facilitate efficient multiple LSPs routing. An analytical model is provided and performance measures are derived to compare the proposed alternatives. Previous works on the performance of multipath routing were only done with simulations [6, 7]. The authors [4] considered the operation of MPLS nodes with regard to the reliability of links in MPLS networks, but did not take into account an aspect of controlling paths.

The paper is organised as follows. We briefly describe MPLS features in Section 2. In Section 3 we present our proposals on the multipath routing operation at ingress nodes. In Section 4 we introduce the analytical performance analysis framework and we provide numerical results to give insights into the proposed multipath routing operation. Finally, we conclude the paper in Section 5.

N.T. Nguyen, T. Van Do, and H.A. Le Thi (Eds.): *ICCSAMA 2013*, SCI 479, pp. 157–165.
DOI: 10.1007/978-3-319-00293-4_13   © Springer International Publishing Switzerland 2013

2   An Overview of MPLS

MPLS is the development of IETF to provide a tool for traffic engineering and management in IP networks [2]. The aims of the MPLS development are to expand the granularity of administrative traffic control (i.e.: Traffic Engineering) [1] for network operators. MPLS is often used in the backbone of IP networks, and it consists of MPLS routers and links between MPLS nodes.

Traffic demands traversing the MPLS domain are conveyed along pipes, or in the MPLS terminology, label switched paths (LSPs). When a packet arrives at the ingress router called Label Edge Router (LER) of the MPLS domain, a short fix length label is appended to it. The packet will be assigned to a specific LSP. The criteria for the assignment is the destination IP address of the incoming packet and some addition considerations concerning the current resource availability in the domain. Afterwards, the packet is forwarded along the specified LSP in the core of the MPLS domain in a rather simple manner. At each core router called Label Switched Router (LSR), the label is simply swapped instead of interrogating IP header, significantly increasing packet forwarding efficiency, which results in tremendous gains in traffic forwarding speed.

Source-based routing is also supported in MPLS beside the shortest path routing based on the routing protocol like Open Shortest Path First (OSPF) [8]. That is, an explicit LSP for a given traffic flow from the ingress LER to the egress LER can be established and maintained based on the operation of constraint based routing (CBR) algorithms and of signalling protocols (e.g. Resource Reservation Protocol–Traffic Engineering, RSVP-TE). These two components allow MPLS to decide the LSP based not only on the link metric (as OSPF does) but also on the currently available resources along the links. By doing in this way, traffic may be routed not along the shortest path but along the most adequate path that has enough resource to meet a given target QoS (e.g. sufficient bandwidth, low delay). Moreover, traffic may also be splitted and routed simultaneously along several LSPs. All of these features make MPLS traffic engineering able to distribute evenly traffic inside the domain.

3   Proposal for the Operation of MPLS Edge Nodes with Multipath Routing

We propose that the ingress nodes perform an action (including a path decision and a path selection procedure) from time to time for each MPLS ingress and egress node-pair. The inter-times between two action points adhere to the exponential distribution with mean $1/x$.

In what follows we concentrate on traffic (referred to as the target traffic from now) supposed to be delivered between a given IE (ingress-egress) node pair of the MPLS domain. Several paths can be defined between the given IE according to some predefined criteria (e.g.: paths with disjoint edges) to carry packets of the target traffic. Assume that N is the maximum number of all possible paths that can be theoretically established between the IE node pair. It is reasonable that only a subset of N paths is used during the operation of the network and the number of paths in use should be made dependent on the traffic offered to the IE node pair (i.e.: the number of actually active paths is somewhere between 1 and N). This is done by varying the number of active paths at each action point with respect to the offered load. The actual number of the active paths depends on the decision (we call path-decision) made up by network operators at each action point with regards to the traffic load condition between the ingress and egress routers. If the offered load is high and there are still some potential paths, then a potential path (or paths) should be taken up to the set of the active paths to increase the throughput of the target traffic. In the contrary, if the offered load is low, an active path (or paths) should be removed from the set of the active paths to avoid wasting service capacity and to increase the resource availability for other traffic in the network.

3.1 Probabilistic Path Decision

The proposed probabilistic path-decision mechanism works as follows at each action point. A path (or paths) is (are) assumed to be taken up with a probability α that is a function of the offered load, expressed in the number j of packets being in the finite buffer of the ingress router plus those being delivered in the paths, i.e. $\alpha = \alpha(j)$. Intuitively, the function $\alpha(j)$ would increase with j, i.e. it has a positive slope. Theoretically, if the number of packets being in the system is too large, $\alpha(j)$ may be set to 1 (i.e. a remained potential path must be taken up with probability 1). But this is done at the expense of serving other traffic flows, since it decreases resource availability from their aspects. Therefore, we suppose that $\alpha(j)$ can only grow to the value α_{max}, ($\alpha_{max} < 1$) when the number of packets being in the system reaches the threshold a. The value of $\alpha(j)$ remains at this maximum constant if the threshold a is passed. With a similar consideration, at each action point an active path is removed with probability $\beta(j)$ which is a decreasing function of j (i.e. a function having a negative slope). Once the number of packets being in the system reaches the value b, $\beta(j)$ becomes constant $\beta(j) = \beta_{min}$, ($\beta_{min} > 0$). The functions $\alpha(j)$ and $\beta(j)$ may have various shape depending on the way traffic flows are interacting in the network.

3.2 LSP Selection Policies

Path-selection addresses the issue of how many and which path(s) should be taken up (removed) if there is a need for such actions. For this aim we introduce a probabilistic selection scenario as follows. Whenever a path-inclusion is decided, a probability u_k is assigned to the event that k paths are taken up simultaneously. Similarly, whenever a path-removal is decided, a predefined probability v_k is assigned to the event that k paths are removed at once. The distribution of u_k and v_k may be chosen in advance or can be adaptively changed during the operation. Note that various policies can be worked out for the identification of the selected k paths. Specifically, one of the alternatives below can be deployed:

- *Largest Removal–Smallest Inclusion* (LR-SI): in this scenario, if a path removal takes place, then from the set of the currently active paths, a path (or paths) with the largest service capacity will be removed first. On the other hand, if a path inclusion takes place, then from the set of the currently inactive paths, a path (or paths) with the smallest service capacity will be taken up first.

– *Smallest Removal–Largest Inclusion* (SR-LI): this scenario works in a contrary way compared to the LR-SI one. If a path removal takes place, then from the set of the currently active paths, a path (or paths) with the smallest service capacity will be removed first. On the other hand, if a path inclusion takes place, then from the set of the currently inactive paths, a path (or paths) with the largest service capacity will be taken up first.

4   Performance Analysis

4.1   The Overview of the Performance Model

The queueing model for an LER in respect of an ingress-egress node pair is illustrated in Fig. 1. The model consists of servers representing active LSPs and a buffer of size K. The service operation of each path between the ingress and egress nodes is considered to have GE (Generalized Exponential) distributed delivery time.

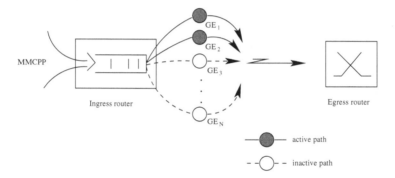

Fig. 1. Model of a multipath routing scenario

The offered traffic is modelled as the aggregation of L independent ON-OFF traffic sources. The ON and OFF periods are exponentially distributed with mean $1/\lambda$ and $1/\nu$, respectively. To keep the generality of modelling perspective, we assume that during its ON period, each source generates traffic which has an inter-arrival pattern according to a GE distribution with parameter pair (σ, θ).

The GE distribution for the inter-arrival times is chosen to capture the bursty property of traffic. The GE distribution choice for the delivery time is motivated by the fact that the GE distribution is a robust two-moment approximation for any service time distribution [9]. Moreover, the use of the GE distribution still leads to the mathematical tractability of the system.

Table 1. Parameters of the GE inter-arrival time distribution of individual sources

Arrival process	
σ (1/sec)	θ
150.669890	0.526471

It is worth emphasizing that our model can be solved in the framework of the Het-Sigma queue [10]. We can derive explicit expressions for the average load in the system, the average number of active paths, the packet loss probability (shortly referred to as loss), the average packet delay (shortly referred to as delay), the average service facility, and the resource utilization, which are the important measures of service quality for the detailed formulas).

4.2   Numerical Results and Discussions

Numerical results are generated with the following input parameters and assumptions.

- There are L sources, each of which has parameters reported in Table 1.
- A set of potential LSPs with different capacities are considered. The capacity of the i-th LSP is set to $(i+1) * 0.5 Mbps$ for $i \geq 1$, i.e. an LSP is chosen to be element of the set $\{1$ Mbps, 1.5 Mbps, 2Mbps, 2.5 Mbps, 3 Mbps, ... $\}$.
- Table 2 shows the relevant parameters of the GE distributed service times for some LSPs.
- If not stated otherwise, the following parameter setting is valid $L = 3$, $\lambda = 0.01$, $y_1 = 3$, $y_2 = 2$, $a = 20$, $b = 5$, $K = 50$. Here

  • L is the number of ON-OFF traffic sources,
  • λ is the transition rate from ON state to OFF state for each individual source,
  • y_1 and y_2 are the upper bounds for batches occurring in the packet arrival process and departure process, respectively,
  • a and b are the thresholds used for the path decisions delineated in Section 3.1,
  • K is the buffer size measured in packets in the model,

The functions $\alpha(j)$ and $\beta(j)$ for the probabilistic path decisions (described in Sub-section 3.1) are chosen to be linear functions. The running parameters are the number of potential paths (N), the mean intervention rates (x) and the ON state probability of the individual sources defined as

$$\frac{E(T_{on})}{E(T_{on}) + E(T_{off})} = \frac{\nu}{\lambda + \nu}.$$

From now on, we focus our attention on the proposed multipath schemes with SR-LI and LR-SI path selection policies, or shortly SR-LI and LR-SI schemes. The distribution

Table 2. Parameters of the GE, service time distribution of LSPs with different capacities

Service time		
LSP bandwidth	μ (1/sec)	ϕ
1 Mbps	256.184596	0.109929
1.5 Mbps	384.276894	0.109929
2 Mbps	512.369192	0.109929
5 Mbps	1280.922980	0.109929
10 Mbps	2561.845960	0.109929

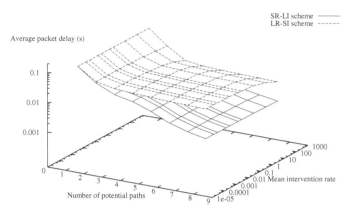

Fig. 2. Delay evolution for different numbers of potential paths and mean intervention rate x. The ON state probability is 0.7.

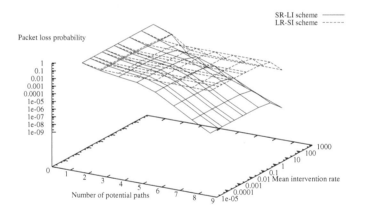

Fig. 3. Packet loss behavior for different numbers of potential paths and mean intervention rate x. The ON state probability is 0.7.

of u_k and v_k for the simultaneous selection of k paths (as introduced in Subsection 3.2) is assumed to be uniform.

In Fig. 2 and Fig. 3 we depict the delay and loss behavior as a function of the mean intervention rate and the number of potential paths. It can be observed that the packet loss probability is more significantly influenced by the intervention rate (especially in the range of 100 - 1000 interventions per second) than the delay. Since the intervention rate is limited by the processing power of the ingress router, it should be chosen with appropriate respect to the QoS performance–router processing capacity tradeoff.

Another observation from Fig. 2 and Fig. 3 is that in all scenario settings the SR-LI scheme achieves better than the LR-SI scheme from the aspect of both loss and delay.

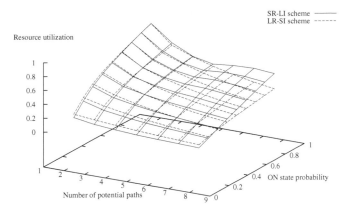

Fig. 4. Resource utilization comparison between SR-LI and LR-SI schemes

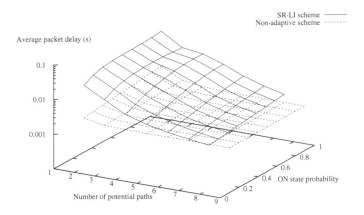

Fig. 5. Delay-related comparison between SR-LI and non-adaptive schemes

In fact, our extensive results in confirm that the performance superiority of the SR-LI scheme is experienced in the whole spectrum of traffic load, size of potential path set and mean intervention rate.

From the aspect of the *resource utilization* metric, which is the ratio between the average carried traffic and the average service capacity, it is expected that the higher the utilization is, the better a given multipath scheme is. The intuitive rationale behind this argument is that the degree of resource wastage for the target traffic between a given ingress-egress pair decreases with the resource utilization.

From Fig. 4, we see that the SR-LI scheme yields better resource utilization than the LR-SI scheme in the regime of moderate and high load and a large number of potential paths. Thus, combining with the previous results, we conclude that the SR-LI is better than the LR-SI scheme from all perspectives including loss, delay and resource utilization.

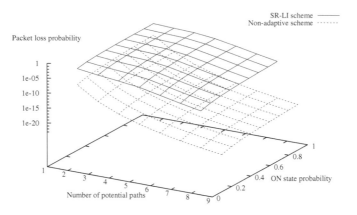

Fig. 6. Loss-related comparison between SR-LI and non-adaptive schemes

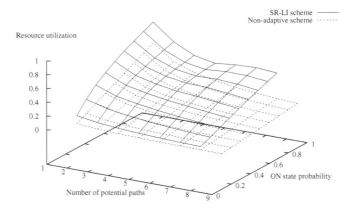

Fig. 7. Resource utilization comparison between SR-LI and non-adaptive schemes

We now compare this SR-LI scheme with the non-adaptive multipath routing scheme where a fix number of LSPs is continuously kept (i.e. there are no interventions like path removals or inclusions). Fig. 5 and Fig. 6 indicate that the non-adaptive scheme achieves better than the SR-LI scheme from both loss and delay perspectives. This is not a surprising fact, because in case of non-adaptive multipath routing the bandwidth of the whole set of potential paths is persistently reserved for the target traffic. Note that increasing the number of potential paths at a given fixed load reduces somewhat the performance bias between the adaptive SR-LI and the non-adaptive schemes, especially in case of delay metric (see Fig. 5).

However, from Fig. 7 we observe that the SR-LI scheme achieves much better performance regarding the resource utilization. For example, in case the ON state probability is 0.6 and we use 8 paths, the utilization gain is approximately 3.5 times. Furthermore, the delay assured by the SR-LI scheme is only 1.37 times worse than that of the non-adaptive scheme. Practically, packet loss is negligible.

5   Conclusions

We have proposed schemes for the multipath operation of ingress nodes in MPLS networks. The idea is to adjust the number of active LSPs according to the actual traffic load. We have developed the performance analysis of the proposed schemes. Numerical results show that the proposed adaptive multipath operation coupled with the SR-LI path selection policy has significant benefits with respect to packet loss, delay and resource utilization.

Acknowledgement. This research was supported by Basic Science Research Program through the National Research Foundation of Korea (NRF) funded by the Ministry of Education, Science and Technology(grant number, 2010-0021215).

References

1. Awduche, D., Malcolm, J., Agogbua, J., O'Dell, M., McManus, J.: Requirements for Traffic Engineering Over MPLS, RFC2702
2. Rosen, E., Viswanathan, A., Callon, R.: Multiprotocol Label Switching Architecture, RFC3031
3. Alouneh, S., En-Nouaary, A., Agarwal, A.: Securing MPLS networks with multi-path routing. In: Fourth International Conference on Information Technology, ITNG 2007, pp. 809–814 (2007)
4. Do, T.V., Papp, D., Chakka, R., Truong, M.X.T.: A Performance Model of MPLS Multipath Routing with Failures and Repairs of the LSPs. In: Kouvatsos, D. (ed.) Performance Modelling and Analysis of Heterogeneous Networks, pp. 27–43. River Publisher (2009)
5. Lin, J.-W., Liu, H.-Y.: Redirection based recovery for MPLS network systems. Journal of Systems and Software 83(4), 609–620 (2010)
6. Villamizar, C.: MPLS Optimized Multipath, Internet-draft (February 1999)
7. Schneider, G.M., Nemeth, T.: A simulation study of the OSPF-OMP routing algorithm. Computer Networks 39, 457–468 (2002)
8. IETF, OSPF Version 2, Internet standard (1998)
9. Kouvatsos, D.D.: MEM for Arbitrary Queueing Networks with Multiple General Servers and Repetitive-Service Blocking. Performance Evaluation 10, 169–195 (1989)
10. Chakka, R., Do The, T.V.: MM $\sum_{k=1}^{K} CPP_k/GE/c/L$ G-Queue with Heterogeneous Servers: Steady state solution and an application to performance evaluation. Performance Evaluation 64, 191–209 (2007)

Part III

Computational Methods for Knowledge Engineering

Meta-concept and Intelligent Data Processing in the Geotechnical Fields

Chi Tran

WM University, Faculty of Technical Sciences, Olsztyn, Poland
tran.chi@uwm.edu.pl

Abstract. Hence, using the counter-concepts *loose* or *consolidation* or *high consolidation* and *intuitive soil* or *non-intuitive soil* to express the knowledge of states of soil consolidation may lead to geotechnical paradoxes. Then, we need a new concept to represent the intelligent data that originates from engineer's intuition and experiences. It requires further a new methodology for representing and information processing.

A new approach presented in this work for this and many practical problems embraces multiple research disciplines based on integration of modern philosophy, logic, mathematics and engineering experiences. They are referred as denotation representing and computing for determining the relative density of sands using CPT data. It helps us to reach unknown reality through *true-false* measures and 'true-false' analysis using the meta-concept, "*true-false*", rather than approximation of the known reality through degrees of truth and 'degree of truth' analysis.

Keywords: Geotechnical engineering, Intelligent data processing, Counter-concept, Meta-concept, True-false measure.

1 Introduction

Usually, the traditional form of the mathematical model based on two-valued logic, in which **counter-concepts** such as true or false, certainty or uncertainty and others, originated from 'being or non-being' philosophy are adequate to solve geotechnical problems. In practice, we want to find out a *true-false adaptive inverse system*, which has some tracking capability in complexity environments, in order to estimate the primary information. This estimation is performed on the basis of data including available data, some a priori knowledge of the real system and engineer's experiences, which we call *intelligent data*. In this approach, however, what underrating and overrating the engineer's experiences in common is that they arise from the binary fallacy, for example, "if not loose, then consolidated. Since soil state is not loose, soil state is consolidated". It boils down to recognizing only 'loose' and 'consolidated', choosing between solely these when there exists a third option that transcends both. In fact, using the counter-concepts *loose* or *consolidation* or *high consolidation* and *intuitive soil* or *non-intuitive soil* to express the knowledge of states of soil consolidation may even lead to geotechnical paradoxes. As, there exist many

N.T. Nguyen, T. Van Do, and H.A. Le Thi (Eds.): *ICCSAMA 2013*, SCI 479, pp. 169–186.
DOI: 10.1007/978-3-319-00293-4_14   © Springer International Publishing Switzerland 2013

options between 'loose' and 'consolidated', which we can study in the framework of other logic based on 'being-non-being' philosophy, rather than 'being or non-being' philosophy, using the meta-concepts: 'loose-consolidated', rather than the counter-concept, 'loose' or 'consolidated'. We need, in this case, not only a deductive system based on the classical law of the excluded middle to recognize the geotechnical world but also a *deductive-inductive* system based on integration of available data (for deductive system) and the human pragmatic competences (for inductive system) using **meta-concepts**, for example, *true-false* rather than *true or false*, modern mathematics without throwing out classical mathematical results. A new approach presented in this work for this and many practical problems embraces multiple research disciplines based on integration of modern philosophy, logic, mathematics and engineering experiences. They are referred, in this paper, as denotation representing and computing for determining the relative density of sands using CPT data. It helps us to reach unknown reality of soil through *true-false* measures and 'true-false' analysis rather than approximation of the known reality through 'degrees of truth' and 'degree of truth' analysis.

2   'Being or Non-being' Philosophy and Counter-Concept

2.1   Counter-Concepts

The term *dualism* was originally created to denote co-eternal binary opposition, a meaning that is preserved in metaphysical and philosophical duality discourse. It creates the first *counter-concepts* such as: mind **or** brain, *being* **or** *non-being, objectivity* **or** *subjectivity* and others. 'Mind or brain' dualism claims that neither the mind nor brain can be reduced to each other in any way. Plato and Aristotle deal with speculation as to the existence of an incorporeal soul that bore the faculties of intelligence and wisdom. They maintained that the faculty of the mind, people's "intelligence", could not be identified with, or explained in terms of, their physical body. In history, the first *counter-concepts* – "*You or others*", "*Good or Evil*", "*Yes or No*" and "*nothing or everything*" appear in Confucius' (551 B.C.E. – 479 B.C.E.) philosophy and are represented as follows:

己 所 不 欲, 勿 施 於 人。

""What you do not wish for yourself, do not do to others."

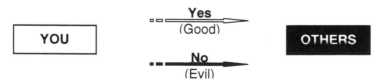

Fig. 1. Confucius' 'binary' philosophy and *counter-concepts*

This philosophy emphasized personal and governmental morality, correctness of social relationships, justice and sincerity. This alternative idea, as well as *counter-concepts* such as: *true or false, certainty or uncertainty* etc. introduced later, creates the so-called binary philosophy. Civilization started when Aristotle (344-322 BC see Barnes, Jonathan) formulated the *'Being or Non-being'* philosophy. He was a brilliant mind for many diverse topics of science and technology. This is shown in the following figure:

"It is impossible for anything to be and not to be at the same time"

"The being"   *"The non-being"*

Fig. 2. Aristotle's philosophy in the ontological sense

"Being" and its opposition, *"Non-being"*, are expressed in the law of the excluded middle, which states the necessity that either an assertion or its negation must be true.

Counter-concepts originating from *'being or non-being'* philosophy are derived from observation, physical testing and logical reasoning based on two-valued logic. It is strictly material reasoning that depends in no way upon the individualized conscious or spiritual state of the observer. Let us return to reasoning derived from observation, for example, beginning with Aristotle's syllogistic, and writing '**Being**' = B and '**Non-being**' = N, we have two alternatives: $B \vee N$. In the framework of binary philosophy, we have two possibilities: 'if not B then N' and vice versa. It is presented in the form:

$$\frac{\begin{array}{l} 1)\ B \vee N \\ 2)\ (\neg B = N) \wedge (\neg N = B) \end{array}}{Consequence: \neg\neg B = \neg N} \tag{1}$$

That is:

$$\neg\neg B = B \tag{2}$$

$$\forall_{p \in \{0,1\}} (p \vee \neg p) = 1 \tag{3}$$

On the other hand, we can represent 'Being' by two possibilities $(B, \neg B)$ and *'Non-being'* by, $(N, \neg N)$. Then, we have *certainty* $(B\neg N$ or $\neg BN)$ and *uncertainty* $(BN$ or $\neg B\neg N)$, i.e. 4 possibilities, which are represented by:

$$\text{'Being or Non-being'} = \{BN, B\neg N, \neg BN, \neg B\neg N\} \tag{4}$$

It is shown in the following network:

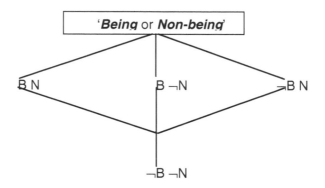

Fig. 3. '*Being* or *Non-being*' perception

Consider, for example, behaviors of explored equipment in the complex environments of soil such as the Cone Penetration Test, CPT. If knowledge about soil properties is obtained from observation of the behavior of a physical "CPT-soil" system, then it is seen as strictly material exploration that depends in no way upon the individual spiritual state of the controller. It is shown in the following figure,

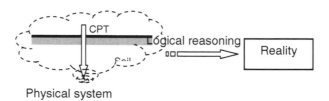

Physical system

Fig. 4. Objective reality derived from available data

2.2   Geotechnical Paradox

A study of the "CPT-soil" system using logical reasoning based on two-valued logic must, in this case, resort to statistical descriptions and what it defines is seen as 'objectivity'. Then, the laws of statistics enable us to understand the behavior of a multitude of disorganized complexities. Hence, we haven't absolute true and absolute false and, the application of binary philosophy with two-valued logic forces us to introduce the concept: certainty and uncertainty in order to evolve a given data towards an output according to certain rules. Many problems originating from different counter-concepts may, in some cases of the geotechnical fields, lead to paradoxes. For example, we can use I_D to indicate the state of denseness (or looseness) of sand soil. Current knowledge on the subject of sand consolidation is limited and qualitative. For example, soil consolidation states, s, are presented by: loose (l – Loose/Low consolidation) for $0 < I_D \leq 0.33$, moderately consolidated (m – medium consolidation) for $0.33 < I_D \leq 0.67$, or highly consolidated (h – high consolidation) for $I_D > 0.67$, which are shown in the following figure:

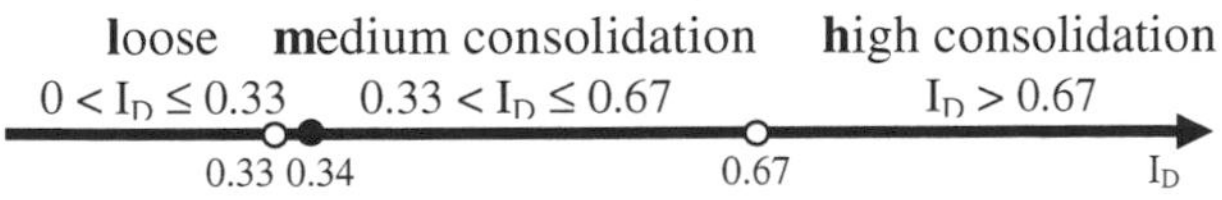

loose   medium consolidation   high consolidation
$0 < I_D \leq 0.33$ $0.33 < I_D \leq 0.67$ $I_D > 0.67$

0.33 0.34   0.67 I_D

Fig. 5. Presentation of soil consolidation states based on 'true or false' philosophy

where I_D stands for the degree of soil consolidation, which is an important parameter in geotechnical engineering. It is represented logically, in the framework of binary philosophy in the form:

$$s \in \{l \lor m \lor h\} \tag{5}$$

where, $\lor$ denotes alternative operator (or) in two-valued logic.

As we can see in the framework of binary philosophy with the *counter-concepts*: *sensitivity* **or** *non-sensitivity* the understandings of the sensitive consolidations of soil (sensitive soil) have in common, that they arise from the binary fallacy:

- *if* change of relative density I_D results in change of consolidation states *then* we have: soil is sensitive to I_D.
- Since change of I_D results in not-change of consolidation states *then* we have: soil is non-sensitive to I_D.

It indicates that the first reason is intuition; whereas the second reason is emotionalism through the binary fallacy, which boils down to recognizing only sensitivity and non-sensitivity, choosing between solely these when exists a third option that transcends both. It leads to contradiction as follows.

With the counter-concepts like *loose* **or** *medium consolidation* and *sensitivity* **or** *non-sensitivity*, we have the paradox comes from the following logical reasoning based on two-valued logic:

(Using *deductive* system)

1. If: $I_D \leq 0.33$ Then: sand is *loose*.
2. If: $0.33 < I_D \leq 0.67$ Then: sand is *medium* consolidated.
3. From 1 and 2 we have: a small changing of I_D: from 0.33 to 0.34 leads to big changing: from *loose* to *medium* consolidated in the states of soil.
4. From 3 we have p_1: it is true that: soil is *sensitive* to changing of I_D, writing: $p_1 =$ true.
5. From 2 we have: big changing of I_D: from 0.33 to 0.64, which results in non-changing of the states of soil: from medium consolidated to *medium* consolidated.
6. From 5 we have p_2: it is true that: soil is *non-sensitive* to changing of I_D, so we write: $p_2 =$ true.
7. Because of p_1: soil is sensitive to changing of I_D; whereas p_2 says: soil is *non-sensitive* to changing of I_D then whereupon we have: $p_2 = \neg p_1$ (in the framework of two valued logic).
8. From 4, $p_1 =$ true, and 6, $p_2 =$ true, from that, we have: $p_1 \land p_2 =$ true.
9. From 7, $p_2 = \neg p_1$, and 8, $p_1 \land p_2 =$ true, we have $p_1 \land \neg p_1 =$ true - **paradox!**

3   Being-Non-being Philosophy and Meta-concept

3.1   Meta-concept

'Affirmation-Denial' Concept: Contrary to dualism, monism does not accept any fundamental divisions between: physical and mental. It is a doctrine saying that *ultimate reality is entirely of 'one' substance*. Monism creates another concept named *meta-concept*. It was found in Indian logic (in the Rigveda 1000 B.C.) in which, we have the four-cornered argumentation including: *affirmation X, denial (negation)* of X, $\neg X$, both *affirmation* and *denial*, $X \wedge \neg X$, *and neither of them* $\neg(X \vee \neg X)$. Here, the joint *affirmation* and *denial* $X \wedge \neg X$ represents a third option that transcends both *affirmation* and *denial*. It we can represent as:

$$X \wedge \neg X = \text{'affirmation-denial'} - \text{is neither } \textit{affirmation} \text{ nor } \textit{denial} \tag{6}$$

The efforts of Indian logic were concentrated on explaining the *unary underlying order* of Hindu philosophy. They want to come to adopt neutral monism, the view that *the possible reality, is of one kind - is both affirmation and its denial.*

'Discrimination-Non-discrimination' Concept : Fundamental undivided nature of reality is found in Trang Chau's *'One'* philosophy, (莊周−*Zhuang Zhou* (Trang Chau) 369-286, in China). It is shown in the figure:

辨 不 辨 不 辨 辨

"Discrimination-Non-discrimination-

Non-discrimination-Discrimination"

Fig. 6. Trang-Chau's *"One"* philosophy

The author claimed that the concept of true and the opposition of it, false, are seen as oneness and according to his dynamic thinking: *"true becomes false and false becomes true ... we can't know where is the starting point, where is the end ... as a **circle**, in which, we can't distinguish big from small, good from evil, discrimination from non-discrimination and true from false"*. Author's idea, *"Discrimination-Non-discrimination-Non-discrimination-Discrimination"* or briefly, *"Discrimination-Non-discrimination"*, represents his philosophy:

"Discrimination-Non-discrimination" is neither *Discrimination* nor *Non-discrimination*

which, according to D.T. Suzuki [4] (鈴木 大拙 貞太郎 *Suzuki Daisetz Teitarō* 1962), is the most deep philosophical paradox.

In fact, in Trang Chau's philosophy, the author claimed that the concept of true and the opposition of it, false, are seen as unity according to his dynamical thinking: *"true becomes false and false becomes true, in which we can't distinguish big from small, good from evil, discrimination from non-discrimination"*. This author's way of thinking we call, as in [5], the Trang Chau's *'One'* philosophy, in which the author wanted to turn away from the *counter-concept, Discrimination* **or** *Non-discrimination*, originating from logical reasoning based on binary philosophy, to devote to the *meta-concept: discrimination-nondiscrimination*, originating from the observer's thoughts and depending rather on his/her feeling - cognitive processing.

'Being-Nonbeing' Concept: The meta-concept: "sac-sac-khong-khong" i.e., *being-being-non-being-non-being* or briefly, *being-nonbeing* is found in the Vietnamese "one" philosophy, as in [5]. That is:

being-non-being, which is neither *being* nor *non-being*

but a third option that transcends both. It is the consequence of both the observers' thinking using cognitive processing and observational results from his/her eye, i.e. a consequence of both material and spiritual phenomena. Nevertheless, its description is beyond the limitations of language expressed by *being* or *non-being*. It suggests, or even forces, us to go beyond the limitations of the counter-concept, being or non-being, in order to discover the other options represented by the *meta-concept: Being-Non-being*. We can represent the ***meta-concept, Being-Non-being*** by the following network:

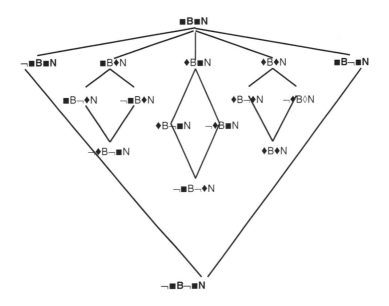

Fig. 7. The *meta-concept*, *'Being-Non-being'*, in the framework of Indian logic

That is, in the framework of Indian logic, using box connective, ■ (affirmation), and diamond connective, ♦ (both ■ and ¬■), for the four-cornered argumentation: ■, ♦, ¬■ and ¬♦. We can represent '*Being*' by 4 symbols, (■*B*, ♦*B*, ¬■*B*, ¬♦*B*), where we have certainty: ■*B*, ¬■*B* (negation), uncertainty: ♦*B* and ¬♦*B* (neither ■*B* nor ¬■*B*); similarly, '*Non-being*' by (■*N*, ♦*N*, ¬■*N*, ¬♦*N*). It is clear that instead of recognizing only being and non-being, choosing between solely these when there exist, in the observer's mind, many thoughts according to many options that transcend both being and non-being. They were presented by the term: *being-non-being*, which we call *meta-concept*. While the *counter-concepts being* or *non-being* are produced by the intercourse of the body senses with objects, originating from the observer's physical perceptions (ordinary perceptions), who can quickly recognize being or non-being by his/her eyes; the *meta-concepts, being-non-being* are individually existing in the observer's spirit in their semantic and pragmatic aspects. They are precisely the conscious development that determines other levels of reality which can be accessed and are beyond the purely physical and deterministic. Saying, it is an ability of the so-called 'psychic power of consciousnesses of the observer's cognitive processing.

'Identity-Different' Concept: The meta-concept '*identity-different*' presented in Georg Wilhelm Friedrich Hegel's philosophy (1770-1831), that is: people do not recognize that "in truth identity is different", i.e.,

Identity-different is neither *identity* nor *different*

'Necessarily_false-Necessarily_true' Concept: The meta-concept: *contingent* (♦) or possibility (◊) (*iff* it is *neither necessarily false nor necessarily true i.e.*, a third option: *necessarily_false-necessarily_true* that transcends both of them) is found in the modal logic. That is:

Necessarily_false-necessarily_true is neither *necessarily false* nor *necessarily true*

'True-false' Concept: The meta-concept *true-false* presented in Łukasiewicz's philosophy (1879–1956) as in [5]: "Truth of future events is neither true nor false", represents either a cognizable value or the 'truth' of judgments, τ, originated from the human competences like engineer's intuition and experiences depending on his/her pragmatic competences. It was created in his multi-valued logic. Author's *true-false* philosophy, in the framework of multi-valued logic, is graphically presented in the following figure:

"...is neither true nor false..."

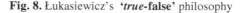

Fig. 8. Łukasiewicz's '*true-false*' philosophy

In Łukasiewicz's logic, the author used negation denoted by: ¬ or ($\overline{\dots}$), and implication, →, as the primitives and defined the other logic operations in term of these two primitives as follows:

$$\begin{cases} a \vee b & = (a \to b) \to b, \\ a \wedge b & = \overline{\overline{a} \vee \overline{b}} \\ a \Leftrightarrow b & = (a \to b) \wedge (b \to a) \end{cases} \tag{7}$$

The primitives of this logic are defined by the following equations:

$$\begin{cases} \overline{a} = 1 - a, \\ a \wedge b = \min(a, \ b), \\ a \vee b = \max(a, \ b) \\ a \to b = \min(1, \ 1 - a + b), \\ a \Leftrightarrow b = 1 - |a - b|. \end{cases} \tag{8}$$

Note that from two-valued logic, ($n = 2$), for m logic variables or functions we have:

$$\Re_2^1 \text{ x } \Re_2^2 \text{ x...x } \Re_2^m \to \Re_2 \tag{9}$$

For $m = 1$ there are ($2^z, z = 2^1$) = 4 truth-functions and for $m = 2$ there are ($2^z, z = 2^2$) = 16 different functions. Thus, we have $\Im$ truth-functions obtained from the following relationship:

$$\Im = n^z, z = n^m \tag{10}$$

To determine the number of possible *true-false* functions with $n = 3$ (multi-valued logic), $m = 2$, we have ($3^z, z = 3^2$) = 19638 possible functions. Let return also to network presented in the figure 7, we can see much more 'being-non-being' cases, depending on the use of n value (n > 4) of multi-valued logic. These results suggest that we should take advantage of the wealth structure of multi-valued logic to human cognitive processing.

'Membership-Non-membership' Concept: The meta-concept *membership-non-membership* is found in Zadeh's theory of fuzzy sets, as in [7]. The *membership grade* of x regarding to set A is shown in the following figure:

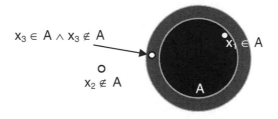

Fig. 9. Zadeh's *'membership-non-membership'* philosophy

It is:

> *membership-non-membership* is neither *membership* nor *non-membership*

It represents an idea:

> x_3 is 'neither *belong* nor *non-belong*' to the set A.

Connectives, in Zadeh's logic, are defined entirely in terms of truth-rules. For any statement, P or Q, we have a truth-value, $\tau(P)$ or $\tau(Q)$ such that:

$$0 \leq \tau(P) \leq 1 \tag{F1}$$

$$\tau(\neg P) = 1 - \tau(P) \tag{F2}$$

$$\tau(P \wedge Q) = \min(\tau(P), \tau(Q)) \tag{F3}$$

$$\tau(P \vee Q) = \max(\tau(P), \tau(Q)) \tag{F4}$$

$$\tau(P \supset Q) = \min(1, 1 - \tau(P) + \tau(Q)) \tag{F5}$$

$$\tau(P \equiv Q) = \min(1 - \tau(P) + \tau(Q), 1 + \tau(P) - \tau(Q)) \tag{F6}$$

3.2   To Go beyond Paradox

The complexity of the "CPT-Soil" system increases when its properties cannot be determined precisely by common measurements such as geometrical, physical measures derived from the behavior of the physical "CPT-Soil" system (crude data) but rather by various types of information with different metrics about noise involved in the behavior of tested equipment in complex environments. This information can be determined depending on what is distinguished by the engineers, i.e. depending on the interaction between the observer and a controlled system, including the behavior of tested equipment, controller's judgment and the individual spiritual state of the scientist. By this way, we can obtain the so called 'meta-knowledge' - '*subjective-objective*' reality. It is presented in the following figure:

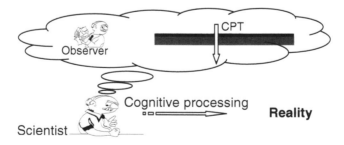

Fig. 10. Meta-knowledge that expresses *objective-subjective* reality

In this example, the soil properties are determined depending on the semantics and pragmatics of a database referred as intelligent data.

Now, a third option between loose and medium (of consolidated states of soil) can be represented by the so called *meta-concept*, as in [5]: *loose-medium* (l-m), which is neither loose nor medium for the consolidated states of sand and *sensitivity-non-sensitivity*, which is neither *sensitivity* nor *non-sensitivity* with truth values, τ, for changing of properties of sand we continue:

(Using *deductive-inductive* system)

10. Since for I_D = 0.34 we have, considering the framework of above traditional principles, a medium consolidated state. However, we have certainly "both in practice and mind" an infinite number of thoughts (denotation of sentences, as in [5]), which can be expressed by an infinite number of *true-false* values. From that, a small changing of I_D from 0.33 to 0.34 leads to a small changing in consolidation states of soil which is expressed with: changing from *loose-medium*|τ(l) = 1.0 $\wedge$ *loose-medium*|τ(m) = 0.0 to *loose-medium*|τ(l) = 0.8 $\wedge$ *loose-medium*|τ(m) = 0.2. Then, we have certainly q_1: it is true that soil is *sensitive-non-sensitive* to changing of I_D, so we write down: q_1 = true.

11. Now, there is a big changing of I_D from 0.33 to 0.64 which results in big changing of consolidation states of soil as well, which is expressed by: changing from *loose-medium*|τ(l) = 1.0 $\wedge$ *loose-medium*| τ(m) = 0.0 to *loose-medium*|τ(l) = 0.0 $\wedge$ *loose-medium*|τ(m) = 1.0. Then, we can write q_2: it is true that: soil is *sensitive-non-sensitive* to changing of I_D also, i.e., q_2 = true.

12. Because of q_1: soil is *sensitive-non-sensitive* to changing of I_D; and q_2: soil is *sensitive-non-sensitive* to changing of I_D thus, we have: $q_2 = q_1$.

13. As a result of 10, 11 and 12 we have: $q_1 \wedge q_2 = q_1 \wedge q_1$ = true – **logically!**

Thus, with this *meta-knowledge* presented by the meta-concept *sensitive-non-sensitive* you can see in the above example, how contradiction occurring in our traditional knowledge disappears by itself. It is why we have to use the geotechnical *deductive-inductive* system.

4   True-False Measures

The knowledge derived from intelligent data is the so called *meta-knowledge*. Its sense is beyond limits of language and is expressed, as in [5], by the so-called *denotation* identified by an infinite number of thoughts or by truth values (according to definition of Friedrich Ludwig Gottlob Frege in his theory of sense and denotation). We can describe, for example, the meta-concept *affirmation-denial* in the *meta-language* form:

affirmation-denial‖Denotation = *affirmation-denial*‖Thoughts/Truth values

Further on, we can use Łukasiewicz's infinite number of *true-false* measures τ according to multi-valued logic to describe the *denotation* of sentences presented in Frege's theory. By this way, we can represent the meta-concept *affirmation-denial* in the conjunctive, $\wedge$, form:

affirmation-denial\thought = affirmation-denial\τ(affirmation) ∧ *affirmation-denial\τ(denial)*

Being-Non-being\thought = Being-Non-being\τ(being) ∧ *Being-Non-being\τ(Non-being)*

likewise,

True-false\thought = True-false\τ(true) ∧ *True-false\τ(false)*

Note, that the symbol: τ, in the framework of the 'being or non-being' philosophy, is called a truth function that expresses the truth according to the degree of truth. However, it, in the framework of the 'being-non-being' philosophy, is called 'true-false' function that expresses the 'true-false' measure (non-additive, fuzzy measure) in respect of its denotation, which can be mathematically represented as follows.

Def. 1. *Borel σ-algebra*: let X be a non-empty set, $X \neq \varnothing$; the family **A** of subsets of the set X is called *σ-algebra* of subsets of X, if:

$$(X \in \mathbf{A}) \wedge (A \in \mathbf{A} \to (X - A) \in \mathbf{A}) \wedge (A_n \in \mathbf{A} \to \overset{\infty}{\underset{n=1}{\cup}} A_n \in \mathbf{A}) \tag{11}$$

The σ-algebra, which is generated by all open intervals in $\mathfrak{R}^n$, is called Borel σ-algebra and denoted as ß. The elements of family ß are called Borel sets.

Def. 2. *Measure*: a function $\tau: \mathbf{A} \Rightarrow \mathfrak{R}_+, \mathfrak{R}_+ = \mathfrak{R} \cup \{+\infty\}$ is called a measure, if:

$$\text{a) } \forall (A \in \mathbf{A}) : \{(\tau(A) \geq 0) \wedge (\tau(\varnothing) = 0)\} \tag{12}$$

$$\text{b) } \forall_{A_i \cap A_j = \varnothing; i \neq j} : \left\{ \tau \left(\overset{\infty}{\underset{i=1}{\cup}} A_i \right) = \overset{\infty}{\underset{i=1}{\sum}} \tau(A_i) \right\} \tag{13}$$

Def. 3. *Measure of a set*: The function $\tau(A)$, $A \in \mathbf{A}$ is called a measure of set A; if $\tau(A) = 0$ then A is called a measurable-zero set, $\tau: \mathbf{A} \to [0, \infty]$ is called a *set function* determined on **A**.

Def. 4. *Measurable set*: Assume, family ℑ composed of elements in the form of: $A = B \cup C$, $B \in \beta$, $\tau(C) = 0$; β is a Borel set; each element of set ℑ is called a measurable set (measurable in the sense defined by Lebesgue), if:

$$\forall_{A \in \mathfrak{I}} : \tau(A) = \tau(B \cup C) = \tau(B) \tag{14}$$

Def. 5. *Measurable function*: mathematically, Let X be a non-empty set, let S be a *σ-field* of subsets of X, let f be a partial function from X to $\overline{\mathfrak{R}}$, $\overline{\mathfrak{R}} = \mathfrak{R} \cup \{-\infty\} \cup \{+\infty\}$ and let A be an element of S. We say that f is measurable on A *iff*: for every real number r holds, as in [1].

$$A \cap LE - dom(f, \overline{\mathfrak{R}}(r)) \text{ is measurable on S} \tag{15}$$

where, the functor LE-dom(f, a) yields a subset of X and is defined by:

$$x \in \text{LE-dom}(f, a) \ \textit{iff}: x \in \text{dom}(f) \wedge \exists y \{y = f(x) \wedge y < a\}, \tag{16}$$

where y denotes an extended real number. By other way, let F be a class of all finite non-negative measurable functions defined on a measurable space, (X, $\mathbf{A}$). Moreover, let the set F_α be called an α-cut of f, $f \in$ F. Function f: X $\Rightarrow \overline{\Re}$ is called a *measurable function iff*: β is a Borel set and:

$$\forall_{B \in \beta} : \left\{ f^{-1}(B) = [x : f(x) \in B] \ belonging \ to \ \sigma - \text{algebra on } X \right\} albo \tag{17}$$

$$\{\forall_{\alpha \in \Re} : F_\alpha = [x : f(x) > \alpha]\} \tag{18}$$

Def. 6. *Quantitative-qualitative evaluation*: Let X be a nonempty set; $\mathbf{A}$ be an σ-algebra on X; τ: $\mathbf{A} \Rightarrow$ [0, 1] be a non-negative, real-valued *set function* defined on $\mathbf{A}$. τ is called a truth/*true-false* measure on (X, $\mathbf{A}$) *iff*:

a. τ is derived from both mathematical results and human "pragmatic" competences.
b. $\tau(\varnothing) = 0$ when $\varnothing \in \mathbf{A}$.
c. $\{(E \in \mathbf{A}, F \in \mathbf{A}) \wedge (E \subset F)\}$ implies $\tau(E) \leq \tau(F)$.
d. $[(\{E_n\} \subset \mathbf{A}; \cup\{E_n\} \in \mathbf{A}) \wedge (\cap\{E_n\} \in \mathbf{A})]$ then

$$[\tau(\cup\{E_n\}) = \max\{\tau(E_n)\}] \wedge [\tau(\cap\{E_n\}) = \min\{\tau(E_n)\}]$$

This number is the *quantitative-qualitative* evaluation corresponding to both quantity and quality factors in X. Hence, set $\varnothing$ has a minimum confidence: 0, i.e., $\tau(\varnothing) = 0$; total space X has a maximum confidence: 1, i.e., $\tau(X) = 1$ (boundary conditions); therefore, for every set of factors E i F, so that $E \subset F$, the confidence of E cannot be greater than the confidence of F. This means that this measure fulfills the conditions: $E \in \mathbf{A}$, $F \in \mathbf{A}$ and $E \subset F$ implies $\tau(E) \leq \tau(F)$ (monotonicity conditions). It is a variable truth coming from one's own intuition or common sense and personal reflection based on his/her experiences.

The measure τ corresponding to above conditions is called a *true-false measure* in the measurable space (X, $\mathbf{A}$), which we call a *non-additive* measure. According to λ-fuzzy measure, let $\lambda \neq 0$ and $\{E_1, E_2, ..., E_n\}$ be a disjoint class of sets in $\mathbf{A}$. So we have:

$$\tau\left(\bigcup_{i=1}^{n} E_i\right) = \frac{1}{\lambda}\left\{\prod_{i=1}^{n}[1 + \lambda\tau(E_i)] - 1\right\} \tag{19}$$

A truth λ-measure is a function τ from 2^X, (i.e., $\{0, 1\}^X$ which is a set of all functions from X to $\{0, 1\}$) to [0, 1] with properties:

$$\tau(X) = 1 \ \wedge \ (E, F \subseteq X \ \wedge \ E \cap F = \varnothing) \rightarrow \tau(E \cup F) = \tau(E) + \tau(F) + \lambda\tau(E)\tau(F) \tag{20}$$

Where λ is determined by:

$$1+\lambda = \left\{\prod_{i=1}^{n}\left[1+\lambda\tau(E_i)\right]\right\}; \quad \lambda \in (-1,\infty) \wedge \lambda \neq 0 \tag{21}$$

An evaluation undertaken by a single expert is always influenced by his/her subjectivity. However, we can imagine that each quality factor x_i of a given object also has its inherent quality index $h(x_i) \in [0, 1]$, $i = 1, 2, ..., n$. That is, we assume the existence of an objective evaluation function $h: X \rightarrow [0, 1]$. The most ideal evaluation, E^*, for the quality of the object is the fuzzy integral: $E^* = \int h \, dg$ of this function h with respect to the importance measure g, which we call the *objective synthetic evaluation*. Wang (1984-1992) introduced a generalization of the *fuzzy integral*. It is non-linear functional, where the integral is defined over measurable sets. Let $A \in \mathbf{A}$, $\mathbf{A}$ is a σ-algebra of sets in $\wp(X)$; $f \in \mathbf{F}$, $\mathbf{F}$ is the class of all finite non-negative measurable functions defined on measurable space $(X, \mathbf{A})$. For any given $f \in \mathbf{F}$ we write $F_\alpha = \{x|f(x) \geq \alpha\}$, $F_{\alpha+} = \{x|f(x) > \alpha\}$, where $\alpha \in [0, \infty]$. Let F_α and $F_{\alpha+}$ be called an α-cut and strict α-cut of f, respectively. Instead of the importance measure g, the *fuzzy integral* of f on A with respect to *true-false* measure, τ, is denoted by $\int_A f \, d\tau$. To simplify the calculation of the fuzzy integral, for a given $(X, \mathbf{A}, \tau), f \in \mathbf{F}$ and $A \in \mathbf{A}$, we have:

$$\int_A f d\tau = \sup_{\alpha \in ALFA} [\alpha \wedge \tau(A \cap F_\alpha)] \tag{22}$$

$$ALFA = \{\alpha \mid \alpha \in [0,\infty], \tau(A \cap F_\alpha) > \tau(A \cap F_\beta), for-any: \beta > \alpha\} \tag{23}$$

5   Determining the Relative Density of Sands Using CPT Data

Let CPT data: the effective overburden stress: $\sigma'_v = 81$ [kPa], the cone-tip resistance: $q_c = 5030$ [kPa], the sleeve friction: $f_s = 3$ [kPa] and the given friction ratio, $r_a = 0.06$ %, are used for determining relative density I_D of sand. Traditionally, interval numbers established depending on the current knowledge of the friction ratio r and the compressibility relationship are used to present three compressibility qualifiers such as low, medium and high. They are: low, L, for $r = [0 \div 0.3]\%$, medium, M, for $r = [0.3 \div 0.7]\%$ and high, H, for $r = [0.7 \div 1.0]\%$, in which $r = f_s/q_c$ [%]. It is represented in the framework of being or non-being philosophy for three levels of compressibility C as low **or** medium **or** high. That is:

$$C \in \{L \vee M \vee H\} \tag{24}$$

It refers as three base correlations, which is graphically presented in the following figure.

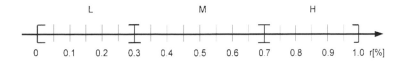

Fig. 11. Sand compressibility in the framework of *being or non-being* philosophy

5.1   Determining Relative Density of Sands Using Fuzzy Philosophy

According to the fuzzy approach, as in [2], there are three compressibility qualifiers: low (L), medium (M) and high (H). They are represented by three fuzzy sets and expressed by three membership functions $\mu(L)$, $\mu(M)$ and $\mu(H)$, which are established on the base of engineering judgments. Those three fuzzy sets Low, Medium and High are shown in the following figure.

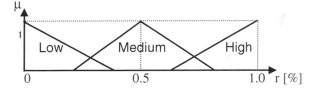

Fig. 12. Three levels of compressibility in the framework of *membership-non-membership* philosophy

To determine I_D, a weighted aggregation technique has been used to combine three base correlations, determined from the correlations, which are defined for sands of low, medium and high compressibility respectively. The relative density I_D obtained from these base correlations, as in [2], is then aggregated as follows:

$$I_D = I_D^L w^L + I_D^M w^M + I_D^H w^H \qquad (25)$$

where, I_D^k, $k = L, M, H$ are relative densities, determined from the correlations defined for sands of low, medium and high compressibility respectively (they usually relate the cone-tip resistance (q_c) to I_D with consideration of effective overburden stress (σ_v) and soil compressibility, as in [2]); w^k, denotes weights which are determined based on a "similarity" measure of three predefined levels of compressibility. The result based on fuzzy logic, as in [2], obtained by Juang et. al. is: $I_D^J = 41\%$.

As, the above aggregation, eq. (25), is based on an implicit assumption, that these effects of three compressibility levels (L, M, H) are viewed as additive. This assumption, however, is not always reasonable as indicated by Wang Z. and Klir G.J. (1992) [6]. On the other hand, although the general idea of degrees of membership agrees well with the general idea of *true-false* philosophy, we have to see in this approach, that the dividing compressibility of sands into three levels, which are described by three fuzzy sets Low, Medium and High separately or partly overlapped, does not agree with the "one" philosophy. In this case, saying as Hegel, *we are constantly using our one reality to subdivide and rearrange new sets of ones.*

5.2   Determining Relative Density of Sands Using True-False Philosophy

Let a set function τ be employed as a *true-false* measure for intelligent data. It is a *qualitative-quantitative* evaluation for the *meta-concept: Low-Medium-High*, L-M-H, which is derived from the engineer's pragmatic competences including both physical and mental perceptions of engineers.

According to Hegel's philosophy: *"reality is saturated in **ones**"*, it refers as appearance of the so called *meta-state* of sand (three states, Low, Medium and High compressibility together, which is not Loose and neither Medium nor High). It is graphically presented for any value, r_i, in the following figure.

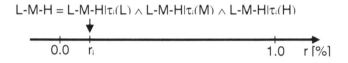

Fig. 13. Three levels of compressibility in the framework of ***true-false*** philosophy

We can distinguish low, medium and high depending on variable *true-false* measures, which is non-additive measure defined by:

$$\tau : R \rightarrow [0, 1] \tag{26}$$

According to the "one" philosophy, it refers as an appearance of the single *meta-condition* L-M-H. That is:

$$\text{L-M-H} = \text{L} \wedge \text{M} \wedge \text{H} \tag{27}$$

As a result, compressibility-level of sands is the '*one*', Low-Medium-High, in which we can recognize low from medium, medium from high and vice versa   through variable *true-false* measures $\tau(k)$, $k = $ L, M, H and *true-false* analysis, as in [5].

Now, determination of relative density (I_D) value for common sands becomes determination of *true-false* measures and *true-false* analysis. For most NC sands, according to Robertson and Campanella (1985), the predefined value of r for medium compressibility $r^*(M)$, is about 0.5%, but for sands of low compressibility $r^*(L) \approx 0\%$ and for sands of high compressibility it is $r^*(H) \approx 1\%$. Hence, the 'difference' measure between r_a and the predefined numbers $r^*(k)$, $k = $ L, M, H is defined by:

$$diff_{r_a}(k) = \left| r_a - r^*(k) \right| \tag{28}$$

It is used as a means of measuring, how close the actual friction ratio r_a is to each of the predefined numbers $r^*(k)$. Smaller distance indicates a higher degree of similarity.

The compressibility measured by friction ratio corresponding to a higher similarity is intuitively assigned a greater value of truth/*true-false* in the general form:

$$\tau(k) = 1 - diff_{r_a}(k), \; k = L, M, H \tag{29}$$

i.e. sand, having the compressibility level k = L, M, H, is intuitively assigned to this *true-false* measure $\tau(k)$. In addition, we support here given evaluations by three experts expressed by: $\tau^*_i(k)$, i = 1, 2, 3. From that we can construct the λ-*true-false* measure for all the other subsets of set X, X = {L ∪ M ∪ H}. Then, we can calculate the λ-*true-false* measures, {$\tau(L ∪ M)$, $\tau(L ∪ H)$ and $\tau(M ∪ H)$} by equations (19, 20, 21) for different subsets {(L ∪ M), (L ∪ H) and (M ∪ H)}, finally, I_D by a cognitive processing with intelligent data using Sugeno (1977) type integral, as in [3], of a *measurable function* $f(k), f(k) \in F_\alpha$ with respect to the true-false measure, τ, in the form:

$$I_D = \int f(k) \, d\tau = \sup_{\alpha \in [0, \infty]} [\alpha \wedge \tau(R \cap F_\alpha)], \; F_\alpha = \{k \mid f(k) \geq \alpha\}, k \subset R; \tag{30}$$

in which, R = {L, M, H}; L, M, H represent the Loose, Medium and High states of sands. The evaluations and obtained results of three experts are shown in the following table.

Table 1. Qualitative-quantitative evaluations and numerical results, I_D^*

Expert	$\tau^*_i(L)$	$\tau^*_i(M)$	$\tau^*_i(H)$	$\tau(L∪M)$	$\tau(L∪H)$	$\tau(M∪H)$	I_D^* [%]
1	0.80	0.30	0.10	0.95	0.85	0.38	41.0
2	0.80	0.50	0.10	0.98	0.84	0.56	42.8
3	0.80	0.30	0.20	0.93	0.88	0.46	42.8

Obviously the changes of results I_D^*, depending on changes of {$\tau^*(L)$, $\tau^*(M)$, $\tau^*(H)$, $\tau(L∪M)$, $\tau(L∪H)$, $\tau(M∪H)$}, confirm the requirement that the relative *true-false* measures, τ, of the compressibility are taken into account in the *true-false*-integral operator. To reduce the influence of qualitative biases of independent experts and to obtain a more reasonable evaluation of I_D^*, we can use an arithmetic average of the results obtained from the three experts. That is:

$$(0.41 + 0.428 + 0.428)/3 = 42.2\%.$$

Although those I_D^* values based on "one" philosophy agree well with the results based on fuzzy logic, as in [2], $I_D^J = 41\%$. This new result depends on the expert's *qualitative-quantitative* evaluations also, which we can improve by a process of inductive thinking. This has been discussed by Łukasiewicz in his paper on matching unknown causes to known effects in order to strengthen the experiences of engineers. A careful presentation of this process will be in another paper.

6   Conclusions

To describe the real systems like soil systems, a formulation of the counter-concept *loose* or *consolidation* based on available data resulted from physical experiments is insufficient. To recognize the real world, we are still confined to the framework of binary philosophy with counter-concepts. It is only a matter of habit, which we should change in the future from dualism to a monist view with different meta-concepts, derived from both physical and mental perceptions of engineer's pragmatic competences through their judgments. Then, we need another concept that transcends limitations of both language and Boolean numbers. For example, the meta-concept: *low-medium-high* represents a third option that transcends both *low*, *medium* and *high*. These expressive needs lead to emergence of the development of new mathematics for denotation computing and their integration with achievements of modern ICT in order to create effective information processing for different fields of sciences.

These areas of research that require more intelligence and contemplation for "the truth will be in the future" should require the meta-concepts derived from being-non-being philosophy and multi-valued logic.

Acknowledgments. I would like to express many thank for supports of the National Center of Sciences in Krakow and WM University in Olsztyn, Poland, in the grant nr. NN 506 215 440.

References

1. Kharazishvili, A.B.: On Almost Measurable Real-valued Functions. Studia Scientiarum Mathematicarum Hungarica 47(2), 257–266 (2010)
2. Juang, C.H., Huang, X.H., Holtz, R.D., Chen, J.W.: Determining Relative Density of Sands from CPT Using Fuzzy Sets. J. of Geo. Eng. 122, 1–5 (1996)
3. Sugeno, M.: Fuzzy measures and Fuzzy integrals: A survey. In: Fuzzy Automata and Decision Processes, Amsterdam, North-Holland, pp. 89–102 (1977)
4. Teitarō, S.D.: C'est ici le paradoxe philosophique le plus profond; L'Essence du Bouddhisme, p. 88 (1962)
5. Tran, C.: Dealing with geotechnical uncertainties by being-non-being philosophy and multi-valued logic. Grant nr. NN 506215440 of National Center of Sciences in Krakow, and WM University, Olsztyn, Poland, pub. WM Univ (2012) (in press)
6. Wang, Z., Klir, G.J.: Fuzzy Measure Theory. Plenum Press, N.Y (1992)
7. Zadeh, L.A.: Fuzzy Sets. Inf. and Contr. 8(3), 338–353 (1965)

Ekman Layers of Rotating Fluids with Vanishing Viscosity between Two Infinite Parallel Plates

Van-Sang Ngo

Laboratoire de Mathématiques Raphaël Salem, UMR 6085 CNRS,
Université de Rouen, France
van-sang.ngo@univ-rouen.fr

Abstract. In this paper, we prove the convergence of weak solutions of fast rotating fluids between two infinite parallel plates towards the two-dimensional limiting system. We also put in evidence the existence of Ekman boundary layers when Dirichlet boundary conditions are imposed on the domain.

Keywords: Navier-Stokes, Rotating fluids, Ekman layer.

1 Introduction

In this paper, we consider a simplified model of geophysical fluids, that is the system of fast rotating, incompressible, homogeneous fluids between two parallel plates, with Dirichlet boundary conditions, as in [16], [18] or [8]

$$
\begin{cases}
\partial_t u^\varepsilon - \nu_h(\varepsilon)\Delta_h u^\varepsilon - \beta\varepsilon\partial_3^2 u^\varepsilon + u^\varepsilon \cdot \nabla u^\varepsilon + \dfrac{e_3 \wedge u^\varepsilon}{\varepsilon} = -\nabla p^\varepsilon & \text{in } \mathbb{R}_+ \times \Omega_h \times [0,1] \\[2mm]
\operatorname{div} u^\varepsilon = 0 & \text{in } \mathbb{R}_+ \times \Omega_h \times [0,1] \\[2mm]
u^\varepsilon\big|_{t=0} = u_0^\varepsilon, & \text{in } \Omega_h \times [0,1].
\end{cases}
\tag{1}
$$

Here, the fluid rotates in the domain $\Omega_h \times [0,1]$, between two "horizontal plates" $\Omega_h \times \{0\}$ and $\Omega_h \times \{1\}$, where Ω_h is a subdomain of $\mathbb{R}^2$. We are interested in the case where Rossby number ε goes to zero and we suppose that $\nu_h(\varepsilon)$ also goes to zero with ε. We emphasize that all along this paper, we always use the index "h" to refer to the horizontal terms and horizontal variables, and the index "v" or "3" to the vertical ones.

The Coriolis force $\varepsilon^{-1}e_3 \wedge u^\varepsilon$ has a very important impact on the behaviors of fast rotating fluids (corresponding to a small Rossby number ε). Indeed, if we suppose that $\overline{u}$ is the formal limit of u^ε when ε goes to zero, we can prove that $\overline{u}$ does not depend on the third space variable x_3. Since the Coriolis force becomes very large as ε becomes small, the "only theorical way" to balance that force is to use the pressure force term $-\nabla p^\varepsilon$. This means that there exists a function φ such that $e_3 \wedge \overline{u} = \nabla\varphi$, or in a equivalent way

$$
\begin{pmatrix} -\overline{u}^2 \\ \overline{u}^1 \\ 0 \end{pmatrix} = \begin{pmatrix} \partial_1 \varphi \\ \partial_2 \varphi \\ \partial_3 \varphi \end{pmatrix}.
$$

N.T. Nguyen, T. Van Do, and H.A. Le Thi (Eds.): *ICCSAMA 2013*, SCI 479, pp. 187–207.
DOI: 10.1007/978-3-319-00293-4_15   © Springer International Publishing Switzerland 2013

Thus, φ (and so $\overline{u}^1$ and $\overline{u}^2$) does not depend on x_3. The incompressibility of the fluid implies that

$$\partial_3 \overline{u}^3 = -\partial_1 \overline{u}^1 - \partial_2 \overline{u}^2 = -\partial_1 \partial_2 \varphi + \partial_1 \partial_2 \varphi = 0,$$

which means the third component $\overline{u}^3$ does not depend on x_3 neither. This independence was justified in the experiment of G.I. Taylor (see [10]), drops of dye injected into a rapidly rotating, homogeneous fluid, within a few rotations, formed perfectly vertical sheets of dyed fluid, known as *Taylor curtains*. In large-scale atmospheric and oceanic flows, the Rossby number is often observed to be very small, and the fluid motions also have a tendency towards columnar behaviors (*Taylor columns*). For example, currents in the western North Atlantic have been observed to extend vertically over several thousands meters without significant change in amplitude and direction ([23]).

The columnar behaviors of the solution of the system (1), in the case where $\nu(\varepsilon) > 0$ is fixed and where the domain has no boundary ($\mathbb{T}^3$ or $\mathbb{R}^3$), were studied by many authors. In the case of periodic domains, Babin, Mahalov and Nicolaenko [1]-[2], Embid and Majda [11], Gallagher [13] and Grenier [15] proved that the weak (and strong) solutions of the system (1) converge to the solution of the limiting system, which is a two-dimensional Navier-Stokes system with three components. In the case of $\mathbb{R}^3$, Chemin, Desjardins, Gallagher and Grenier proved in [6] and [7] that if the initial data are in $\mathbf{L}^2(\mathbb{R}^3)$ then the limiting system is zero. If the initial data are of the form

$$u_0 = \overline{u}_0 + v_0, \tag{2}$$

where

$$\overline{u}_0 = \left(\overline{u}_0^1(x_1, x_2), \overline{u}_0^2(x_1, x_2), \overline{u}_0^3(x_1, x_2) \right)$$

is a divergence-free vector field, independent of x_3 and

$$v_0 = \left(v_0^1(x_1, x_2, x_3), v_0^2(x_1, x_2, x_3), v_0^3(x_1, x_2, x_3) \right),$$

the limiting system is also proved to be a two-dimensional Navier-Stokes system with three components. The case where $\nu_h(\varepsilon) \to 0$ as $\varepsilon \to 0$ and the domain is $\mathbb{R}^3$ was studied by the author of this paper in [19] and [20]. We also refer to [14] in which, Gallagher and Saint-Raymond proved the convergence of the weak solutions of the system (1) to the solution of the two-dimensional limiting system in the more general case where the axis of rotation is not fixed to be e_3.

Things are very different in the case where the domain is $\Omega_h \times [0,1]$ with Dirichlet boundary conditions. Indeed, when the rotation goes to infinity, the *Taylor columns* are only formed in the interior of the domain. Near the boundary, Ekman boundary layers exist. The behaviors of the fluid become very complex and the friction slows the fluid down in a way that the velocity is zero on the boundary. In the works of Grenier and Masmoudi [16] ($\Omega_h = \mathbb{T}^2$) and Chemin *et al.* [8] ($\Omega_h = \mathbb{R}^2$), it was proved that, in the limiting system, we obtain an additional damping term of the form $\sqrt{2\beta}\overline{u}$. This phenomenon is well known in fluid mechanics as the Ekman pumping.

Since the viscosity is positive in all three directions ($\nu_h = \nu_h(\varepsilon) > 0$ and $\nu_v = \beta\varepsilon > 0$), the system (1) possesses a weak Leray solution

$$u^\varepsilon \in \mathbf{L}^\infty(\mathbb{R}_+, \mathbf{L}^2(\Omega_h \times [0,1])) \cap \mathbf{L}^2(\mathbb{R}_+, \dot{\mathbf{H}}^1(\Omega_h \times [0,1])).$$

In the case where $\nu_h > 0$ is fixed and where the initial data are well prepared, i.e.

$$\lim_{\varepsilon\to 0} u_0^\varepsilon = \overline{u}_0 = (\overline{u}_0^1(x_1, x_2), \overline{u}_0^2(x_1, x_2), 0) \quad \text{in } \mathbf{L}^2(\mathbb{R}_h^2 \times [0,1])$$

and $\overline{u}_0$ is a divergence-free two-dimensional vector field in $\mathbf{H}^\sigma(\mathbb{R}_h^2)$, $\sigma > 2$, it was proved by Grenier and Masmoudi in [16] ($\Omega_h = \mathbb{T}_h^2$) and by Chemin *et al.* in [8] ($\Omega_h = \mathbb{R}_h^2$) that, when ε goes to zero, u^ε converges to the solution of the following limiting system in $\mathbf{L}^\infty(\mathbb{R}_+, \mathbf{L}^2(\mathbb{R}^3))$

$$\begin{cases} \partial_t \overline{u}^h - \nu_h \Delta_h \overline{u}^h + \overline{u}^h \cdot \nabla_h \overline{u}^h + \sqrt{2\beta}\,\overline{u}^h = -\nabla_h \overline{p} \\ \partial_t \overline{u}^3 - \nu_h \Delta_h \overline{u}^3 + \overline{u}^h \cdot \nabla_h \overline{u}^3 + \sqrt{2\beta}\,\overline{u}^3 = 0 \\ \operatorname{div}_h \overline{u}^h = 0 \\ \partial_3 \overline{u} = 0 \\ \overline{u}|_{t=0} = \overline{u}_0. \end{cases} \quad (3)$$

The case of ill-prepared data, where $\overline{u}_0 = (\overline{u}_0^1(x_1, x_2), \overline{u}_0^2(x_1, x_2), \overline{u}_0^3(x_1, x_2))$ has all the three components different from 0, was studied in [8].

In this paper, we consider the system (1) in the case where $\Omega_h = \mathbb{R}^2$, where $\nu_h(\varepsilon) \to 0$ as $\varepsilon \to 0$ and where the data are well prepared. The limiting system is the following

$$\begin{cases} \partial_t \overline{u}^h + \overline{u}^h \cdot \nabla_h \overline{u}^h + \sqrt{2\beta}\,\overline{u}^h = -\nabla_h \overline{p} \\ \partial_t \overline{u}^3 + \overline{u}^h \cdot \nabla_h \overline{u}^3 + \sqrt{2\beta}\,\overline{u}^3 = 0 \\ \operatorname{div}_h \overline{u}^h = 0 \\ \partial_3 \overline{u} = 0 \\ \overline{u}|_{t=0} = \overline{u}_0. \end{cases} \quad (4)$$

We want to remark that in this case where the data are well prepared, as $\overline{u}_0^3 = 0$, the third component $\overline{u}^3 = 0$ for any $t > 0$. In [16] and [8], it was proved that, in the case where $\nu_h \to 0$ as $\varepsilon \to 0$, the weak solutions of the system (1) converge to the solution of the limiting system (4), but the convergence is only local with respect to the time variable. In this paper, we show the exponential decay of the solution of the system (4) in appropriate Sobolev norms, and we improve the result of [16] and [8]. More precisely, we prove the uniform convergence (with respect to the time variable) of (1) towards (4).

Theorem 1. *Suppose that*

$$\lim_{\varepsilon\to 0} \nu_h(\varepsilon) = 0 \quad and \quad \lim_{\varepsilon\to 0} \frac{\varepsilon^{\frac{1}{2}}}{\nu_h(\varepsilon)} = 0.$$

Let $u_0^\varepsilon \in \mathbf{L}^2(\mathbb{R}_h^2 \times [0,1])$ be a family of initial data such that

$$\lim_{\varepsilon \to 0} u_0^\varepsilon = \overline{u}_0 = (\overline{u}_0^1(x_1,x_2), \overline{u}_0^2(x_1,x_2),0) \quad in \quad \mathbf{L}^2(\mathbb{R}_h^2 \times [0,1]),$$

where $\overline{u}_0$ is a divergence-free two-dimensional vector field in $\mathbf{H}^\sigma(\mathbb{R}_h^2)$, $\sigma > 2$. Let $\overline{u}$ be the solution of the limiting system (4) with initial data $\overline{u}_0$ and, for each $\varepsilon > 0$, let u^ε be a weak solution of (1) with initial data u_0^ε. Then

$$\lim_{\varepsilon \to 0} \|u^\varepsilon - \overline{u}\|_{\mathbf{L}^\infty(\mathbb{R}_+, \mathbf{L}^2(\mathbb{R}_h^2 \times [0,1]))} = 0.$$

2   Preliminaries

In this section, we briefly recall the properties of dyadic decompositions in the Fourier space and give some elements of the Littlewood-Paley theory. Using dyadic decompositions, we redefine some classical function spaces, which will be used in this paper. In what follows, we always denote by (c_q) (respectively (d_q)) a square-summable (respectively summable) sequence, with $\sum_q c_q^2 = 1$ (respectively $\sum_q d_q = 1$), of positive numbers (which can depend on several parameters). We also remarque that, in order to simplify the notations, we use the bold character $\mathbf{X}$ to indicate the space of vector fields, each component of which belongs to the space X.

We recall that $\mathcal{F}$ and $\mathcal{F}^{-1}$ are the Fourier transform and its inverse, and that we also write $\widehat{u} = \mathcal{F}u$. For any $d \in \mathbb{N}^*$ and $0 < r < R$, we denote $B_d(0,R) = \{\xi \in \mathbb{R}^d \mid |\xi| \le R\}$, and $C_d(r,R) = \{\xi \in \mathbb{R}^d \mid r \le |\xi| \le R\}$. The following Bernstein lemma gives important properties of a distribution u when its Fourier transform is well localized. We refer the reader to [5] for the proof of this lemma.

Lemma 2. Let $k \in \mathbb{N}$, $d \in \mathbb{N}^*$ and $r_1, r_2 \in \mathbb{R}$ satisfy $0 < r_1 < r_2$. There exists a constant $C > 0$ such that, for any $a,b \in \mathbb{R}$, $1 \le a \le b \le +\infty$, for any $\lambda > 0$ and for any $u \in L^a(\mathbb{R}^d)$, we have

$$supp \ (\widehat{u}) \subset B_d(0, r_1\lambda) \quad \Longrightarrow \quad \sup_{|\alpha|=k} \|\partial^\alpha u\|_{L^b} \le C^k \lambda^{k+d\left(\frac{1}{a}-\frac{1}{b}\right)} \|u\|_{L^a}, \quad (5)$$

and

$$supp \ (\widehat{u}) \subset C_d(r_1\lambda, r_2\lambda) \quad \Longrightarrow \quad C^{-k}\lambda^k \|u\|_{L^a} \le \sup_{|\alpha|=k} \|\partial^\alpha u\|_{L^a} \le C^k \lambda^k \|u\|_{L^a}. \quad (6)$$

Let ψ be an even smooth function in $C_0^\infty(\mathbb{R})$, whose support is contained in the ball $B_1(0, \frac{4}{3})$, such that ψ is equal to 1 on a neighborhood of the ball $B_1(0, \frac{3}{4})$. Let

$$\varphi(z) = \psi\left(\frac{z}{2}\right) - \psi(z).$$

Then, the support of φ is contained in the ring $C_1(\frac{3}{4}, \frac{8}{3})$, and φ is identically equal to 1 on the ring $C_1(\frac{4}{3}, \frac{3}{2})$. The functions ψ and φ allow us to define a dyadic partition of $\mathbb{R}^d$, $d \in \mathbb{N}^*$, as follows

$$\forall z \in \mathbb{R}, \quad \psi(z) + \sum_{j \in \mathbb{N}} \varphi(2^{-j} z) = 1.$$

Moreover, this decomposition is almost orthogonal, in the sense that, if $|j - j'| \geq 2$, then

$$\text{supp } \varphi(2^{-j}(\cdot)) \cap \text{supp } \varphi(2^{-j'}(\cdot)) = \emptyset.$$

We introduce the following dyadic frequency cut-off operators. We refer to [3] and [5] for more details.

Definition 3. *For any $d \in \mathbb{N}^*$ and for any tempered distribution $u \in \mathcal{S}'(\mathbb{R}^d)$, we set*

$$\begin{aligned}
\Delta_q u &= \mathcal{F}^{-1}\left(\varphi(2^{-q}|\xi|)\widehat{u}(\xi)\right), & \forall q \in \mathbb{N}, \\
\Delta_{-1} u &= \mathcal{F}^{-1}\left(\psi(|\xi|)\widehat{u}(\xi)\right), & \\
\Delta_q u &= 0, & \forall q \leq -2, \\
S_q u &= \sum_{q' \leq q-1} \Delta_{q'} u, & \forall q \geq 1.
\end{aligned}$$

Using the properties of ψ and φ, one can prove that for any tempered distribution $u \in \mathcal{S}'(\mathbb{R}^d)$, we have

$$u = \sum_{q \geq -1} \Delta_q u \quad \text{in} \quad \mathcal{S}'(\mathbb{R}^d),$$

and the (isotropic) nonhomogeneous Sobolev spaces $H^s(\mathbb{R}^d)$, with $s \in \mathbb{R}$, can be characterized as follows

Proposition 4. *Let $d \in \mathbb{N}^*$, $s \in \mathbb{R}$ and $u \in H^s(\mathbb{R}^d)$. Then,*

$$\|u\|_{H^s} := \left(\int_{\mathbb{R}^d} (1 + |\xi|^2)^s |\widehat{u}(\xi)|^2 \, d\xi\right)^{\frac{1}{2}} \sim \left(\sum_{q \geq -1} 2^{2qs} \|\Delta_q u\|_{L^2}^2\right)^{\frac{1}{2}}$$

Moreover, there exists a square-summable sequence of positive numbers $\{c_q(u)\}$ with $\sum_q c_q(u)^2 = 1$, such that

$$\|\Delta_q u\|_{L^2} \leq c_q(u) 2^{-qs} \|u\|_{H^s}.$$

The decomposition into dyadic blocks also gives a very simple characterization of Hölder spaces.

Definition 5. *Let $d \in \mathbb{N}^*$ and $r \in \mathbb{R}_+ \setminus \mathbb{N}$.*

1. *If $r \in]0, 1[$, we denote $C^r(\mathbb{R}^d)$ the set of bounded functions $u : \mathbb{R}^d \to \mathbb{R}$ such that there exists $C > 0$ satisfying*

$$\forall\, (x, y) \in \mathbb{R}^d \times \mathbb{R}^d, \qquad |u(x) - u(y)| \le C\,|x - y|^r.$$

2. *If $r > 1$ is not an integer, we denote $C^r(\mathbb{R}^d)$ the set of $[r]$ times differentiable functions u such that $\partial^\alpha u \in C^{r-[r]}(\mathbb{R}^d)$, for any $\alpha \in \mathbb{N}^d$, $|\alpha| \le [r]$, where $[r]$ is the largest integer smaller than r.*

One can prove that the set $C^r(\mathbb{R}^d)$, endowed with the norm

$$\|u\|_{C^r} := \sum_{|\alpha| \le [r]} \left(\|\partial^\alpha u\|_{L^\infty} + \sup_{x \ne y} \frac{|\partial^\alpha u(x) - \partial^\alpha u(y)|}{|x - y|^{r-[r]}} \right)$$

is a Banach space. Moreover, we have the following result, the proof of which can be found in [5].

Proposition 6. *There exists a constant $C > 0$ such that, for any $r \in \mathbb{R}_+ \setminus \mathbb{N}$ and for any $u \in C^r(\mathbb{R}^d)$, we have*

$$\sup_q 2^{qr} \|\Delta_q u\|_{L^\infty} \le \frac{C^{r+1}}{[r]!} \|u\|_{C^r}.$$

Conversely, if the sequence $\left(2^{qr} \|\Delta_q u\|_{L^\infty} \right)_{q \ge -1}$ is bounded, then

$$\|u\|_{C^r} \le C^{r+1} \left(\frac{1}{r - [r]} + \frac{1}{[r] + 1 - r} \right) \sup_q 2^{qr} \|\Delta_q u\|_{L^\infty}.$$

Finally, we need the following results (for a proof, see [21]). Let $[.,.]$ denote the usual commutator.

Lemma 7. *Let $d \in \mathbb{N}^*$. There exists a constant $C > 0$ such that, for any tempered distributions u, v in $\mathcal{S}'(\mathbb{R}^d)$, we have*

$$\|[\Delta_q, u]\,v\|_{L^2} = \|\Delta_q(uv) - u\Delta_q v\|_{L^2} \le C 2^{-q} \|\nabla u\|_{L^\infty} \|v\|_{L^2}.$$

3   Estimates for the Limiting System

In this section, we give useful auxiliary results concerning the 2D limiting system (4). Throughout this paper, for any vector field $\overline{u} = (\overline{u}^1, \overline{u}^2, \overline{u}^3)$ independent of the vertical variable x_3, we denote by $\overline{w}$ the associated horizontal vorticity, $\overline{w} = \partial_1 \overline{u}^2 - \partial_2 \overline{u}^1$. For the sake of the simplicity, let $\gamma = \sqrt{2\beta}$. The first result of this section is the following lemma

Lemma 8. *Let* $\overline{u}_0 = (\overline{u}_0^1(x_1, x_2), \overline{u}_0^2(x_1, x_2), \overline{u}_0^3(x_1, x_2)) \in \mathbf{L}^2(\mathbb{R}_h^2)$ *be a divergence-free vector field, the horizontal vorticity of which*

$$\overline{w}_0 = \partial_1 \overline{u}_0^2 - \partial_2 \overline{u}_0^1 \in L^2(\mathbb{R}_h^2) \cap L^\infty(\mathbb{R}_h^2).$$

Then, the system (4), *with initial data* $\overline{u}_0$, *has a unique, global solution*

$$\overline{u} \in \mathbf{C}(\mathbb{R}_+, \mathbf{L}^2(\mathbb{R}_h^2)) \cap \mathbf{L}^\infty(\mathbb{R}_+, \mathbf{L}^2(\mathbb{R}_h^2)).$$

Moreover,

(i) There exists a constant $C > 0$ *such that, for any* $p \geq 2$ *and for any* $t > 0$, *we have*

$$\left\| \nabla_h \overline{u}^h(t) \right\|_{\mathbf{L}^p(\mathbb{R}_h^2)} \leq C M p \, e^{-\gamma t}, \tag{7}$$

$$\left\| \overline{u}^h(t) \right\|_{\mathbf{L}^p(\mathbb{R}_h^2)} \leq C M e^{-\gamma t}, \tag{8}$$

where

$$M = \max \left\{ \left\| \overline{u}_0^h \right\|_{\mathbf{L}^2(\mathbb{R}_h^2)}, \left\| \overline{w}_0 \right\|_{L^2(\mathbb{R}_h^2)}, \left\| \overline{w}_0 \right\|_{L^\infty(\mathbb{R}_h^2)} \right\}.$$

(ii) For any $p \geq 2$, *if* $\overline{u}_0^3 \in \mathbf{L}^p(\mathbb{R}_h^2)$, *then,*

$$\left\| \overline{u}^3(t) \right\|_{L^p(\mathbb{R}_h^2)} \leq \left\| \overline{u}_0^3 \right\|_{L^p(\mathbb{R}_h^2)} e^{-\gamma t}. \tag{9}$$

To prove of Lemma 8, we remark that in (4), the first two components of $\overline{u}$ verify a two-dimensional Euler system with damping term. Then, according to the Yudovitch theorem [25] (see also [5]), this system has a unique solution $\overline{u}^h \in \mathbf{C}\left(\mathbb{R}_+, \mathbf{L}^2(\mathbb{R}_h^2)\right) \cap \mathbf{L}^\infty\left(\mathbb{R}_+, \mathbf{L}^2(\mathbb{R}_h^2)\right)$ such that the horizontal vorticity $\overline{w} \in \mathbf{L}^\infty(\mathbb{R}_+, \mathbf{L}^2(\mathbb{R}_h^2)) \cap \mathbf{L}^\infty(\mathbb{R}_+, \mathbf{L}^\infty(\mathbb{R}_h^2))$. Since the third component $\overline{u}^3$ satisfies a linear transport-type equation, then we can deduce the existence and uniqueness of the solution $\overline{u}$ of the limiting system (4). Then, Inequalities (7)-(9) can be deduced from classical $\mathbf{L}^p$ estimates for Euler equations and transport equations. $\square$

Next, we need the following Brezis-Gallouet type inequality. For a proof of Lemma 9 below, see [5] or [20]. We also refer to ([4]).

Lemma 9. *Let* $r > 1$. *Under the hypotheses of Lemma 8 and the additional hypothesis that* $\overline{w}_0 \in H^r(\mathbb{R}_h^2)$, *there exists a positive constant* C_r *such that*

$$\left\| \nabla_h \overline{u}^h \right\|_{\mathbf{L}^\infty(\mathbb{R}_h^2)} \leq C_r \left\| \overline{w} \right\|_{L^\infty(\mathbb{R}_h^2)} \ln \left(e + \frac{\left\| \overline{w} \right\|_{H^r(\mathbb{R}_h^2)}}{\left\| \overline{w} \right\|_{L^\infty(\mathbb{R}_h^2)}} \right). \tag{10}$$

Now, we can give a $\mathbf{L}^\infty$-estimate of $\nabla_h \overline{u}^h$ in the following lemma.

Lemma 10. *Under the hypotheses of Lemma 9, there exist positive constants* C_1, C_2, *depending on* γ, $\left\| \overline{w}_0 \right\|_{H^r(\mathbb{R}^2)}$, *such that*

$$\left\| \overline{w}(t) \right\|_{H^r(\mathbb{R}_h^2)} \leq C_1 e^{-\gamma t}, \tag{11}$$

and

$$\left\| \nabla_h \overline{u}^h(t) \right\|_{\mathbf{L}^\infty(\mathbb{R}_h^2)} \leq C_2 e^{-\gamma t}. \tag{12}$$

Proof of Lemma 10

First of all, there exist a constant $C > 0$ and a summable sequence of positive numbers $(d_q)_{q \geq -1}$ such that (see [20])

$$\left| \left\langle \Delta_q^h (\overline{u}^h \cdot \nabla_h \overline{w}) \mid \Delta_q^h \overline{w} \right\rangle \right| \leq C d_q 2^{-2qr} \left(\left\| \nabla_h \overline{u}^h \right\|_{\mathbf{L}^\infty} + \left\| \overline{w} \right\|_{L^\infty} \right) \left\| \overline{w} \right\|_{H^r}. \quad (13)$$

Then, for any $r > 1$, we get the following energy estimate in Sobolev H^r-norm:

$$\frac{1}{2} \frac{d}{dt} \left\| \overline{w}(t) \right\|_{H^r}^2 + \gamma \left\| \overline{w}(t) \right\|_{H^r}^2 \leq C \left(\left\| \overline{w}(t) \right\|_{L^\infty} + \left\| \nabla_h \overline{u}^h(t) \right\|_{\mathbf{L}^\infty} \right) \left\| \overline{w}(t) \right\|_{H^r}^2. \quad (14)$$

Taking into account Estimate (10), we rewrite (14) as follows

$$\frac{d}{dt} \left\| \overline{w}(t) \right\|_{H^r} + \gamma \left\| \overline{w}(t) \right\|_{H^r} \leq C \left\| \overline{w} \right\|_{L^\infty} \left(1 + \ln \left(e + \frac{\left\| \overline{w} \right\|_{H^r}}{\left\| \overline{w} \right\|_{L^\infty}} \right) \right) \left\| \overline{w}(t) \right\|_{H^r}. \quad (15)$$

Since $\overline{w}$ is solution of a linear transport equation, it is easy to prove that $\left\| \overline{w}(t) \right\|_{L^p} \leq CM e^{-\gamma t}$, where C is a positive constant and

$$M = \max \left\{ \left\| \overline{u}_0^h \right\|_{\mathbf{L}^2(\mathbb{R}_h^2)}, \left\| \overline{w}_0 \right\|_{L^2(\mathbb{R}_h^2)}, \left\| \overline{w}_0 \right\|_{L^\infty(\mathbb{R}_h^2)} \right\}.$$

Since C and M do not depend on p, we have

$$\left\| \overline{w}(t) \right\|_{L^\infty} \leq CM e^{-\gamma t}.$$

Therefore, considering $y(t) = \left\| \overline{w}(t) \right\|_{H^r} e^{\gamma t}$, we can deduce from (15) that

$$\frac{d}{dt} y(t) \leq CM e^{-\gamma t} y(t) \left[1 + \ln \left(e + y(t) \right) \right]. \quad (16)$$

Integrating (16) with respect to t, we obtain the existence of $C_1 > 0$ such that

$$\left\| \overline{w}(t) \right\|_{H^r} \leq C_1 e^{-\gamma t}.$$

Combining the above estimate with (10) and using the fact that $x \ln \left(e + \frac{\alpha}{x} \right)$ is an increasing function, we obtain the existence of a positive constant C_2, depending on γ and $\left\| \overline{w}_0 \right\|_{H^r}$, such that

$$\left\| \nabla_h \overline{u}^h(t) \right\|_{\mathbf{L}^\infty} \leq C_2 e^{-\gamma t}. \qquad \square$$

In what follows, we wish to prove an estimate similar to (12) for the third component $\overline{u}^3$ of the solution $\overline{u}$ of the system (4).

Lemma 11. *Let $2 < r < 3$ and $\overline{u}(t,x)$ be a solution of (4), with initial data $\overline{u}_0$ in $\mathbf{H}^r(\mathbb{R}_h^2)$. Then, there exist a positive constant C_3, depending on γ and $\left\| \overline{u}_0 \right\|_{\mathbf{H}^r(\mathbb{R}_h^2)}$ such that, for any $t \geq 0$,*

$$\left\| \overline{u}^3(t) \right\|_{H^r(\mathbb{R}_h^2)} \leq C_3 e^{-\gamma t}. \quad (17)$$

Proof of Lemma 11

Differentiating two times the equation verified by $\overline{u}^3$, for any $i, j \in \{1, 2\}$, we have

$$\partial_t \partial_i \partial_j \overline{u}^3 + \gamma \partial_i \partial_j \overline{u}^3 + (\partial_i \partial_j \overline{u}^h) \cdot \nabla_h \overline{u}^3 + (\partial_i \overline{u}^h) \cdot \nabla_h \partial_j \overline{u}^3$$
$$+ (\partial_j \overline{u}^h) \cdot \nabla_h \partial_i \overline{u}^3 + \overline{u}^h \cdot \nabla_h \partial_i \partial_j \overline{u}^3 = 0.$$

Taking the H^{r-2} scalar product of the above equation with $\partial_i \partial_j \overline{u}^3$, we get

$$\frac{1}{2} \frac{d}{dt} \left\| \partial_i \partial_j \overline{u}^3 \right\|_{H^{r-2}}^2 + \gamma \left\| \partial_i \partial_j \overline{u}^3 \right\|_{H^{r-2}}^2 \qquad (18)$$
$$\leq \left| \langle (\partial_i \partial_j \overline{u}^h) \cdot \nabla_h \overline{u}^3 \mid \partial_i \partial_j \overline{u}^3 \rangle_{H^{r-2}} \right| + \left| \langle (\partial_i \overline{u}^h) \cdot \nabla_h \partial_j \overline{u}^3 \mid \partial_i \partial_j \overline{u}^3 \rangle_{H^{r-2}} \right|$$
$$+ \left| \langle (\partial_j \overline{u}^h) \cdot \nabla_h \partial_i \overline{u}^3 \mid \partial_i \partial_j \overline{u}^3 \rangle_{H^{r-2}} \right| + \left| \langle \overline{u}^h \cdot \nabla_h \partial_i \partial_j \overline{u}^3 \mid \partial_i \partial_j \overline{u}^3 \rangle_{H^{r-2}} \right|.$$

The divergence-free property allow us to write

$$(\partial_i \partial_j \overline{u}^h) \cdot \nabla_h \overline{u}^3 = \partial_i \left((\partial_j \overline{u}^h) \cdot \nabla_h \overline{u}^3 \right) - (\partial_j \overline{u}^h) \cdot \nabla_h \partial_i \overline{u}^3$$
$$= \partial_i \left((\partial_j \overline{u}^h) \cdot \nabla_h \overline{u}^3 \right) - \mathrm{div}_h \left(\partial_i \overline{u}^3 \partial_j \overline{u}^h \right).$$

Then, using the Cauchy-Schwarz inequality, classical estimates for Sobolev spaces (see [[5], Theorem 2.4.1]) and the Sobolev embedding $H^{r-1}(\mathbb{R}_h^2) \hookrightarrow L^\infty(\mathbb{R}_h^2)$, we obtain

$$\left| \langle (\partial_i \partial_j \overline{u}^h) \cdot \nabla_h \overline{u}^3 \mid \partial_i \partial_j \overline{u}^3 \rangle_{H^{r-2}} \right|$$
$$\leq \left\| (\partial_j \overline{u}^h) \cdot \nabla_h \overline{u}^3 \right\|_{H^{r-1}} \left\| \partial_i \partial_j \overline{u}^3 \right\|_{H^{r-2}} + \left\| \partial_i \overline{u}^3 \partial_j \overline{u}^h \right\|_{\mathbf{H}^{r-1}} \left\| \partial_i \partial_j \overline{u}^3 \right\|_{H^{r-2}}$$
$$\leq \left(\left\| \partial_j \overline{u}^h \right\|_{\mathbf{L}^\infty} \left\| \overline{u}^3 \right\|_{H^r} + \left\| \nabla_h \overline{u}^3 \right\|_{L^\infty} \left\| \partial_j \overline{u}^h \right\|_{\mathbf{H}^{r-1}} \right) \left\| \overline{u}^3 \right\|_{H^r}$$
$$+ \left(\left\| \partial_i \overline{u}^3 \right\|_{L^\infty} \left\| \partial_j \overline{u}^h \right\|_{\mathbf{H}^{r-1}} + \left\| \partial_j \overline{u}^h \right\|_{\mathbf{L}^\infty} \left\| \overline{u}^3 \right\|_{H^r} \right) \left\| \overline{u}^3 \right\|_{H^r}$$
$$\leq C \left\| \overline{w} \right\|_{H^{r-1}} \left\| \overline{u}^3 \right\|_{H^r}^2.$$

The same arguments imply

$$\left| \langle (\partial_i \overline{u}^h) \cdot \nabla_h \partial_j \overline{u}^3 \mid \partial_i \partial_j \overline{u}^3 \rangle_{H^{r-2}} \right|$$
$$\leq \left\| \mathrm{div}_h \left(\partial_j \overline{u}^3 \partial_i \overline{u}^h \right) - \partial_j \overline{u}^3 \partial_i (\mathrm{div}_h \overline{u}^h) \right\|_{H^{r-2}} \left\| \partial_i \partial_j \overline{u}^3 \right\|_{H^{r-2}}$$
$$\leq \left(\left\| \partial_j \overline{u}^3 \right\|_{L^\infty} \left\| \partial_i \overline{u}^h \right\|_{\mathbf{H}^{r-1}} + \left\| \partial_i \overline{u}^h \right\|_{\mathbf{L}^\infty} \left\| \overline{u}^3 \right\|_{H^r} \right) \left\| \overline{u}^3 \right\|_{H^r}$$
$$\leq C \left\| \overline{w} \right\|_{H^{r-1}} \left\| \overline{u}^3 \right\|_{H^r}^2,$$

and likewise,

$$\left| \langle (\partial_j \overline{u}^h) \cdot \nabla_h \partial_i \overline{u}^3 \mid \partial_i \partial_j \overline{u}^3 \rangle_{H^{r-2}} \right| \leq C \left\| \overline{w} \right\|_{H^{r-1}} \left\| \overline{u}^3 \right\|_{H^r}^2.$$

For the last term of (18), since $2 < r < 3$, a slightly different version of Estimate (13) yields

$$\left| \langle \overline{u}^h \cdot \nabla_h \partial_i \partial_j \overline{u}^3 \mid \partial_i \partial_j \overline{u}^3 \rangle_{H^{r-2}} \right| \leq C \left\| \nabla_h \overline{u}^h \right\|_{\mathbf{L}^\infty} \left\| \partial_i \partial_j \overline{u}^3 \right\|_{H^{r-2}}^2 \leq C \left\| \nabla_h \overline{u}^h \right\|_{\mathbf{L}^\infty} \left\| \overline{u}^3 \right\|_{H^r}^2.$$

Multiplying (18) by $e^{\gamma t}$, then integrating the obtained equation with respect to time and using Lemma 10, we get

$$\left\|\overline{u}^3(t)\right\|_{H^r} e^{\gamma t} \leq \left\|\overline{u}_0^3\right\|_{H^r} + C \int_0^t \left(\left\|\nabla_h \overline{u}^h(\tau)\right\|_{\mathbf{L}^\infty} + \left\|\overline{w}(\tau)\right\|_{H^{r-1}}\right) \left\|\overline{u}^3(\tau)\right\|_{H^r} e^{\gamma \tau} d\tau$$

$$\leq \left\|\overline{u}_0^3\right\|_{H^r} + C(C_1 + C_2) \int_0^t \left(\left\|\overline{u}^3(\tau)\right\|_{H^r} e^{\gamma \tau}\right) e^{-\gamma \tau} d\tau.$$

Thus, the Gronwall lemma allow us to obtain (17).   □

In the next paragraphs, we will not directly compare the system (1) with the limiting system (4) because of technical difficulties. Instead of (4), we consider the following system

$$\begin{cases} \partial_t \overline{u}^{\varepsilon,h} - \nu_h(\varepsilon) \Delta_h \overline{u}^{\varepsilon,h} + \gamma \overline{u}^{\varepsilon,h} + \overline{u}^{\varepsilon,h} \cdot \nabla_h \overline{u}^{\varepsilon,h} = -\nabla \overline{p}^\varepsilon \\ \partial_t \overline{u}^{\varepsilon,3} - \nu_h(\varepsilon) \Delta_h \overline{u}^{\varepsilon,3} + \gamma \overline{u}^{\varepsilon,3} + \overline{u}^{\varepsilon,h} \cdot \nabla_h \overline{u}^{\varepsilon,3} = 0 \\ \operatorname{div}_h \overline{u}^{\varepsilon,h} = 0 \\ \partial_3 \overline{u}^\varepsilon = 0 \\ \overline{u}^\varepsilon \big|_{t=0} = \overline{u}_0 \end{cases} \tag{19}$$

with $\lim_{\varepsilon \to 0} \nu_h(\varepsilon) = 0$.

Proposition 12. *Like the system (4), the system (19) has a unique, global solution*

$$\overline{u}^\varepsilon \in \mathbf{C}\left(\mathbb{R}_+, \mathbf{L}^2(\mathbb{R}_h^2)\right) \cap \mathbf{L}^\infty\left(\mathbb{R}_+, \mathbf{L}^2(\mathbb{R}_h^2)\right) \cap \mathbf{L}^2\left(\mathbb{R}_+, \dot{\mathbf{H}}^1(\mathbb{R}_h^2)\right),$$

which also satisfies Lemmas 8, 9, 10 and 11.

In the following lemma, we will prove the convergence of $\overline{u}^\varepsilon$ towards $\overline{u}$ when ε goes to zero.

Lemma 13. *Suppose that $\nu_h(\varepsilon)$ converges to 0 when ε goes to 0 and that $\overline{u}_0 \in \mathbf{H}^\sigma(\mathbb{R}_h^2)$, $\sigma > 2$. Then, $\overline{u}^\varepsilon$ converges towards the solution $\overline{u}$ of (4) in $\mathbf{L}^\infty(\mathbb{R}_+, \mathbf{L}^2(\mathbb{R}_h^2))$, as ε goes to 0.*

Proof of Lemma 13

Using the previously proved results of this section, for any $t > 0$, we have

$$\|\overline{u}(t)\|_{\mathbf{L}^2(\mathbb{R}_h^2)} \leq M e^{-\gamma t} \quad \text{and} \quad \|\overline{u}^\varepsilon(t)\|_{\mathbf{L}^2(\mathbb{R}_h^2)} \leq M e^{-\gamma t}. \tag{20}$$

Thus, for fixed $\mu > 0$, there exists $T_\mu > 0$ such that, for any $t \geq T_\mu$,

$$\|\overline{u}^\varepsilon(t)\|_{\mathbf{L}^2(\mathbb{R}_h^2)} + \|\overline{u}(t)\|_{\mathbf{L}^2(\mathbb{R}_h^2)} \leq \frac{\mu}{2}.$$

On the interval $[0, T_\mu]$, let $v^\varepsilon = \overline{u}^\varepsilon - \overline{u}$. Then, v^ε is a solution of the following system

$$\begin{cases} \partial_t v^{\varepsilon,h} - \nu_h(\varepsilon)\Delta_h v^{\varepsilon,h} + \gamma v^{\varepsilon,h} + \overline{u}^{\varepsilon,h} \cdot \nabla_h v^{\varepsilon,h} + v^{\varepsilon,h} \cdot \nabla_h \overline{u}^h = \nu_h \Delta_h \overline{u}^h - \nabla \tilde{p}, \\ \partial_t v^{\varepsilon,3} - \nu_h(\varepsilon)\Delta_h v^{\varepsilon,3} + \gamma v^{\varepsilon,3} + \overline{u}^{\varepsilon,h} \cdot \nabla_h v^{\varepsilon,3} + v^{\varepsilon,h} \cdot \nabla_h \overline{u}^3 = \nu_h \Delta_h \overline{u}^3, \\ \mathrm{div}_h v^{\varepsilon,h} = 0, \\ \partial_3 v^\varepsilon = 0, \\ v^\varepsilon|_{t=0} = 0. \end{cases}$$

Taking the $\mathbf{L}^2$-scalar product of the first two equations of the above system with $v^{\varepsilon,h}$ and $v^{\varepsilon,3}$ respectively, we get

$$\frac{1}{2}\frac{d}{dt}\|v^\varepsilon\|_{\mathbf{L}^2}^2 + \nu_h(\varepsilon)\|\nabla_h v^\varepsilon\|_{\mathbf{L}^2}^2 + \gamma\|v^\varepsilon\|_{\mathbf{L}^2}^2 \le \nu_h(\varepsilon)\|\nabla_h \overline{u}\|_{\mathbf{L}^2}\|\nabla_h v^\varepsilon\|_{\mathbf{L}^2} + \left|\left\langle v^{\varepsilon,h} \cdot \nabla_h \overline{u}|v^\varepsilon\right\rangle\right|.$$

Hence,

$$\frac{1}{2}\frac{d}{dt}\|v^\varepsilon\|_{\mathbf{L}^2}^2 + \gamma\|v^\varepsilon\|_{\mathbf{L}^2}^2 \le \nu_h(\varepsilon)\|\nabla_h \overline{u}\|_{\mathbf{L}^2}^2 + \|\nabla_h \overline{u}\|_{\mathbf{L}^\infty}\|v^\varepsilon\|_{\mathbf{L}^2}^2.$$

Integrating the obtained inequality, we come to

$$\|v^\varepsilon(t)\|_{\mathbf{L}^2}^2 \le \frac{\nu_h(\varepsilon)}{\gamma}\|\nabla_h \overline{u}\|_{\mathbf{L}^\infty([0,T_\mu],\mathbf{L}^2)}^2 + \int_0^t \|\nabla_h \overline{u}(\tau)\|_{\mathbf{L}^\infty}\, e^{-\gamma(t-\tau)}\|v^\varepsilon(\tau)\|_{\mathbf{L}^2}^2\, d\tau.$$

Then, the Gronwall Lemma proves that, for any $0 < t < T_\mu$,

$$\|v^\varepsilon(t)\|_{\mathbf{L}^2}^2 \le C\nu_h(\varepsilon)M^2 T_\mu \exp\left\{\int_0^t \|\nabla_h \overline{u}(\tau)\|_{\mathbf{L}^\infty}\, d\tau\right\}.$$

Combining with (20), this above estimate implies that

$$\lim_{\varepsilon \to 0}\|\overline{u}^\varepsilon - \overline{u}\|_{\mathbf{L}^\infty(\mathbb{R}_+,\mathbf{L}^2(\mathbb{R}_h^2))} = 0. \qquad \square$$

4   Ekman Boundary Layers

As mentioned in the introduction, when ε goes to 0, the fluid has the tendency to have a two-dimensional behavior. In the interior part of the domain, far from the boundary, the fluid moves in vertical columns, according to the Taylor-Proudman theorem. Near the boundary, the Taylor columns are destroyed and thin boundary layers are formed. The movements of the fluid inside the layers are very complex and the friction stops the fluid on the boundary. The goal of this paragraph is to briefly recall the mathematical construction of these boundary layers. More precisely, we will "correct" the solution of the limiting system (4) (which is a divergence-free vector field, independent of x_3) by adding a "boundary layer term" $\mathcal{B}$ such that $\overline{u} + \mathcal{B}$ is a divergence-free vector field which vanishes on the boundary.

In order to construct such boundary layers, a typical approach consists in looking for the approximate solutions of the system (1) in the following form (the Ansatz):

$$
\begin{aligned}
u_{app}^{\varepsilon} &= u_{0,int} + u_{0,BL} + \varepsilon u_{1,int} + \varepsilon u_{1,BL} + \cdots \\
p_{app}^{\varepsilon} &= \frac{1}{\varepsilon} p_{-1,int} + \frac{1}{\varepsilon} p_{-1,BL} + p_{0,int} + p_{0,BL} + \cdots,
\end{aligned}
\tag{21}
$$

where the terms with the index "int" stand for the "interior" part, which is smooth functions of (x_h, x_3) and the index "BL" refers to the boundary layer part, which is smooth functions of the form

$$
(x_h, x_3) \to F_0(t, x_h, \frac{x_3}{\delta}) + F_1(t, x_h, \frac{1 - x_3}{\delta}),
$$

where $F_0(x_h, \zeta)$ and $F_1(x_h, \zeta)$ rapidly decrease in ζ at infinity. The quantity $\delta > 0$, which goes to zero as ε goes to zero, represents the size of the boundary layers. It is proved that δ is of the same order as ε (see [16], [18], [8] and [9]). In this paper, we simply choose $\delta = \varepsilon$.

Let $E = 2\beta\varepsilon^2$ be the Ekman number and $\bar{u}$ be the solution of the limiting system (4). We recall that the third component $\bar{u}^3 = 0$ and we pose $\mathrm{curl}(\bar{u}) = \partial_1 \bar{u}^2 - \partial_2 \bar{u}^1$. In [16], [18] and [8], by studying carefully the Ansatz (21), the authors proved that we can write the boundary layer part in the following form

$$
\mathcal{B} = \mathcal{B}^1 + \mathcal{B}^2 + \mathcal{B}^3 + \mathcal{B}^4,
$$

where $\mathcal{B}^i$, $i \in \{1, 2, 3, 4\}$, are defined as follows.
1. The term $\mathcal{B}^1$ is defined by

$$
\mathcal{B}^1 = \begin{pmatrix} \tilde{w}_1 + \breve{w}_1 \\ \tilde{w}_2 + \breve{w}_2 \\ \sqrt{\frac{E}{2}} \, \mathrm{curl}(\bar{u}) \, G(x_3) \end{pmatrix}
$$

where

$$
\tilde{w}_1 = -e^{-\frac{x_3}{\sqrt{E}}} \left(\bar{u}^1 \cos\left(\frac{x_3}{\sqrt{E}}\right) + \bar{u}^2 \sin\left(\frac{x_3}{\sqrt{E}}\right) \right),
$$

$$
\tilde{w}_2 = -e^{-\frac{x_3}{\sqrt{E}}} \left(\bar{u}^2 \cos\left(\frac{x_3}{\sqrt{E}}\right) - \bar{u}^1 \sin\left(\frac{x_3}{\sqrt{E}}\right) \right),
$$

$$
\breve{w}_1 = -e^{-\frac{1-x_3}{\sqrt{E}}} \left(\bar{u}^1 \cos\left(\frac{1 - x_3}{\sqrt{E}}\right) + \bar{u}^2 \sin\left(\frac{1 - x_3}{\sqrt{E}}\right) \right),
$$

$$
\breve{w}_2 = -e^{-\frac{1-x_3}{\sqrt{E}}} \left(\bar{u}^2 \cos\left(\frac{1 - x_3}{\sqrt{E}}\right) - \bar{u}^1 \sin\left(\frac{1 - x_3}{\sqrt{E}}\right) \right),
$$

$$
G(x_3) = -e^{-\frac{x_3}{\sqrt{E}}} \sin\left(\frac{x_3}{\sqrt{E}} + \frac{\pi}{4}\right) + e^{-\frac{1-x_3}{\sqrt{E}}} \sin\left(\frac{1 - x_3}{\sqrt{E}} + \frac{\pi}{4}\right).
$$

2. The terms $\mathcal{B}^2$ and $\mathcal{B}^3$ are defined by

$$\mathcal{B}^2 = \begin{pmatrix} \sqrt{E}\,\bar{u}^2 \\ -\sqrt{E}\,\bar{u}^1 \\ \sqrt{E}\,\mathrm{curl}(\bar{u})\,\left(\frac{1}{2} - x_3\right) \end{pmatrix}$$

$$\mathcal{B}^3 = e^{-\frac{1}{\sqrt{E}}}\cos\left(\frac{1}{\sqrt{E}}\right)\begin{pmatrix} \bar{u}^1 \\ \bar{u}^2 \\ 0 \end{pmatrix}.$$

3. Finally,

$$\mathcal{B}^4 = f(x_3)\begin{pmatrix} \bar{u}^2 \\ -\bar{u}^1 \\ 0 \end{pmatrix} + g(x_3)\begin{pmatrix} 0 \\ 0 \\ \mathrm{curl}(\bar{u}) \end{pmatrix},$$

where

$$f(x_3) = a\left(e^{-\frac{x_3}{\sqrt{E}}} + e^{-\frac{1-x_3}{\sqrt{E}}}\right) + b,$$

$$g(x_3) = -\sqrt{\frac{E}{2}}\,e^{-\frac{1}{\sqrt{E}}}\sin\left(\frac{1}{\sqrt{E}} + \frac{\pi}{4}\right) - \int_0^{x_3} f(s)\,ds,$$

and where (a, b) is the solution of the linear system

$$\begin{cases} a\left(1 + e^{-\frac{1}{\sqrt{E}}}\right) + b = -\sqrt{E} + e^{-\frac{1}{\sqrt{E}}}\sin\left(\frac{1}{\sqrt{E}}\right) \\[2mm] 2a\sqrt{E}\left(1 - e^{-\frac{1}{\sqrt{E}}}\right) + b = \sqrt{2E}e^{-\frac{1}{\sqrt{E}}}\sin\left(\frac{1}{\sqrt{E}} + \frac{\pi}{4}\right). \end{cases} \tag{22}$$

We remark that the determinant of the system (22) is

$$D = 1 + e^{-\frac{1}{\sqrt{E}}} - 2\sqrt{E}\left(1 - e^{-\frac{1}{\sqrt{E}}}\right).$$

Thus, for $\varepsilon > 0$ small enough, we have $D > \frac{1}{2}$ and (22) always has the following solution

$$a = \frac{J_E - K_E}{D} \quad \text{and} \quad b = \frac{K_E\left(1 + e^{-\frac{1}{\sqrt{E}}}\right) - 2J_E\sqrt{E}\left(1 - e^{-\frac{1}{\sqrt{E}}}\right)}{D},$$

where

$$J_E = -\sqrt{E} + e^{-\frac{1}{\sqrt{E}}}\sin\left(\frac{1}{\sqrt{E}}\right),$$

$$K_E = \sqrt{2E}e^{-\frac{1}{\sqrt{E}}}\sin\left(\frac{1}{\sqrt{E}} + \frac{\pi}{4}\right).$$

It is easy to prove that when $\varepsilon > 0$ is small enough, then

$$|a| < 4(\beta + \sqrt{\beta})\varepsilon \quad \text{and} \quad |b| < 32\beta\varepsilon^2.$$

With the previously defined boundary layer term $\mathcal{B}$, we can verify that

$$\text{div}\,(\overline{u} + \mathcal{B}) = 0 \qquad \text{and} \qquad (\overline{u} + \mathcal{B})_{|\{x_3=0\}} = (\overline{u} + \mathcal{B})_{|\{x_3=1\}} = 0.$$

Now, let

$$B_0(x_3) = \begin{bmatrix} -e^{-\frac{x_3}{\sqrt{E}}}\cos\frac{x_3}{\sqrt{E}} - e^{-\frac{1-x_3}{\sqrt{E}}}\cos\frac{1-x_3}{\sqrt{E}} & -e^{-\frac{x_3}{\sqrt{E}}}\sin\frac{x_3}{\sqrt{E}} - e^{-\frac{1-x_3}{\sqrt{E}}}\sin\frac{1-x_3}{\sqrt{E}} \\[2mm] e^{-\frac{x_3}{\sqrt{E}}}\sin\frac{x_3}{\sqrt{E}} + e^{-\frac{1-x_3}{\sqrt{E}}}\sin\frac{1-x_3}{\sqrt{E}} & -e^{-\frac{x_3}{\sqrt{E}}}\cos\frac{x_3}{\sqrt{E}} - e^{-\frac{1-x_3}{\sqrt{E}}}\cos\frac{1-x_3}{\sqrt{E}} \end{bmatrix}$$

Then, we can write $\mathcal{B}$ in the following form

$$\mathcal{B} = \mathcal{M}(x_3)A(t,x_1,x_2),$$

where

$$A(t,x_1,x_2) = {}^t\big(\overline{u}^1, \overline{u}^2, \text{curl}(\overline{u})\big) \qquad \text{and} \qquad \mathcal{M}(x_3) = \begin{bmatrix} M(x_3) & 0 \\ 0 & m(x_3) \end{bmatrix}$$

with $M(x_3)$ and $m(x_3)$ defined by

$$M(x_3) = B_0(x_3) + \left(\sqrt{E} + f(x_3)\right)\begin{pmatrix} 0 & 1 \\ -1 & 0 \end{pmatrix} + e^{-\frac{1}{\sqrt{E}}}\cos\frac{1}{\sqrt{E}}\begin{pmatrix} 1 & 0 \\ 0 & 1 \end{pmatrix},$$

$$m(x_3) = \sqrt{\frac{E}{2}}\,G(x_3) + \sqrt{E}\left(\frac{1}{2} - x_3\right) + g(x_3).$$

We can also prove the existence of a constant $C > 0$ such that, for any $p \geq 1$, we have

$$\begin{cases} \|\mathcal{M}(\cdot)\|_{\mathbf{L}^p_{x_3}} \leq C\varepsilon^{\frac{1}{p}}, \quad \|\mathcal{M}(\cdot)\|_{\mathbf{L}^\infty_{x_3}} \leq C, \quad \|\mathcal{M}'(\cdot)\|_{\mathbf{L}^p_{x_3}} \leq C\varepsilon^{\frac{1}{p}-1}, \\[2mm] \qquad\qquad \|m(\cdot)\|_{L^\infty_{x_3}} \leq C\varepsilon, \quad \|m(\cdot)\|_{L^p_{x_3}} \leq C\varepsilon \\[2mm] \displaystyle\sup_{x_3\in[0,\frac{1}{2}]}\big|x_3^2 M'(x_3)\big| \leq C\varepsilon \quad \text{and} \quad \sup_{x_3\in[\frac{1}{2},1]}\big|(1-x_3)^2 M'(x_3)\big| \leq C\varepsilon. \end{cases} \tag{23}$$

5 Convergence to the Limiting System

In this paragraph, we provide a priori estimates needed and a sketch the proof of Theorem 14. These a priori estimates can be justified by a classical approximation by smooth fonctions (see for instance [9]). For any $\varepsilon > 0$, we consider the following 2D damped Navier-Stokes system with three components:

$$\begin{cases} \partial_t \overline{u}^{\varepsilon,h} - \nu_h(\varepsilon)\Delta_h \overline{u}^{\varepsilon,h} + \sqrt{2\beta}\,\overline{u}^{\varepsilon,h} + \overline{u}^{\varepsilon,h}\cdot\nabla_h \overline{u}^{\varepsilon,h} = -\nabla_h \overline{p}^{\varepsilon} \\ \partial_t \overline{u}^{\varepsilon,3} - \nu_h(\varepsilon)\Delta_h \overline{u}^{\varepsilon,3} + \sqrt{2\beta}\,\overline{u}^{\varepsilon,3} + \overline{u}^{\varepsilon,h}\cdot\nabla_h \overline{u}^{\varepsilon,3} = 0 \\ \text{div}_h\overline{u}^{\varepsilon,h} = 0 \\ \partial_3\overline{u}^{\varepsilon} = 0 \\ \overline{u}^{\varepsilon}_{|t=0} = \overline{u}_0. \end{cases} \tag{24}$$

Then, Lemma 13 implies that Theorem 1 is a corollary of the following theorem

Theorem 14. *Suppose that*

$$\lim_{\varepsilon \to 0} \nu_h(\varepsilon) = 0 \quad and \quad \lim_{\varepsilon \to 0} \frac{\varepsilon^{\frac{1}{2}}}{\nu_h(\varepsilon)} = 0.$$

Let $u_0^\varepsilon \in \mathbf{L}^2(\Omega)$ be a family of initial data such that

$$\lim_{\varepsilon \to 0} u_0^\varepsilon = \overline{u}_0 = (\overline{u}_0^1(x_1, x_2), \overline{u}_0^2(x_1, x_2), 0) \quad in \quad \mathbf{L}^2(\mathbb{R}_h^2 \times [0,1]),$$

where $\overline{u}_0$ is a divergence-free two-dimensional vector field in $\mathbf{H}^\sigma(\mathbb{R}_h^2)$, $\sigma > 2$. Let $\overline{u}^\varepsilon$ be the solution of the system (24) *with initial data $\overline{u}_0$ and for each $\varepsilon > 0$, let u^ε be a weak solution of* (1) *with initial data u_0^ε. Then*

$$\lim_{\varepsilon \to 0} \|u^\varepsilon - \overline{u}^\varepsilon\|_{\mathbf{L}^\infty\left(\mathbb{R}_+, \mathbf{L}^2(\mathbb{R}_h^2 \times [0,1])\right)} = 0.$$

Proof of Theorem 14

We first remark that we can construct the boundary layers term $\mathcal{B}^\varepsilon$ for the system (24) in the same way as we did to construct $\mathcal{B}$, with $\overline{u}$ being replaced by $\overline{u}^\varepsilon$. It is easy to prove that $\mathcal{B}^\varepsilon$ is small, *i.e.*, $\mathcal{B}^\varepsilon$ goes to 0 in $\mathbf{L}^\infty\left(\mathbb{R}_+, \mathbf{L}^2(\mathbb{R}_h^2 \times [0,1])\right)$ as ε goes to 0. Then, our goal is to prove that $v^\varepsilon = u^\varepsilon - \overline{u}^\varepsilon - \mathcal{B}^\varepsilon$ converge to 0 in $\mathbf{L}^\infty\left(\mathbb{R}_+, \mathbf{L}^2(\mathbb{R}_h^2 \times [0,1])\right)$ as ε goes to 0.

We recall that a two-dimensional divergence-free vector field (independant of x_3) belongs to the kernel of the operator $\mathbb{P}(e_3 \wedge \cdot)$, where $\mathbb{P}$ is the Leray projection of $\mathbf{L}^2(\mathbb{R}^3)$ onto the subspace of divergence-free vector fields. As a consequence, $e_3 \wedge \overline{u}^\varepsilon$ is a gradient term. Replacing u^ε by $v^\varepsilon + \overline{u}^\varepsilon + \mathcal{B}^\varepsilon$ in the system (1), we deduce that v^ε satisfied the following equation

$$\partial_t v^\varepsilon - \nu_h(\varepsilon)\Delta_h v^\varepsilon - \beta\varepsilon\partial_3^2 v^\varepsilon + L_1 + u^\varepsilon \cdot \nabla v^\varepsilon + \mathcal{B}^\varepsilon \cdot \nabla \mathcal{B}^\varepsilon$$

$$+ \mathcal{B}^\varepsilon \cdot \nabla \overline{u}^\varepsilon + v^\varepsilon \cdot \nabla \mathcal{B}^\varepsilon + v^\varepsilon \cdot \nabla \overline{u}^\varepsilon - L_2 + \frac{e_3 \wedge v^\varepsilon}{\varepsilon} = -\nabla \widetilde{p}^\varepsilon, \quad (25)$$

where

$$L_1 = \partial_t \mathcal{B}^\varepsilon - \nu_h(\varepsilon)\Delta_h \mathcal{B}^\varepsilon + \overline{u}^\varepsilon \cdot \nabla \mathcal{B}^\varepsilon$$

$$L_2 = \beta\varepsilon\partial_3^2 \mathcal{B}^\varepsilon - \frac{e_3 \wedge \mathcal{B}^\varepsilon}{\varepsilon} + \sqrt{2\beta}\,\overline{u}^\varepsilon.$$

Taking the $\mathbf{L}^2$ scalar product of (25) with v^ε, then integrating by parts the obtained equation and taking into account the fact that v^ε satisfies the Dirichlet boundary condition, we get

$$\frac{1}{2}\frac{d}{dt}\|v^\varepsilon\|_{\mathbf{L}^2}^2 + \nu_h(\varepsilon)\|\nabla_h v^\varepsilon\|_{\mathbf{L}^2}^2 + \beta\varepsilon\|\partial_3 v^\varepsilon\|_{\mathbf{L}^2}^2$$

$$= -\langle L_1, v^\varepsilon\rangle - \langle u^\varepsilon \cdot \nabla v^\varepsilon, v^\varepsilon\rangle - \langle \mathcal{B}^\varepsilon \cdot \nabla \mathcal{B}^\varepsilon, v^\varepsilon\rangle - \langle \mathcal{B}^\varepsilon \cdot \nabla \overline{u}^\varepsilon, v^\varepsilon\rangle$$

$$- \langle v^\varepsilon \cdot \nabla \mathcal{B}^\varepsilon, v^\varepsilon\rangle - \langle v^\varepsilon \cdot \nabla \overline{u}^\varepsilon, v^\varepsilon\rangle + \langle L_2, v^\varepsilon\rangle. \quad (26)$$

In what follows, we will seperately estimate the seven terms on the right-hand side of Inequation (26). Using the same notations as in [16] and [18], we denote B_1, B_2 and b (V_1, V_2 and v respectively) the three components of $\mathcal{B}^\varepsilon$ (v^ε respectively) and we write $B = (B_1, B_2)$ et $V = (V_1, V_2)$.

1. Applying the operator curl to the first two equations of the system (24) (we recall that in this paper, curl only acts on the horizontal components and we already defined $\mathrm{curl}(\overline{u}) = \partial_1 \overline{u}^2 - \partial_2 \overline{u}^1$), we obtain

$$\partial_t (\mathrm{curl}\ \overline{u}^\varepsilon) - \nu_h(\varepsilon)\Delta_h (\mathrm{curl}\ \overline{u}^\varepsilon) + \sqrt{2\beta}\,(\mathrm{curl}\ \overline{u}^\varepsilon) + \overline{u}^\varepsilon \cdot \nabla \mathrm{curl}\ (\overline{u}^\varepsilon) = 0.$$

We recall that $A(t, x_1, x_2) = {}^t(\overline{u}_1^\varepsilon, \overline{u}_2^\varepsilon, \mathrm{curl}\ (\overline{u}^\varepsilon))$. So combining the above equation with the first two equations of (24), we deduce that

$$\partial_t A - \nu_h(\varepsilon)\Delta_h A + \sqrt{2\beta}\, A + \overline{u}^\varepsilon \cdot \nabla A = -(\nabla_h \overline{p}^\varepsilon, 0).$$

Since $\overline{u}^{\varepsilon,3} = 0$, div $v^\varepsilon = 0$, $\partial_3 \overline{p} = 0$ and $\mathcal{B}^\varepsilon = \mathcal{M}(x_3)A(t, x_1, x_2)$, we can write

$$|\langle L_1, v^\varepsilon \rangle| = |\langle \mathcal{M}(x_3)\left(\partial_t A - \nu_h(\varepsilon)\Delta_h A + \overline{u}^\varepsilon \cdot \nabla A\right), v^\varepsilon \rangle|$$
$$\leq \sqrt{2\beta}\ \|\mathcal{M}(\cdot)\|_{\mathbf{L}^2_{x_3}} \|A\|_{\mathbf{L}^2_{x_h}} \|v^\varepsilon\|_{\mathbf{L}^2}.$$

Then, Estimate (23), Lemma 8 and Young's inequality imply

$$|\langle L_1, v^\varepsilon \rangle| \leq C(\overline{u}_0)\, \varepsilon^{\frac{1}{2}} e^{-t\sqrt{2\beta}} \left(1 + \|v^\varepsilon\|_{\mathbf{L}^2}^2\right). \tag{27}$$

2. For the second term, using the divergence-free property of u^ε, we simply have

$$\langle u^\varepsilon \cdot \nabla v^\varepsilon, v^\varepsilon \rangle = 0. \tag{28}$$

3. We decompose the third term into two parts:

$$\langle \mathcal{B}^\varepsilon \cdot \nabla \mathcal{B}^\varepsilon, v^\varepsilon \rangle = \langle B \cdot \nabla_h \mathcal{B}^\varepsilon, v^\varepsilon \rangle + \langle b \partial_3 \mathcal{B}^\varepsilon, v^\varepsilon \rangle.$$

Using an integration by parts, the "horizontal" part can be bounded as follows

$$|\langle B \cdot \nabla_h \mathcal{B}^\varepsilon, v^\varepsilon \rangle| \leq |\langle (\mathrm{div}_h B)\mathcal{B}^\varepsilon, v^\varepsilon \rangle| + |\langle B \otimes \mathcal{B}^\varepsilon, \nabla_h v^\varepsilon \rangle| = |\langle B \otimes \mathcal{B}^\varepsilon, \nabla_h v^\varepsilon \rangle|.$$

Hence, Hölder's inequality, Estimates (23), Lemma 8 and Young's inequality yield

$$|\langle B \cdot \nabla_h \mathcal{B}^\varepsilon, v^\varepsilon \rangle| \leq \|B\|_{\mathbf{L}^4} \|\mathcal{B}^\varepsilon\|_{\mathbf{L}^4} \|\nabla_h v^\varepsilon\|_{\mathbf{L}^2} \tag{29}$$
$$\leq \|\mathcal{M}(\cdot)\|_{\mathbf{L}^4_{x_3}}^2 \left(\|\overline{u}^\varepsilon\|_{\mathbf{L}^4_{x_h}}^2 + \|\nabla_h \overline{u}^\varepsilon\|_{\mathbf{L}^4_{x_h}}^2\right) \|\nabla_h v^\varepsilon\|_{\mathbf{L}^2}$$
$$\leq C(\overline{u}_0)\, \varepsilon^{\frac{1}{2}} \nu_h(\varepsilon)^{-1} e^{-t\sqrt{2\beta}} + \frac{\nu_h(\varepsilon)}{16} \|\nabla_h v^\varepsilon\|_{\mathbf{L}^2}^2.$$

Likewise, we have the following estimate for the vertical part:

$$|\langle b \partial_3 \mathcal{B}^\varepsilon, v^\varepsilon \rangle| \leq \|b\|_{L^\infty} \|\partial_3 \mathcal{B}^\varepsilon\|_{\mathbf{L}^2} \|v^\varepsilon\|_{\mathbf{L}^2} \tag{30}$$
$$\leq C(\overline{u}_0)\, e^{-t\sqrt{2\beta}} \|m(\cdot)\|_{L^\infty_{x_3}} \|\mathcal{M}'(\cdot)\|_{\mathbf{L}^2_{x_3}} \|v^\varepsilon\|_{\mathbf{L}^2}$$
$$\leq C(\overline{u}_0)\, \varepsilon^{\frac{1}{2}} e^{-t\sqrt{2\beta}} \left(1 + \|v^\varepsilon\|_{\mathbf{L}^2}^2\right).$$

4. For the fourth term, taking into account the fact that $\overline{u}^\varepsilon$ is independent of x_3, Estimates (23) and Lemma 8 imply

$$|\langle \mathcal{B}^\varepsilon \cdot \nabla \overline{u}^\varepsilon, v^\varepsilon \rangle| \leq \|\mathcal{B}^\varepsilon\|_{\mathbf{L}^4_{x_h}\,\mathbf{L}^2_{x_3}} \|\nabla_h \overline{u}^\varepsilon\|_{\mathbf{L}^4_{x_h}} \|v^\varepsilon\|_{\mathbf{L}^2} \tag{31}$$

$$\leq \|\mathcal{M}(\cdot)\|_{\mathbf{L}^2_{x_3}} \left(\|\overline{u}^\varepsilon\|^2_{\mathbf{L}^4} + \|\nabla_h \overline{u}^\varepsilon\|^2_{\mathbf{L}^4} \right) \|v^\varepsilon\|_{\mathbf{L}^2}$$

$$\leq C(\overline{u}_0)\, \varepsilon^{\frac{1}{2}} \mathrm{e}^{-t\sqrt{2\beta}} \left(1 + \|v^\varepsilon\|^2_{\mathbf{L}^2} \right).$$

5. The fifth term is the most difficult to treat. First, we decompose this term as follows

$$\langle v^\varepsilon \cdot \nabla \mathcal{B}^\varepsilon, v^\varepsilon \rangle = \langle V \cdot \nabla_h B, V \rangle + \langle V \cdot \nabla_h b, v \rangle + \langle v\, \partial_3 b, v \rangle + \langle v\, \partial_3 B, V \rangle.$$

For the first term on the right-hand side, Hölder inequality implies that

$$|\langle V \cdot \nabla_h B, V \rangle| \leq C\, \|V\|_{\mathbf{L}^2} \|\nabla_h B\|_{\mathbf{L}^\infty} \|V\|_{\mathbf{L}^2} \leq C\, \|\mathcal{M}(\cdot)\|_{\mathbf{L}^\infty_{x_3}} \|\nabla_h \overline{u}^\varepsilon\|_{\mathbf{L}^\infty_{x_h}} \|v^\varepsilon\|^2_{\mathbf{L}^2}.$$

Then, using Estimates (23) and Lemma 8, we obtain

$$|\langle V \cdot \nabla_h B, V \rangle| \leq C(\overline{u}_0)\, \mathrm{e}^{-t\sqrt{2\beta}} \|v^\varepsilon\|^2_{\mathbf{L}^2}. \tag{32}$$

Next, by integrating by parts and using Hölder's inequality, we deduce that

$$|\langle V \cdot \nabla_h b, v \rangle| \leq C\, \|\nabla_h v^\varepsilon\|_{\mathbf{L}^2} \|b\|_{\mathbf{L}^\infty} \|v^\varepsilon\|_{\mathbf{L}^2}.$$

So, Estimates (23), Lemmas 8 and 10 and Young's inequality imply

$$|\langle V \cdot \nabla_h b, v \rangle| \leq C\, \|m(\cdot)\|_{\mathbf{L}^\infty_{x_3}} \|\mathrm{curl}\,\overline{u}^\varepsilon\|_{\mathbf{L}^\infty_{x_h}} \|\nabla_h v^\varepsilon\|_{\mathbf{L}^2} \|v^\varepsilon\|_{\mathbf{L}^2} \tag{33}$$

$$\leq C(\overline{u}_0)\, \varepsilon^2 \nu_h(\varepsilon)^{-1} \mathrm{e}^{-t\sqrt{2\beta}} \|v^\varepsilon\|^2_{\mathbf{L}^2} + \frac{\nu_h(\varepsilon)}{16} \|\nabla_h v^\varepsilon\|^2_{\mathbf{L}^2}.$$

Performing an integration by parts, we can control the third term in the same way as the second one:

$$|\langle v\, \partial_3 b, v \rangle| = 2\, |\langle bv, \partial_3 v \rangle| = 2\, |\langle bv, \mathrm{div}\,_h V \rangle|$$

$$\leq C(\overline{u}_0)\, \varepsilon^2 \nu_h(\varepsilon)^{-1} \mathrm{e}^{-t\sqrt{2\beta}} \|v^\varepsilon\|^2_{\mathbf{L}^2} + \frac{\nu_h(\varepsilon)}{16} \|\nabla_h v^\varepsilon\|^2_{\mathbf{L}^2}. \tag{34}$$

In order to estimate the last term of the right-hand side, we decompose it into two parts, the first part corresponding to the boundary layer near $\{x_3 = 0\}$ and the other corresponding to the one near $\{x_3 = 1\}$:

$$\langle v\, \partial_3 B, V \rangle = \int_{\mathbb{R}^2_h \times [0,\frac{1}{2}]} (v\, \partial_3 B) \cdot V\, dx + \int_{\mathbb{R}^2_h \times [\frac{1}{2},1]} (v\, \partial_3 B) \cdot V\, dx$$

For the fisrt part, since v^ε vanishes on $\{x_3 = 0\}$, using Hölder's inequality and Hardy-Littlewood inequality, we get

$$\left| \int_{\mathbb{R}^2_h \times [0,\frac{1}{2}]} (v\, \partial_3 B) \cdot V\, dx \right| \leq \sup_{x_3 \in [0,\frac{1}{2}]} |x_3^2 M'(x_3)|\, \|\overline{u}^\varepsilon\|_{\mathbf{L}^\infty_{x_h}} \left\| \frac{v}{x_3} \right\|_{L^2} \left\| \frac{V}{x_3} \right\|_{\mathbf{L}^2}$$

$$\leq \sup_{x_3 \in [0,\frac{1}{2}]} |x_3^2 M'(x_3)|\, \|\overline{u}^\varepsilon\|_{\mathbf{L}^\infty_{x_h}} \|\partial_3 v\|_{L^2} \|\partial_3 V\|_{\mathbf{L}^2}.$$

We recall that $\partial_3 v = -\mathrm{div}_h V$. Then, Lemmas 8 and 10, Estimates (23) and Young's inequality imply

$$\left| \int_{\mathbb{R}_h^2 \times [0,\frac{1}{2}]} (v\,\partial_3 B) \cdot V\,dx \right| \leq C(\bar{u}_0)\,\varepsilon\,\mathrm{e}^{-t\sqrt{2\beta}}\,\|\mathrm{div}_h v\|_{\mathbf{L}^2}\,\|\partial_3 V\|_{\mathbf{L}^2} \qquad (35)$$

$$\leq C(\bar{u}_0)\,\varepsilon\,\|\nabla_h v^\varepsilon\|_{\mathbf{L}^2}^2 + \frac{\beta\varepsilon}{4}\,\|\partial_3 v^\varepsilon\|_{\mathbf{L}^2}^2\,.$$

For the second part concerning the boundary layer near $\{x_3 = 1\}$, since $v^\varepsilon = (V, v)$ vanishes on $\{x_3 = 1\}$, Hardy-Littlewood inequality implies that

$$I_v = \int_{\mathbb{R}_h^2} \left(\int_{\frac{1}{2}}^{1} \left| \frac{v(x_h, x_3)}{1 - x_3} \right|^2 dx_3 \right) dx_h = \int_{\mathbb{R}_h^2} \left(\int_{0}^{\frac{1}{2}} \left| \frac{v(x_h, 1 - x_3)}{x_3} \right|^2 dx_3 \right) dx_h$$

$$\leq C \int_{\mathbb{R}_h^2} \left(\int_{0}^{\frac{1}{2}} |\partial_3 v(x_h, 1 - x_3)|^2\,dx_3 \right) dx_h$$

$$\leq C\,\|\partial_3 v\|_{L^2}^2 = C\,\|\mathrm{div}_h V\|_{L^2}^2\,.$$

Likewise,

$$I_V = \int_{\mathbb{R}_h^2} \left(\int_{\frac{1}{2}}^{1} \left| \frac{V(x_h, x_3)}{1 - x_3} \right|^2 dx_3 \right) dx_h \leq C\,\|\partial_3 V\|_{L^2}^2\,.$$

Thus, using Hölder inequality, we get

$$\left| \int_{\mathbb{R}_h^2 \times [\frac{1}{2}, 1]} (v\,\partial_3 B) \cdot V\,dx \right| \leq \sup_{x_3 \in [\frac{1}{2}, 1]} \left| (1 - x_3)^2 M'(x_3) \right| \|\bar{u}^\varepsilon\|_{\mathbf{L}_{x_h}^\infty}\,\sqrt{I_v}\sqrt{I_V} \qquad (36)$$

$$\leq C(\bar{u}_0)\,\varepsilon\,\mathrm{e}^{-t\sqrt{2\beta}}\,\|\mathrm{div}_h v\|_{\mathbf{L}^2}\,\|\partial_3 V\|_{\mathbf{L}^2}$$

$$\leq C(\bar{u}_0)\,\varepsilon\,\|\nabla_h v^\varepsilon\|_{\mathbf{L}^2}^2 + \frac{\beta\varepsilon}{4}\,\|\partial_3 v^\varepsilon\|_{\mathbf{L}^2}^2\,.$$

6. The sixth term on the right-hand side of (26) can be treated using Hölder inequality and Lemma 10. We have

$$|\langle v^\varepsilon \cdot \nabla \bar{u}^\varepsilon, v^\varepsilon \rangle| \leq C\,\|\nabla_h \bar{u}^\varepsilon\|_{\mathbf{L}^\infty(\mathbb{R}_h^2)}\,\|v^\varepsilon\|_{\mathbf{L}^2}^2 \leq C(\bar{u}_0)\,\mathrm{e}^{-t\sqrt{2\beta}}\,\|v^\varepsilon\|_{\mathbf{L}^2}^2 \qquad (37)$$

7. We will evaluate the seventh term as in [16] or [18]. We have

$$\langle L_2, v^\varepsilon \rangle = \left\langle \beta\varepsilon\partial_3^2 \mathcal{B}^\varepsilon, v^\varepsilon \right\rangle - \left\langle \frac{e_3 \wedge \mathcal{B}^\varepsilon}{\varepsilon}, v^\varepsilon \right\rangle + \left\langle \sqrt{2\beta}\,\bar{u}^\varepsilon, v^\varepsilon \right\rangle\,.$$

We recall that

$$\mathcal{B}^\varepsilon = \mathcal{B}^{\varepsilon,1} + \mathcal{B}^{\varepsilon,2} + \mathcal{B}^{\varepsilon,3} + \mathcal{B}^{\varepsilon,4},$$

and for any $i \in \{1, 2, 3, 4\}$, we set $\mathcal{B}^{\varepsilon,i} = (B^i, b^i)$, where B^i and b^i denote the horizontal and vertical components of $\mathcal{B}^{\varepsilon,i}$ respectively. Then, the following identities are immediate

$$\partial_3^2 B^3 = 0,$$

$$\beta\varepsilon\partial_3^2 B^1 - \frac{e_3 \wedge \mathcal{B}^1}{\varepsilon} = 0,$$

$$\beta\varepsilon\partial_3^2 B^2 - \frac{e_3 \wedge \mathcal{B}^2}{\varepsilon} + \sqrt{2\beta}\,\overline{u}^\varepsilon = 0.$$

For the remaining terms, we have

$$
\begin{aligned}
\beta\varepsilon\left|\langle\partial_3^2 b, v\rangle\right| &\leq \beta\varepsilon\,\|\partial_3 b\|_{\mathbf{L}^2}\,\|\partial_3 v\| && (38)\\
&\leq \beta\varepsilon\,\|\nabla_h B\|_{\mathbf{L}^2}\,\|\nabla_h v^\varepsilon\|_{\mathbf{L}^2}\\
&\leq \beta\varepsilon\,\|M(x_3)\|_{\mathbf{L}^2_{x_3}}\,\|\nabla_h\overline{u}^\varepsilon\|_{\mathbf{L}^2_{x_h}}\,\|\nabla_h v^\varepsilon\|_{\mathbf{L}^2}\\
&\leq C(\overline{u}_0)\,\varepsilon^3\nu_h(\varepsilon)^{-1}e^{-t\sqrt{2\beta}} + \frac{\nu_h(\varepsilon)}{16}\,\|\nabla_h v^\varepsilon\|_{\mathbf{L}^2}^2\,;
\end{aligned}
$$

$$\left|\left\langle\frac{e_3 \wedge \mathcal{B}^3}{\varepsilon}, V\right\rangle\right| \leq C\varepsilon^{-1}e^{-\frac{1}{\varepsilon}}\,\|\overline{u}^\varepsilon\|_{\mathbf{L}^2}\,\|v^\varepsilon\|_{\mathbf{L}^2} \leq C(\overline{u}_0)\,\varepsilon\,e^{-t\sqrt{2\beta}}\left(1 + \|v^\varepsilon\|_{\mathbf{L}^2}^2\right). \tag{39}$$

We recall that

$$B^4 = f(x_3)\begin{pmatrix} \overline{u}^{\varepsilon,2} \\ -\overline{u}^{\varepsilon,1} \end{pmatrix},$$

where

$$f(x_3) = a\left(e^{-\frac{x_3}{\sqrt{E}}} + e^{-\frac{1-x_3}{\sqrt{E}}}\right) + b,$$

and where $E = 2\beta\varepsilon^2$ is the Ekman number. We also recall that, if $\varepsilon > 0$ is small enough, we have

$$|a| < 4(\beta + \sqrt{\beta})\varepsilon \quad \text{and} \quad |b| < 32\beta\varepsilon^2.$$

Then,

$$
\begin{aligned}
\beta\varepsilon\left|\langle\partial_3^2 B^4, V\rangle\right| &= \beta\varepsilon\left|\langle f''(x_3)\overline{u}^\varepsilon, V\rangle\right| && (40)\\
&\leq C\varepsilon^{\frac{1}{2}}\,\|\overline{u}^\varepsilon\|_{\mathbf{L}^2}\,\|v^\varepsilon\|_{\mathbf{L}^2}\\
&\leq C(\overline{u}_0)\,\varepsilon^{\frac{1}{2}}e^{-t\sqrt{2\beta}}\left(1 + \|v^\varepsilon\|_{\mathbf{L}^2}^2\right).
\end{aligned}
$$

Finally, we have

$$\left|\left\langle\frac{e_3 \wedge \mathcal{B}^4}{\varepsilon}, V\right\rangle\right| \leq C\left[\left(\int_0^1\left|e^{-\frac{x_3}{\sqrt{E}}} + e^{-\frac{1-x_3}{\sqrt{E}}}\right|^2 dx_3\right)^{\frac{1}{2}} + \beta\varepsilon\right]\|\overline{u}^\varepsilon\|_{\mathbf{L}^2}\,\|V\|_{\mathbf{L}^2} \tag{41}$$

$$\leq C(\overline{u}_0)\,\varepsilon^{\frac{1}{2}}e^{-t\sqrt{2\beta}}\left(1 + \|v^\varepsilon\|_{\mathbf{L}^2}^2\right).$$

End of the proof: Summing all the inequalities from (27) to (41), we deduce from (26) that

$$\frac{d}{dt}\|v^\varepsilon\|_{\mathbf{L}^2}^2 + \nu_h(\varepsilon)\,\|\nabla_h v^\varepsilon\|_{\mathbf{L}^2}^2 + \beta\varepsilon\,\|\partial_3 v^\varepsilon\|_{\mathbf{L}^2}^2$$
$$\leq C(\overline{u}_0)\,\varepsilon^{\frac{1}{2}}\nu_h(\varepsilon)^{-1}\mathrm{e}^{-t\sqrt{2\beta}} + C(\overline{u}_0)\,\mathrm{e}^{-t\sqrt{2\beta}}\,\|v^\varepsilon\|_{\mathbf{L}^2}^2 + C(\overline{u}_0)\,\varepsilon\,\|\nabla_h v^\varepsilon\|_{\mathbf{L}^2}^2\,.$$

Since

$$\lim_{\varepsilon\to 0}\frac{\varepsilon^{\frac{1}{2}}}{\nu_h(\varepsilon)} = 0,$$

there exists $\varepsilon_0 = \varepsilon_0(\overline{u}_0) \in]0,1[$ such that, for any $0 < \varepsilon < \varepsilon_0$, we have $C(\overline{u}_0)\,\varepsilon < \nu_h(\varepsilon)$. Therefore, for any $\varepsilon \in]0,\varepsilon_0[$ small enough, by integrating the above inequality with respect to the time variable, we get

$$\|v^\varepsilon(t)\|_{\mathbf{L}^2}^2 \leq \|v^\varepsilon(0)\|_{\mathbf{L}^2}^2 + \overline{C}(\overline{u}_0)\varepsilon^{\frac{1}{2}}\nu_h(\varepsilon)^{-1} + C(\overline{u}_0)\int_0^t \mathrm{e}^{-s\sqrt{2\beta}}\,\|v^\varepsilon(s)\|_{\mathbf{L}^2}^2\,ds. \quad (42)$$

We recall that $v^\varepsilon = u^\varepsilon - \overline{u}^\varepsilon - \mathcal{B}^\varepsilon$. Thus,

$$\|v^\varepsilon(0)\|_{\mathbf{L}^2}^2 \leq \|u^\varepsilon(0) - \overline{u}^\varepsilon(0)\|_{\mathbf{L}^2}^2 + \|\mathcal{B}^\varepsilon(0)\|_{\mathbf{L}^2}^2$$
$$\leq \|u_0^\varepsilon - \overline{u}_0\|_{\mathbf{L}^2}^2 + \|\mathcal{M}(\cdot)\|_{\mathbf{L}^2_{x_3}}^2\,\|\overline{u}_0\|_{\mathbf{L}^2}^2 \leq \|u_0^\varepsilon - \overline{u}_0\|_{\mathbf{L}^2}^2 + C\varepsilon^{\frac{1}{2}}\,\|\overline{u}_0\|_{\mathbf{L}^2}^2\,.$$

According to Gronwall lemma, it follows from (42) that

$$\|v^\varepsilon(t)\|_{\mathbf{L}^2}^2 \leq \left(\|u_0^\varepsilon - \overline{u}_0\|_{\mathbf{L}^2}^2 + C\varepsilon^{\frac{1}{2}}\,\|\overline{u}_0\|_{\mathbf{L}^2}^2 + \overline{C}(\overline{u}_0)\varepsilon^{\frac{1}{2}}\nu_h(\varepsilon)^{-1}\right)\exp\left\{\frac{\overline{C}(\overline{u}_0)}{\sqrt{2\beta}}\right\}.$$

Using the hypotheses that

$$\lim_{\varepsilon\to 0}\frac{\varepsilon^{\frac{1}{2}}}{\nu_h(\varepsilon)} = 0 \quad\text{and}\quad \lim_{\varepsilon\to 0}\|u_0^\varepsilon - \overline{u}_0\|_{\mathbf{L}^2} = 0,$$

we obtain

$$\lim_{\varepsilon\to 0}\|v^\varepsilon\|_{\mathbf{L}^\infty(\mathbb{R}_+,\mathbf{L}^2)} = 0,$$

and Theorem 14 is proved. $\square$

References

1. Babin, A., Mahalov, A., Nicolaenko, B.: Global Splitting, Integrability and Regularity of 3D Euler and Navier-Stokes Equations for Uniformly Rotating Fluids. European Journal of Mechanics 15, 291–300 (1996)
2. Babin, A., Mahalov, A., Nicolaenko, B.: Global regularity of 3D rotating Navier-Stokes equations for resonant domains. Indiana University Mathematics Journal 48, 1133–1176 (1999)
3. Bony, J.-M.: Calcul symbolique et propagation des singularités pour les équations aux dérivées partielles non linéaires. Annales de l'École Normale Supérieure 14, 209–246 (1981)

4. Brezis, H., Gallouet, T.: Nonlinear Schrödinger evolution equations. Nonlinear Analysis 4(4), 677–681 (1980)
5. Chemin, J.-Y.: Fluides parfaits incompressibles. Astérisque 230 (1995)
6. Chemin, J.-Y., Desjardins, B., Gallagher, I., Grenier, E.: Anisotropy and dispersion in rotating fluids. Nonlinear Partial Differential Equations and their Application, Collège de France Seminar, Studies in Mathematics and its Applications 31, 171–191 (2002)
7. Chemin, J.-Y., Desjardins, B., Gallagher, I., Grenier, E.: Fluids with anisotropic viscosity, Special issue for R. Temam's 60th birthday. M2AN. Mathematical Modelling and Numerical Analysis 34(2), 315–335 (2000)
8. Chemin, J.-Y., Desjardins, B., Gallagher, I., Grenier, E.: Ekman boundary layers in rotating fluids. ESAIM Controle Optimal et Calcul des Variations, A tribute to J.-L. Lions 8, 441–466 (2002)
9. Chemin, J.-Y., Desjardins, B., Gallagher, I., Grenier, E.: Mathematical Geophysics: An introduction to rotating fluids and to the Navier-Stokes equations. Oxford University Press (2006)
10. Cushman-Roisin, B.: Introduction to geophysical fluid dynamics. Prentice-Hall (1994)
11. Embid, E., Majda, A.: Averaging over fast gravity waves for geophysical flows with arbitrary potential vorticity. Communications in Partial Differential Equations 21, 619–658 (1996)
12. Fujita, H., Kato, T.: On the Navier-Stokes initial value problem I. Archiv for Rational Mechanic Analysis 16, 269–315 (1964)
13. Gallagher, I.: Applications of Schochet's Methods to Parabolic Equation. Journal de Mathématiques Pures et Appliquées 77, 989–1054 (1998)
14. Gallagher, I., Saint-Raymond, L.: Weak convergence results for inhomogeneous rotating fluid equations. Journal d'Analyse Mathématique 99, 1–34 (2006)
15. Grenier, E.: Oscillatory perturbations of the Navier-Stokes equations. Journal de Mathématiques Pures et Appliquées 76, 477–498 (1997)
16. Grenier, E., Masmoudi, N.: Ekman layers of rotating fluid, the case of well prepared initial data. Communications in Partial Differential Equations 22(5-6), 953–975 (1997)
17. Leray, J.: Essai sur le mouvement d'un liquide visqueux emplissant l'espace. Acta Matematica 63, 193–248 (1933)
18. Masmoudi, N.: Ekman layers of rotating fluids: the case of general initial data. Communications on Pure and Applied Mathematics 53(4), 432–483 (2000)
19. Ngo, V.-S.: Rotating Fluids with small viscosity. International Mathematics Research Notices IMRN (10), 1860–1890 (2009)
20. Ngo, V.-S.: Rotating fluids with small viscosity - The case of ill-prepared data (submitted)
21. Paicu, M.: Equation anisotrope de Navier-Stokes dans des espaces critiques. Revista Matemática Iberoamericana 21(1), 179–235 (2005)
22. Pedlosky, J.: Geophysical fluid dynamics. Springer (1979)
23. Schmitz, W.J.: Weakly depth-dependent segments of the North Atlantic circulation. Journal of Marine Research 38, 111–133
24. Schochet, S.: Fast singular limits of hyperbolic PDEs. Journal of Differential Equations 114, 476–512 (1994)
25. Yudovitch, V.: Non stationnary flows of an ideal incompressible fluid. Zh. Vych. Math. 3, 1032–1066 (1963)

Parallelization of the Fast Multipole Method for Molecular Dynamics Simulations on Multicore Computers

Nguyen Hai Chau

Faculty of Information Technology
VNUH University of Engineering and Technology
chaunh@vnu.edu.vn, nhchau@gmail.com

Abstract. We have parallelized the fast multipole method (FMM) on multicore computers using OpenMP programming model. The FMM is the one of the fastest approximate force calculation algorithms for molecular dynamics simulations. Its computational complexity is linear. Parallelization of FMM on multicore computers using OpenMP has been reported since the multicore processors become increasingly popular. However the number of those FMM implementations is not large. The main reason is that those FMM implementations have moderate or low parallel efficiency for high expansion orders due to sophisticated formulae of the FMM. In addition, parallel efficiency of those implementations for high expansion orders rapidly drops to 40% or lower as the number of threads increases to 8 or higher. Our FMM implementation on multicore computers using a combination approach as well as a newly developed formula and a computational procedure (A2P) solved the above issues. Test results of our FMM implementation on a multicore computer show that our parallel efficiency with 8 threads is at least 70% for moderate and high expansion orders $p = 4, 5, 6, 7$. Moreover, the parallel efficiency for moderate and high expansion orders gradually drops from 96% to 70% as the number of threads increases.

Keywords: molecular dynamics simulations, fast multipole method, multicore, OpenMP, parallelization.

1 Introduction

Molecular dynamics (MD) simulation methods [1] are orthodox means for studying large-scale physical/chemical systems. The methods were originally proposed in 1950s but they only began to use widely in the mid-1970s when digital computers became powerful and affordable. Nowadays MD methods are being continued to use widely for studying physical/chemical systems [2,3,4].

MD is simply stated that ones numerically solve the N-body problems of classical mechanics (Newton mechanics). Solving N-body problems for a large number of particles N in the considered systems requires a great amount of

N.T. Nguyen, T. Van Do, and H.A. Le Thi (Eds.): *ICCSAMA 2013*, SCI 479, pp. 209–224.
DOI: 10.1007/978-3-319-00293-4_16   © Springer International Publishing Switzerland 2013

time and it needs powerful computers as well as efficient algorithms. Calculation of Coulombic interaction force is the most dominated task in solving N-body problem. The calculation often dominates 90-95% total calculation time of MD simulations [1,2,3,4].

The simplest and most accurate algorithm for calculation force is the direct summation (DS) algorithm. The DS calculates interaction force between every pair of particles in the system using Coulombic force formula:

$$\boldsymbol{F}_{ij} = k_e \frac{q_i q_j}{||\boldsymbol{r}_{ij}||^2} \cdot \frac{\boldsymbol{r}_{ij}}{||\boldsymbol{r}_{ij}||}, \tag{1}$$

where $\boldsymbol{F}_{ij}$ is the Coulombic force between two particles located at $\boldsymbol{r}_i$ and $\boldsymbol{r}_i$ with charges are q_i and q_j, respectively; $k_e = \frac{1}{4\pi\epsilon_0} = \frac{c_0^2 \mu_0}{4\pi} = c_0^2 10^{-7} \text{H m}^{-1}$ is the Coulomb constant and $\boldsymbol{r}_{ij} = \boldsymbol{r}_j - \boldsymbol{r}_i$. Here c_0 is the speed of light in vacuum, H is the Henry unit and m is the meter. The potential due to q_j at position $\boldsymbol{r}_i$ is

$$\Phi_{ij} = k_e \frac{q_j}{||\boldsymbol{r}_{ij}||}. \tag{2}$$

The DS has $O(N^2)$ computational complexity, where N is the number of particles in the system. However DS is only applicable for small particle systems where N is up to 10^5. Using DS for larger systems will consume a huge amount of time. To reduce force calculation cost for N-body simulations, fast algorithms such as Barnes-Hut treecode (BH) [5,6] and fast multipole method (FMM) [7,8,9] has been developed. The computational complexity of the algorithms are $O(N \log N)$ and $O(N)$, respectively.

FMM has a broad range of applications in many fields of research: large-scale molecular dynamics simulations [10], accelerating boundary elements methods [11], vortex methods [12] etc. Large-scale N-body or MD simulations those have a very large N or where periodic boundary condition is not applicable are good examples of FMM's applications.

Because of the wide usability of FMM, there are many efforts to parallelize FMM for achieving high performance simulations. There are different approaches of FMM parallelization have been done so far. The first and most popular one is parallelization of FMM on distributed memory platforms using message passing interface (MPI). The second one is usage of special-purpose hardware for parallelization of FMM, including graphics processing units (GPU), GRAPE (GRAvity piPE) computer family and others. The third one is using OpenMP programming model for multiprocessor or multicore platforms. The last one combines the mentioned above approaches.

Parallelization of FMM using OpenMP programming model has been reported since the multicore processors become increasingly popular. However, there are is a small numbers of FMM implementations for OpenMP. The reason is as follows. Due to FMM's sophisticated formulae and data structures, existing OpenMP implementations of FMM have moderate or low parallel efficiency. The parallel efficiency often drops down for high expansions orders those needed by high accuracy applications such as molecular dynamics simulations. As an example, parallel efficiency for 8 threads of a multi-level FMM implementation using OpenMP is about 40% and drops to less than 25% with 16 threads [13]. Parallel efficiency of another FMM implementation using OpenMP on a single node of the Kraken supercomputer is relatively high (78%) for a low expansion order $p = 3$ [14]. However such a low expansion order $p = 3$ is suitable for N-body simulations in astrophysics but not for molecular dynamics simulations. Note that molecular dynamics simulations often require high expansion orders to achieve higher accuracy than astrophysics simulations do. The parallel efficiency of this implementation for higher expansion orders is not reported in details and drops down rapidly. The main drawback of the authors in [14] is that the parallelism of the M2L kernel in their implementation becomes finer with high order expansions. This would affect parallel efficiency of the implementation favourably.

In this paper we describe our approach for implementation of FMM on multicore computers using OpenMP programming model [15] to overcome drawbacks of the existing implementations. We develop a new formula for L2L stage and a computational procedure to simplify thus speed up the far field force stage of the FMM. The main advantages our approach is its simplicity and parallel efficiency.

The rest parts of this paper are as follows. In section 2, we describe the original FMM and its variations. The implementation of FMM on multicore computers is presented in section 3. Section 4 describes experimental results and section 5 concludes.

2   The Fast Multipole Method and Its Variations

In this section, we describe briefly the fast multipole method (section 2.1), and the most relevant algorithms to our work: the Anderson's method (section 2.2) and the pseudoparticle multipole method (section 2.3) by Makino.

2.1   Fast Multipole Method

The FMM is an $O(N)$ approximate algorithm to calculate forces among particles. The $O(N)$ scaling is achieved by approximation of the forces using the multipole and local expansion techniques. The algorithm is applicable for both two-dimensional [7] and three-dimensional [8,9] particle systems.

Figure 1 shows schematic idea of force approximation in the FMM. The force from a group of distant particles are approximated by a multipole expansion (M2M). At an observation point, the multipole expansion is converted to local expansion (M2L). The local expansion is then evaluated by each particle around the observation point (L2L).

A hierarchical tree structure (the octree) is used for grouping the particles. In all FMM implementations, the particle system is assume to locate inside a cube refering as the root cell. The root cell is then subdivided into eight equal subcells and the subdivision process for subcells continues until certain criteria is met. The root cell and subcells form an octree and the subdivision process is the octree construction. The octree construction stops when the number of

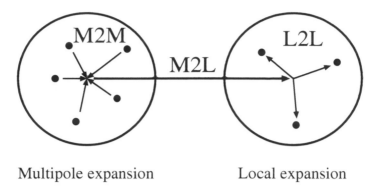

Multipole expansion   Local expansion

Fig. 1. Schematic idea of force approximation in FMM

octree levels reach a predefined number, often defined by the required accuracy of the simulations. In original FMM and its variations, the octree constructions are simple and similar. We refer readers to Greengard and Rokhlin's paper for details of the octree construction [7,8].

After the completion of the octree construction, the FMM goes to its main stages: multipole expansions to multipole expansions transition (M2M), multipoles expansion to local expansions conversion (M2L), local expansions to local expansions transition (L2L) and finally force evaluation. Among them, the M2L stage in the most computationally time consuming. The force evaluation stage contains two parts: near field force evaluation and far field force evaluation. All the variations of the FMM follow exactly stages in the original FMM: M2M, M2L, L2L and force evalution. The only difference is that each variation uses different mathematical formulae for the stages.

In the rest part of this section (2.1) we briefly describe the mathematical formulae for M2M, M2L and L2L by Greengard, Cheng and Rokhlin [9].

Spherical Harmonics. We begin by the definition of the spherical harmonics function of degree n and order m by the formula

$$Y_n^m(\theta, \phi) = \sqrt{\frac{(n - |m|)!}{(n + |m|)!}} P_n^{|m|} cos(\theta) e^{im\phi}. \tag{3}$$

Here, P_n^m is the assosiated Legendre functions, defined by Rodrigues' formula

$$P_n^m(x) = (-1)^m (1 - x^2)^{m/2} \frac{d^m}{dx^m} P_n(x), \tag{4}$$

where $P_n(x)$ denotes the Legendre polynomials of degree n.

Multipole Expansion. Given the definition of the spherical harmonics function, the multipole expansion is defined as

$$\Phi(X) = \sum_{n=0}^{\infty} \sum_{m=-n}^{n} \frac{M_n^m}{r^{n+1}} Y_n^m(\theta, \phi),\qquad(5)$$

where $X_1, X_2, ..., X_n$ are positions of N charges of strength $q_1, q_2, ..., q_n$ with spherical coordinates $(\rho_1, \alpha_1, \beta_1)$, $(\rho_2, \alpha_2, \beta_2)$, ..., $(\rho_n, \alpha_n, \beta_n)$, respectively. Assume that $X_1, X_2, ..., X_n$ are inside a sphere of radius a centered at the origin. The point X's spherical coordinate is (r, ϕ, θ). The M_n^m is defined as

$$M_n^m = \sum_{i=1}^{N} q_i \rho_i^n Y_n^{-m}(\alpha_i, \beta_i).\qquad(6)$$

Local Expansion. The local expansion is defined as

$$\Phi(X) = \sum_{j=0}^{\infty} \sum_{k=-j}^{j} L_j^k Y_j^k(\theta, \phi) r^j,\qquad(7)$$

where

$$L_j^k = \sum_{l=1}^{N} q_l \frac{Y_j^{-k}(\alpha_l, \beta_l)}{\rho_l^{j+1}}.\qquad(8)$$

Here we have N charges of strength $q_1, q_2, ..., q_n$ located at $X_1, X_2, ..., X_n$ with spherical coordinates $(\rho_1, \alpha_1, \beta_1)$, $(\rho_2, \alpha_2, \beta_2)$, ..., $(\rho_n, \alpha_n, \beta_n)$, respectively and all the points $X_1, X_2, ..., X_n$ are outside of a sphere radius a centered at the origin. The point X's spherical coordinate is (r, ϕ, θ).

Transition of a Multipole Expansion (M2M). Assume that N charges of strength $q_1, q_2, ..., q_n$ are located inside a sphere D of radius a centered at $X_0 = (\rho, \alpha, \beta)$. Suppose that for any point $X = (r, \phi, \theta) \in \mathbb{R}^3 \setminus D$, the potential due to these charges is given by the multipole expansion

$$\Phi(X) = \sum_{n=0}^{\infty} \sum_{m=-n}^{n} \frac{O_n^m}{r'^{n+1}} Y_n^m(\theta', \phi'),\qquad(9)$$

where (r', θ', ϕ') are the spherical coordinates of the vector $X - X_0$. Then for any point $X = (r, \theta, \phi)$ outside a sphere $D1$ of radius $a + \rho$ center at the origin,

$$\Phi(X) = \sum_{j=0}^{\infty} \sum_{k=-j}^{j} \frac{M_j^k}{r^{j+1}} Y_j^k(\theta, \phi),\qquad(10)$$

where

$$M_j^k = \sum_{n=0}^{j} \sum_{m=-n}^{n} \frac{O_{j-n}^{k-m} i^{|k|-|m|-|k-m|} A_{j-n}^{k-m} A_n^m \rho^n Y_n^{-m}(\alpha, \beta)}{A_j^k},\qquad(11)$$

with A_n^m defined by

$$A_n^m = \frac{(-1)^n}{\sqrt{(n-m)!(n+m)!}} \qquad (12)$$

Conversion of a Multipole Expansion to a Local Expansion (M2L).
Suppose that N charges of strengths $q_1, q_2, ..., q_n$ are located inside the sphere D_{X_0} of radius a centered at the point $X_0 = (\rho, \alpha, \beta)$, and that $\rho > (c+1)a$ for some $c > 1$. Then the corresponding multipole (13) converges inside the sphere D_0 of radius a centered at the origin. For any point $X \in D_0$ with coordinates (r, θ, ϕ), the potential due to the charges $q_1, q_2, ..., q_n$ is described by the local expansion

$$\Phi(X) = \sum_{j=0}^{\infty} \sum_{k=-j}^{j} L_j^k Y_j^k(\theta, \phi) r^j, \qquad (13)$$

where

$$L_j^k = \sum_{n=0}^{\infty} \sum_{m=-n}^{n} \frac{O_n^m i^{|k-m|-|k|-|m|} A_n^m A_j^k Y_{j+n}^{m-k}(\alpha, \beta)}{(-1)^n A_{j+n}^{m-k} \rho^{j+n+1}}, \qquad (14)$$

with A_n^m defined by (12).

Translation of a Local Expansion (L2L). Suppose that X, X_0 are a pair of points in $\mathbb{R}^3$ with spherical coordinates $(\rho, \alpha, \beta), (r, \theta, \phi)$, respectively, and (r', θ', ϕ') are the spherical coordinates of the vector $X - X_0$ and p is a natural number. Let X_0 be the center of a p-th order local expansion with p finite; its expression at the point X is given by

$$\Phi(X) = \sum_{n=0}^{p} \sum_{m=-n}^{n} O_n^m Y_n^m(\theta', \phi') r'^n. \qquad (15)$$

Then

$$\Phi(X) = \sum_{j=0}^{p} \sum_{k=-j}^{j} L_j^k Y_j^k(\theta, \phi) r^j, \qquad (16)$$

everywhere in $\mathbb{R}^3$, with

$$L_j^k = \sum_{n=j}^{p} \sum_{m=-n}^{n} \frac{O_n^m i^{|m|-|m-k|-|k|} A_{n-j}^{m-k} A_j^k Y_{n-j}^{m-k}(\alpha, \beta) \rho^{n-j}}{(-1)^{n+j} A_n^m}, \qquad (17)$$

and A_n^m defined by (12).

The M2M, M2L and L2L stages of the FMM are depicted in Figure 1. In sections 2.2 and 2.3, we describe two variations of the FMM those are the most relevant to our approach: Anderson's and Makino's methods.

2.2 Anderson's Method

Anderson [16] proposed a variant of the FMM using a new formulation of the multipole and local expansions. The advantage of his method is its simplicity. Anderson's method makes the implementation of the FMM significantly simple. The following is a brief description of Anderson's method.

Anderson's method is based on the Poisson's formula. This formula gives solution of the boundary value problem of the Laplace equation. When the potential on the surface of a sphere of radius a is given, the potential Φ at position $\boldsymbol{r} = (r, \phi, \theta)$ is expressed as

$$\Phi(\boldsymbol{r}) = \frac{1}{4\pi} \int_S \sum_{n=0}^{\infty} (2n+1) \left(\frac{a}{r}\right)^{n+1} P_n \left(\frac{\boldsymbol{s} \cdot \boldsymbol{r}}{r}\right) \Phi(a\boldsymbol{s}) ds \qquad (18)$$

for $r \geq a$, and

$$\Phi(\boldsymbol{r}) = \frac{1}{4\pi} \int_S \sum_{n=0}^{\infty} (2n+1) \left(\frac{r}{a}\right)^{n} P_n \left(\frac{\boldsymbol{s} \cdot \boldsymbol{r}}{r}\right) \Phi(a\boldsymbol{s}) ds \qquad (19)$$

for $r \leq a$. Note that here we use a spherical coordinate system. Here, $\Phi(a\boldsymbol{s})$ is the given potential on the sphere surface. The area of the integration S covers the surface of the unit sphere centered at the origin. The function P_n denotes the n-th Legendre polynomial.

In order to use these formulae as replacements of the multipole and local expansions, Anderson proposed a discrete version of them, i.e., he truncated the right-hand side of the Eq. (18)–(19) at a finite n, and replaced the integrations over S with numerical ones using a spherical t-design. Hardin and Sloane define the spherical t-design [17] as follows.

A set of K points $\wp = \{P_1, ..., P_K\}$ on the unit sphere $\Omega_d = S^{d-1} = \{x = (x_1, ..., x_d) \in R^d : x \cdot x = 1\}$ forms a spherical t-design if the identity

$$\int_{\Omega_d} f(x) d\mu(x) = \frac{1}{K} \sum_{i=1}^{K} f(P_i) \qquad (20)$$

(where μ is uniform measure on Ω_d normalized to have total measure 1) holds for all polynomials f of degree $\leq t$ [17].

Note that the optimal set, i.e., the smallest set of the spherical t-design is not known so far for general t. In practice we use spherical t-designs as empirically found by Hardin and Sloane. Examples of such t-designs are available at http://www.research.att.com/~njas/sphdesigns/.

Using the spherical t-design, Anderson obtained the discrete versions of (18) and (19) as follows:

$$\Phi(\boldsymbol{r}) \approx \sum_{i=1}^{K} \sum_{n=0}^{p} (2n+1) \left(\frac{a}{r}\right)^{n+1} P_n \left(\frac{\boldsymbol{s}_i \cdot \boldsymbol{r}}{r}\right) \Phi(a\boldsymbol{s}_i) w_i \qquad (21)$$

for $r \geq a$ (outer expansion) and

$$\Phi(\boldsymbol{r}) \approx \sum_{i=1}^{K} \sum_{n=0}^{p} (2n+1) \left(\frac{r}{a}\right)^{n} P_{n}\left(\frac{\boldsymbol{s}_{i} \cdot \boldsymbol{r}}{r}\right) \Phi(a\boldsymbol{s}_{i}) w_{i} \qquad (22)$$

for $r \leq a$ (inner expansion). Here w_i is constant weight value and p is the number of untruncated terms.

Anderson's method uses Eq. (21) for M2M, M2L stages and (22) for L2L transitions. The procedures of other stages are the same as that of the original FMM.

2.3   Pseudoparticle Multipole Method

Makino [18] proposed the pseudoparticle multipole method (P²M²) – yet another formulation of the multipole expansion. The advantage of his method is that the expansions can be evaluated using simple equations Eq. (1) or Eq. (2).

The basic idea of P²M² is to use a small number of pseudoparticles to express the multipole expansions. In other words, this method approximates the potential field of physical particles by the field generated by a small number of pseudoparticles. This idea is very similar to that of Anderson's method. Both methods use discrete quantities to approximate the potential field of the original distribution of the particles. The difference is that P²M² uses the distribution of point charges, while the Anderson's method uses potential values. In the case of P²M², the potential is expressed by point charges using Eq. (2).

In the following, we describe the formulation procedure of P²M². The distribution of pseudoparticles is determined so that it correctly describes the coefficients of a multipole expansion. A naive approach to obtain the distribution is to directly invert the multipole expansion formula. For relatively small expansion order, say $p \leq 2$, we can solve the inversion formula, and obtain the optimal distribution with minimum number of pseudoparticles [19].

However, it is rather difficult to solve the inversion formula for higher p, since the formula is nonlinear. For solution with $p > 2$, Makino fixed the pseudoparticles positions given by the spherical t-design [17], and only their charges can change. This makes the formula linear, although the necessary number of pseudoparticles increases. The degree of freedom assigned to each pseudoparticle is then reduced from four to one.

Makino's approach gives the solution of the inversion formula as follows:

$$Q_{j} = \sum_{i=1}^{N} q_{i} \sum_{l=0}^{p} \frac{2l+1}{K} \left(\frac{r_{i}}{a}\right)^{l} P_{l}(\cos\gamma_{ij}), \qquad (23)$$

where Q_j is charge of pseudoparticle, $\boldsymbol{r}_i = (r_i, \phi, \theta)$ is position of physical particle, γ_{ij} is angle between $\boldsymbol{r}_i$ and position vector $\boldsymbol{R}_j$ of the j-th pseudoparticle. For the derivation procedure of Eq. (23), see [18].

3   Implementation of the FMM on Multicore Computers Using OpenMP

3.1   A New Calculation Procedure for L2L Stage

As described above, the FMM has five stages including the octree construction, M2M transition, M2L conversion, L2L transition and force evaluation. The force evaluation stage contains two parts: near field and far field force evaluation. Anderson's method uses outer expansion (Eq. (21)) for M2M and M2L stages and inner expansion (Eq. (22)) for L2L stage. The P^2M^2 method by Makino is only applicable for M2M stage. However using Makino's method the M2L stage is simplied significantly by using direct pair-wise interaction given in Eq. (2).

We have done two implementations of the FMM for the special-purpose computer GRAPE so far. In the first implementation (hereafter FMMGRAPE1) [20], we combined Anderson's method and Makino's method. The P^2M^2 formula is used for M2M stage, then Eq. (2) is used for M2L. Next, the inner expansion by Anderson in Eq. (22) is used for L2L. With this approach, M2L is simplified using Eq. (2) and speeded up thanks to the special-purpose computer GRAPE. Another advantage of the combination of Anderson's and Makino's methods is that it is easy to parallelize the M2L stage for multicore architecture. However a new computational bottleneck appears in far-field force calculation as follow.

Using Eq. (22), the far field potential on a particle at position r can be calculated from the set of potential values of the leaf cell that contains the particle. Consequently, the far field force is calculated using derivative of Eq. (22) [20]:

$$-\nabla \Phi(r) = \sum_{i=1}^{K} \sum_{n=0}^{p} \left(nr P_n(u) + \frac{ur - s_i\, r}{\sqrt{1-u^2}} \nabla P_n(u) \right) (2n+1) \frac{r^{n-2}}{a^n} g(as_i) w_i, \quad (24)$$

where $u = s_i \cdot r/r$. Force calculation using Eq. (24) is complicated and hard to parallelize efficiently. In FMMGRAPE1, calculation of Eq. (24) dominates a significant part of the total calculation [20].

In the second implementation (hereafter FMMGRAPE2) [21], we fixed the bottleneck by developing a new formula and a new conversion procedure named A2P. We first developed a new formula for inner expansion using pseudoparticles. Eq. (23) by Makino gives the solution for outer expansion. We followed a similar approach [22] and proved that the solution for inner expansion is

$$Q_j = \sum_{i=1}^{N} q_i \sum_{l=0}^{p} \frac{2l+1}{K} \left(\frac{a}{r_i} \right)^{l+1} P_l(\cos \gamma_{ij}). \quad (25)$$

The A2P conversion is used to obtain a distribution of pseudoparticles that reproduces the potential field given by Anderson's inner expansion. Once the distribution of pseudoparticles is obtained, L2L stage can be performed using formula (Eq. (25)), and then the force evaluation stage is totally done using the simple Eq. (1). Hereafter we describe the A2P procedure indetails.

For the first step, we distribute pseudoparticles on the surface of a sphere with radius b using the spherical t-design. Here, b should be larger than the radius of the sphere a on which Anderson's potential values $g(a\boldsymbol{s}_i)$ are defined. According to Eq. (25), it is guaranteed that we can adjust the charge of the pseudoparticles so that $g(a\boldsymbol{s}_i)$ are reproduced. Therefore, the relation

$$\sum_{j=1}^{K} \frac{Q_j}{|\boldsymbol{R}_j - a\boldsymbol{s}_i|} = \varPhi(a\boldsymbol{s}_i) \qquad (26)$$

should be satisfied for all $i = 1..K$. Using a matrix $\mathcal{R} = \{1/|\boldsymbol{R}_j - a\boldsymbol{s}_i|\}$ and vectors $\boldsymbol{Q} = {}^T[Q_1, Q_2, ..., Q_K]$ and $\boldsymbol{P} = {}^T[\varPhi(a\boldsymbol{s}_1), \varPhi(a\boldsymbol{s}_2), ..., \varPhi(a\boldsymbol{s}_K)]$, we can rewrite Eq. (26) as

$$\mathcal{R}\boldsymbol{Q} = \boldsymbol{P}. \qquad (27)$$

In the next step, we solve the linear equation (27) to obtain charges Q_j. For a given cell with edge length is 1.0, the radius a and b for outer expansion and inner expansion are 0.75 and 6.0, respectively. Because of that, solving the linear equations system (27) is simply performing matrix-vector multiplication of $\mathcal{R}^{-1}$ and $\boldsymbol{P}$. Once solution of Q_j is obtained, far-field force calculated in Eq. (24) is replaced by the calculation pairwise interactions with Q_j using Eq. (1) that is much simpler. Note that the calculation of $\mathcal{R}^{-1}$ is simple and takes a negligible amount of time.

We have done numerical tests for accuracy of potential and force calculation performed with Eq. (25) [21]. In the tests, we approximate force and potential exerted from a particle q to a point L using Eq. (25) and the A2P procedure and compare results with potential and force calculated using Eq. (1) and Eq. (2). We change the distance r from q to L in a range of [1,10] and calculate relative error for both potential and force exerted from q to L. The test results shows that for expansion orders $p = 1$ to 5, potential error scales as $r^{-(p+2)}$ and force error scales as $r^{-(p+1)}$ as theoretically expected. For $p = 6$, potential and force error scales as $r^{-(p+2)}$ and $r^{-(p+1)}$ for $r < 6$, respectively and slowly descreasing for $r \geq 6$. The test results show that the Eq. (25) gives similar numerical accuracy of Anderson's given in Eq. (22).

Tables 1 and 2 compare formulae in original FMM to its variations and describe the mathematical formulae we have used in stages of our own FMM implementations.

3.2   Parallelization of FMM Using OpenMP

As shown in Tables 1 and 2, we combine methods of Anderson and Makino and our new calculation procedure A2P to implement the FMM. We apply the same approach to implement FMM on multicore computer. Hereafter we refer this implementation as FMMOpenMP. The main advantages of this combination is that we are able to use very simple mathematical formulae for stages of the FMM. In computational aspect, we have four kernels of calculation. The first one is the Eq. (1) used for near and far field force calculation. The second one is

Table 1. Mathematical formulae used in different variations of FMM

Stages	Original FMM	Anderson's method	Makino's method
M2M	Eq. (9), (10)	Eq. (21)	Eq. (23)
M2L	Eq. (13), (14)	Eq. (21)	Eq. (2)
L2L	Eq. (15), (16)	Eq. (22)	*Not available*
Near field force	Eq. (1)	Eq. (1)	Eq. (1)
Far field force	Evaluation of local expansions	Eq. (24)	*Not available*

the Eq. (2) used for M2L stage. The third one is the Eq. (23) used for M2M stage and the last one (Eq. (22)) used for L2L stage. Eq. (25) and Eq. (26) do not dominate computationally. However, Eq. (25) and Eq. (26) help us to calculate the far field force using the simple Eq. (1).

Table 2. Mathematical formulae used in our implementations of FMM

Stages	FMMGRAPE1	FMMGRAPE2	FMMOpenMP
M2M	Eq. (23)	Eq. (23)	Eq. (23)
M2L	Eq. (2)	Eq. (2)	Eq. (2)
L2L	Eq. (22)	Eq. (22)	Eq. (22)
Near field force	Eq. (1)	Eq. (1)	Eq. (1)
Far field force	Eq. (24)	Eq. (1), Eq. (25), Eq. (26)	Eq. (1), Eq. (25), Eq. (26)

Parallelization of the FMM using OpenMP becomes easier with our combination method. We need to parallelize the loops for potential/force pairwise interaction and the loops that used Eq. (23) and Eq. (22). We have developed a flops counter to find optimal level of the octree. As reported from the counter, number of flops due to pairwise interactions takes at least 90% number of FMM floating point operation. Therefore parallelization of the pairwise interaction is the most important task. Thanks to Eq. (1), Eq. (2) for their simplicity, the parallelization pseudocode of Eq. (1) (for near and far field force calculation) and Eq. (2) (for M2L stage) is straightforward as follows:

```
#pragma omp parallel for default(shared) private(i,j,...)
for (j=0;j<k;j++) { // For each destination particle
    for (i=0;i<n;i++) { // For each source particle
        // Calculate distance between particle i and particle j
                ...
        // Calculate Coulombic force (Eq. (1)) or
        // Coulombic potential (Eq. (2))
                ...
    }
}
```

Parallelization of M2M and L2L stages are also simple. The following is the parallelization pseudocode of the M2M calculation:

```
for (l=levels-1;l>=0;l--) { // Traverse octree from leaf to root
   #pragma omp parallel for default(shared) private(node,...)
   for (node=0;node<num;node++) { // For every node in this level
      // if the current node is a leaf then calculate
      // pseudoparticles masses based on postions and masses
      // of real particles inside the node,
      // otherwise calculate pseudoparticles masses based on
      // positions and masses of the pseudoparticles
      // of its children nodes.
   }
}
```

and the parallelization pseudocode of L2L is

```
#pragma omp parallel for default(shared) private(i,j,...)
   for (i=0;i<nbchild;i++) { // For every child of the current node
      // if the child contains no real particles then ignore,
      // otherwise perform L2L using inner expansion formula
   }.
```

We see that the parallelization of the FMM using OpenMP is relatively simple using our calculation scheme. Our calculation scheme has been implemented for the special-purpose computer GRAPE and achieved a speedup from 3-60 depending on accuracy of force calculation. Since our method uses simple mathematical formulae, it is also simple to parallelize FMM on multicore computers as shown above. We will show our experimental results in the next section.

4   Experimental Results

The FMMOpenMP is tested on a multiprocessor computer. The computer is equiped with four dual-core Intel(R) Xeon(R) CPU X5355 2.66 GHz processors and 4 GB RAM so that it is able to run 8 threads in parallel. The computer runs x86_64 Ubuntu operating system with 2.6.32-21 Linux kernel. We use gcc 4.4.3 compiler with POSIX threading model enabled. We develop the FMMOpenMP using C++ programming language.

We performed tests on performance and parallel efficiency of the FMM OpenMP. In all the tests, we distributed particles uniformly in a unit cube centered at the origin and evaluated force on all particles. The number of particles is from 64K to 8M, where K denotes 1024 and M denotes 1024×1024. The accuracy of FMM force calculation for uniform distribution of particles is described in table 3. Since the accuracy of force calculation with $p = 1$ and $p = 2$ is not enough for production runs, we perform tests for $p \geq 3$. Figure 2 shows FMM performance versus that of direct summation. We can see that the direct summation algorithm scales as $O(N^2)$ while FMM scales as $O(N)$. When N is

Table 3. Accuracy of FMM force calculation

Expansion order	Force RMS relative error
$p = 1$	1.6×10^{-1}
$p = 2$	2.1×10^{-2}
$p = 3$	4.9×10^{-3}
$p = 4$	7.4×10^{-4}
$p = 5$	1.6×10^{-4}
$p = 6$	5.7×10^{-5}
$p = 7$	1.7×10^{-5}

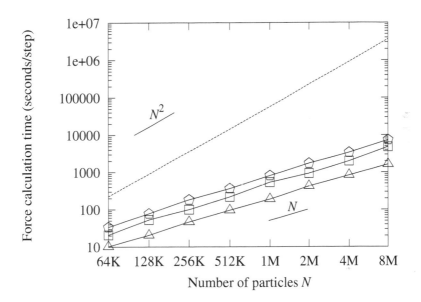

Fig. 2. Performance of direct summation and FMM algorithms. The dashed and solid curves represent the direct summation's and FMM's performance, respectively. Triangles, squares and pentagons denote performance of FMM with $p = 3$, $p = 5$ and $p = 7$, respectively.

8M, FMM runs faster than direct summation from 497 times to 2253 times when p runs from 7 to 3.

In the next tests, we show speedup and parallel efficiency of the FMMOpenMP on the computer described above. If P is the number of threads, T_1 is the execution time of the sequential FMM and T_P is the execution time of FMMOpenMP with P threads then speedup is

$$S_p = \frac{T_1}{T_P} \qquad (28)$$

and parallel efficiency is

$$E_P = \frac{S_P}{P}. \qquad (29)$$

Table 4. Speed up S_P of FMMOpenMP for N=8M

P	$p = 3$	$p = 4$	$p = 5$	$p = 6$	$p = 7$
2	1.52	1.91	1.91	1.93	1.91
4	2.23	3.59	3.60	3.62	3.62
8	2.36	5.47	5.52	5.60	5.53

Table 5. Parallel efficiency E_P of FMMOpenMP for N=8M

P	$p = 3$	$p = 4$	$p = 5$	$p = 6$	$p = 7$
2	76.1%	95.5%	95.7%	96.4%	95.4%
4	55.7%	89.7%	90.1%	90.5%	90.5%
8	29.5%	68.3%	69.0%	70.0%	69.1%

Tables 4 and 5 show speedup and parallel efficiency of FMMOpenMP for a test with 8M uniform distribution particles. The test results are similar for other values of N. Test results shows that speedup S_P of FMMOpenMP is low with expansion order $p = 3$ and becomes much higher with moderate and high order expansions $p = 4, 5, 6, 7$.

As a result, the parallel efficiency E_P of the FMMOpenMP rapidly drops from 76.1% to 29.5% when the number of threads increases from 2 to 8. However the parallel efficiency gradually drops from 96% to 70% with moderate and high expansions orders $p = 4, 5, 6, 7$. We can see that FMMOpenMP efficiency is better than that of Pan et. al. [13] which equal or lower than 40% for 8 threads.

As shown in Table 5, for a given number of threads the FMMOpenMP's parallel efficiency increases or unchanges as expansion order p increases. This behaviour is opposite to that of Yokota et. al. [14]. The parallel efficiency of Yokota's implementation is relatively high (78%) for low expansion order $p = 3$ but becomes low with high expansion orders [14]. The behaviour of parallel efficiency in our FMMOpenMP implementation shows that our approach is better than that of Yokota for applications require high accuracy of force calculation.

Reasons to explain our FMMOpenMP's behaviours include the usage of combination approach and simple mathematical formulae thanks to the A2P procedure. With the combination approach and A2P, the easy-to-parallelize pairwise interaction of FMMOpenMP, that has high parallel efficiency, dominates the total calculation. As expansion order p increases, the number of pairwise interaction increases accordingly and so does the parallel effciency.

5   Conclusions

We have successfully implemented the fast multipole method for multicore computers using OpenMP programming model. Test results show that it is simple to parallelize our FMM code using OpenMP thanks to the combination approach and A2P procedure. Our implementation's parallel efficiency for moderate and high expansion orders p are higher than those of Pan et. al. [13] and Yokota et. al. [14]. This behaviour makes our approach is suitable for high accuracy demand applications. The parallel efficiency of FMMOpenMP drops gradually for moderate and high expansion orders. This also shows another advantage of our implementation over those of Pan et. al. and Yokota et. al.

We still have room for improvements since the parallelization of M2M and L2L kernels is not optimized yet. The speedup and parallel efficiency of FMMOpenMP will become higher once the issues for improvements have been solved.

Acknowledgement. This work is partly supported by research project No. QG.2013 "Acceleration of the fast multipole method using graphics processing units and multicore computers for molecular dynamics simulations" granted by Vietnam National University, Hanoi.

References

1. Haile, M.: Molecular dynamics simulation: Elementary methods. Wiley-Interscience (1997)
2. Rappaport, D.C.: The art of molecular dynamics simulation, 2nd edn. Cambridge University Press (2004)
3. Satou, A.: Introduction to Practice of Molecular Simulation: Molecular Dynamics, Monte Carlo, Brownian Dynamics, Lattice Boltzmann and Dissipative Particle Dynamics. Elsevier (2010)
4. Griebel, M.: Numerical Simulation in Molecular Dynamics: Numerics, Algorithms, Parallelization, Applications. Springer (2010)
5. Barnes, J.E., Hut, P.: A hierarchical $O(N \log N)$ force calculation algorithm. Nature 324, 446 449 (1986)
6. Barnes, J.E.: A modified tree code: Don't laugh, it runs. Journal of Computational Physics 87, 161–170 (1990)
7. Greengard, L., Rokhlin, V.: A fast algorithm for particle simulations. Journal of Computational Physics 73, 325–348 (1987)
8. Greengard, L., Rokhlin, V.: A new version of the fast multipole method for the Laplace equation in three dimensions. Acta Numerica 6, 229–269 (1997)
9. Cheng, H., Greengard, L., Rokhlin, V.: A fast adaptive multipole algorithm in three dimensions. Journal of computational physics 155, 468–498 (1999)
10. Lupo, J.A., Wang, Z.Q., McKenney, A.M., Pachter, R., Mattson, W.: A large scale molecular dynamics simulation code using the fast multipole algorithm (FMD): performance and application. Journal of Molecular Graphics and Modelling 21, 89–99 (2002)
11. Gumerov, N.A., Duraiswami, R., Zotkin, D.N., Fantalgo, M.D.: Fast Multipole Accelerated Boundary Elements for Numerical Computation of the Head Related Transfer Function. In: IEEE International Conference on Acoustics, Speech and Signal Processing, ICASSP 2007 (2007)
12. Gumerov, N.A., Duraiswami, R.: Efficient FMM accelerated vortex methods in three dimensions via the Lamb-Helmholtz decomposition. Submitted to Journal of Computational Physics, arXiv:1201.5430
13. Pan, X.M., Pi, W.C., Sheng, X.Q.: On OpenMP parallelization of the multilevel fast multipole algorithm. Progress in Electromagnetics Research 112, 199–213 (2011)

14. Yokota, R., Barba, L.: A Tuned and Scalable Fast Multipole Method as a Preeminent Algorithm for Exascale Systems, arXiv:1106.2176v2 [cs.NA] (2011)
15. http://www.openmp.org
16. Anderson, C.R.: An implementation of the fast multipole method without multipoles. SIAM Journal on Scientific and Statistical Computing 13(4), 923–947 (1992)
17. Hardin, R.H., Sloane, N.J.A.: McLaren's improved snub cube and other new spherical design in three dimensions. Discrete and Computational Geometry 15, 429–441 (1996)
18. Makino, J.: Yet another fast multipole method without multipoles - pseudoparticle multipole method. Journal of Computational Physics 151, 910–920 (1999)
19. Kawai, J.M.: Pseudoparticle multipole method: A simple method to implement a high-accuracy treecode. The Astrophysical Journal 550, L143–L146 (2001)
20. Chau, N.H., Kawai, A., Ebisuzaki, T.: Implementation of fast multipole algorithm on special-purpose computer MDGRAPE-2. In: Proceedings of the 6th World Multiconference on Systemics, Cybernetics and Informatics, SCI 2002, Orlando, Colorado, USA, July 14-18, pp. 477–481 (2002)
21. Chau, N.H., Kawai, A., Ebisuzaki, T.: Acceleration of fast multipole method using special-purpose computer GRAPE. International Journal of High Performance Computing Applications 22(2), 194–205 (2008)
22. Chau, N.H.: A new formulation for fast calculation of far field force in molecular dynamics simulations. Journal of Science (Mathematics-Physics) 23(1), 1–8 (2007)

An Adaptive Neuro-Fuzzy Inference System for Seasonal Forecasting of Tropical Cyclones Making Landfall along the Vietnam Coast

Trong Hai Duong[1], Duc Cuong Nguyen[1], Sy Dung Nguyen[2], and Minh Hien Hoang[3]

[1] International University – VNU-HCM, Vietnam
[2] Department of Mechanical Engineering, Ho Chi Minh University of Industry, HUI, Vietnam
[3] Disaster Management Center, Ministry of Agriculture and Rural Development, Vietnam
{haiduongtrong,ndcuong69}@gmail.com, nsidung@yahoo.com,
hmh@netnam.vn

Abstract. The regression is a causal forecasting method that fits curves to the entire data set to minimize the forecasting errors. It should be noted that the linear statistic-based regression models does not support nonlinear in forecasting. According to literature, Bayesian- and Neural Network-based regression for seasonal typhoon activity forecasting is more effective than the traditional regression models. In this paper, a conjunct space cluster-based adaptive neuro-fuzzy inference system (ANFIS) is applied for seasonal forecasting of tropical cyclones making landfall along the Vietnam coast. The experimental results indicated that the conjunct space cluster-based ANFIS for seasonal forecasting of tropical cyclones is an effective approach with high accuracy.

1 Introduction

Vietnam has more than 3000 km of coastline which is directly affected by tropical storms and tropical depressions derived from the North-West Pacific or the South China Sea. Tropical storms extremely caused damage to people and property, especially when they were making landfall. Annual average of 5 to 6 tropical storms and tropical depressions made landfall along the Vietnam coast. Years 1964 and 1973, the number of tropical storms and tropical depressions made landfall in Vietnam is up to 10 attacks. Topical storm, especially when combined with a cold air front often causes heavy rains in Vietnam, flooding in a large area, extensive damage of life and property. With the advancement of climate science, weather and extreme weather event such as tropical cyclones, have considerably improved. However, at present in Vietnam, the forecasting of weather and tropical cyclones mainly applied for short-term or extremely short-term warning. The purpose of planning preventive and proactive mitigation offer more long-term warnings about our capabilities and activities. To develop the strategic plans to be more appropriate for the warning services, economic development, and the progress of Vietnam, the development of seasonal forecast of extreme weather events such as tropical cyclones are essential. This motivation leading

N.T. Nguyen, T. Van Do, and H.A. Le Thi (Eds.): *ICCSAMA 2013*, SCI 479, pp. 225–236.
DOI: 10.1007/978-3-319-00293-4_17   © Springer International Publishing Switzerland 2013

us set up a research for seasonal forecasting of tropical cyclones making landfall along with the Vietnam coast.

The most previous works used the regression-based model for seasonal tropical forecasting. The regression is a causal forecasting method that fits curves to the entire data set to minimize the forecasting errors. It should be noted that as higher order polynomial models are used, the overall degree of error will be reduced. However, the seasonal tropical forecasting requires high-dimensional data, so the forecasting ability also is reduced for higher order polynomial models. In addition, the regression-based linear statistical models does not support nonlinear in forecasting. According to Chu [4-7], Bayesian-based model for seasonal typhoon activity forecasting is more effective than regression-based models. Another aspect, Azizi [1] shown that ANFIS model was proven as more efficient and provides better forecasting accuracy compared with Bayesian model. Hence, ANFIS model is recommended to be used for production estimation under random uncertainties. According our study, the ANFIS has been not yet applied for seasonal tropical forecasting up to this research. Another importance to choose ANFIS model is that it can be used to combine all predictor factors for forecasting, while, other approaches only use several factors by transforming high-dimensional data to low-dimensional data.

The aim at this research is to offer usefully realistic supports for seasonal forecast of tropical cyclone activities in the region. We applied our proposed conjunct space cluster-based ANFIS for seasonal forecasting of tropical cyclones making landfall along the Vietnam coast using the ENSO, atmospheric and oceanographic data related to formation conditions and activity of ENSO factors and tropical cyclone activity in the study area in the 61-year period (1951-2011).

2   Related Works

According to our studies, we found that there are two issues to be successful in seasonal forecasting of tropical cyclones making landfall: forecasting factors and forecasting methods.

2.1   Forecasting Factors

The anomaly of tropical cyclone (TC) activity bears relation to the fluctuations of global climate. Nicholls [18] documents that TC activity in the Australian region has an interannual variation and related to ENSO phenomenon. Gray [13, 14] and Shapiro [22] found the relationship between TC activity in the Atlantic and the stratospheric quasi-biennial oscillation (QBO) phenomena. Afterward, methods for forecasting Australian and the Atlantic seasonal TC activity have established by Nicholls [19] and Gray [14]. Therefore, the seasonal variability and forecasting of TC activity over each of the ocean basins has received considerable attention. Over the Western North Pacific (WNP), Chan [2] found that TC activity is very much related to the ENSO phenomenon. Dong [8] also showed that this activity is correlated with the sea-surface temperature (SST) over the eastern equatorial Pacific that can be considered to be a proxy of ENSO. Chu [7] identified five parameters (sea surface temperature, sea level pressure, perceptible water, low-level relative vorticity, and vertical wind shear) as predictor datasets to predict the seasonal tropical cyclone count in the vicinity of Taiwan. Chan [3] pointed out that the correlation between the number of landfalling TCs over China, the ENSO and QBO phenomena should be possible to make the predictions. In addition to ENSO, Landsea [16] shown that other global factors such as the stratospheric Quasi-Biennial Oscillation and local factors such as sea surface temperature, monsoon intensity and rainfall, sea level pressures and tropospheric vertical shear can also help modulate tropical cyclone variability. Goh and Chan [10, 11] shown seven factors can be divided into two groups: steering factors (500 hPa geopotential height and zonal wind) and genesis factors (850 hPa geopotential height and relative vorticity, 200 hPa divergence, 200-850 hPa vertical wind shear, and 1000–500 hPa moist static energy) over Korean peninsula and Japan. Therefore, the seven factors from Goh and Chan [10, 11], along with the ENSO and PDO indices were the basis of investigation for a study of Goh and Chan [12].

2.2   Forecasting Methods

According to literature, forecasting methods can be categorized into qualitative and quantitative approaches. The former usually based on the opinions of people, which refers to a long or medium forecast by asking a group of knowledgeable experts for their opinions with regard to future values of the things being forecasted. The well-known method called Delphi involving a group of experts who eventually reach to a consensus of a forecast. In ecommerce, a firm often uses customer survey to forecast on the customers' purchasing plans. The later refers to quantitative, mathematical formulations or statistical forecasting, which includes time series models and casual models. The regression, a causal forecasting model that fits curves to the entire data set to minimize the forecasting errors, is often applied for the seasonal forecasting of TCs.

William Gray and his team pioneered the seasonal hurricane prediction enterprise using regression-based linear statistical models [14]. In a study, Chu [7], Fan and Wang [9] presented a multivariate linear regression model applied to predict the seasonal tropical cyclone count in the vicinity of Taiwan. The model is based on the least absolute deviation so that regression estimates are more resistant than those derived from the ordinary least square method. Kim H. S., et al. [15] used least absolute deviation (LAD) regression and the Poisson regression method. Poisson model is being slightly more skillful than the LAD model. Goh and Chan [12] presented an improved prediction scheme for the number of TCs making landfall on the coast of south China. The schemes for the early, late, and JD seasons all provide reasonable results. Chu and Zhao [5, 6] applied a hierarchical Bayesian change point analysis to detect abrupt shifts in the TC time series over the central North Pacific (CNP). Chu [4] extended the probabilistic Bayesian framework suggested in the prior works from the CNP [6], with a particular focus toward the vicinity of the Taiwan area. Different from prior studies, he adopts a feature classification approach based on the fuzzy clustering analysis of TC tracks.

3   Conjunct Space Clustering-Based ANFIS

We consider a black-box-typed model expressing a mathematical relationship between input and output spaces of a system based on a given data set as follows:

$$(\overline{x}_i, y_i), \overline{x}_i = [x_{i1} \ x_{i2}...x_{in}], \quad i = 1...P \tag{1}$$

where $\overline{x}_i$ is the ith input vector of the data set and y_i is the output; P is number of samples. This work can be seen as system-identifying process, in which the model works as a mathematical function f expressed by a mapping as follows:

$$f: \ \mathfrak{R}^n \to \mathfrak{R}^1$$
$$\overline{x}_i \mapsto y_i \mid y_i = f(\overline{x}_i)$$

In this paper, the above mathematical model is expressed by the ANFIS.

ANFIS is one of the most popular types of fuzzy neural network. The clustering techniques are commonly used to create fuzzy rules of ANFIS networks based on clustering a training set of numerical examples of the unknown mapping to be approximated. Panella et. al [20] analyze various clustering methods adopting for ANFIS, including clustering in input space, clustering in output space and clustering in input-output space (conjunct space). The clustering based on the data set only in the input space, which assumed that points potentially belonging to the same cluster in the input space are mapped into points potentially belonging to the same cluster in the output space. Its disadvantage that the output clusters could not reflect the real structure of the mapping in the output space.   In the other the hand, the clustering method considers only output space, which can be ensured that the possibility to discover the real structure of the mapping in the output space. Unfortunately, there can be contradictory rules having similar input MFs but different output coefficients, which is unacceptable in ANFIS networks. To overcome these problems, M. Panella [20, 21] and Dung [17] presented a clustering method in conjunct space for ANFIS networks construction. The clustering in input-output space mentioning in the previous works combines a linear cluster (e.g. hyperplane cluster) and Simpson's min–max models for classification (min-max classification).

The hyperplane clustering algorithm is expressed as follows:

Initialization: The C-Means algorithm is used to initialize hyperplanes by clustering the input space into M clusters $\Gamma^{(k)}, k = 1..M$. The correspondence between such clusters and initialized hyperplanes is based on following criterion: If an input pattern $\overline{x}_i, i = 1..P$ belongs to the cluster $\Gamma^{(q)}, 1 \le q \le M$, then the corresponding input-output pair $(\overline{x}_i, y_i)$ is assigned to the hyperplane A_q.

Step 1: The coefficients of each kth hyperplane, $y_t = \sum_{j=1}^{n} a_j^{(k)} x_{tj} + a_0^{(k)}, k = 1..M$, is updated using the pairs assigned to either in the *initialization* or in the successive *step 2*, where index t spans all the pairs assigned to the kth hyperplane using suited least-squares techniques.

Step 2: Each pair $(\bar{x}_i, y_i)$ assigned to a hyperplane A_q, which has the minimum orthogonal distance d_i from it. The stop condition is determined using a convergence quantity $\sigma = (|D - D^{(old)}|/D^{(old)})$ where D is the current approximation error defined by $D = \frac{1}{p}\sum_{i=1}^{P} d_i$ and $D^{(old)}$ is the previous approximation error. The algorithm will be stopped if it satisfy $\sigma \leq$ a predefined threshold ε, otherwise, it goes back to step 1.

The previous algorithm is a linear clustering that only yields the linear consequent of Sugeno rules. According to [21], several clusters of the input space could be associated with the same hyperplane. To solve this problem, the well-known Simpson's min–max models for classification (min-max classification) were applied by M. Panella [20, 21] and Dung [17]. The combination of the hyperplane clustering and the max-min classification on the input space supports to effectively determine the ANFIS network for a given number of rules. The min–max classification technique uses hyperboxes (HBs) which have boundary hyperplanes parallel to the coordinate axes of the patterns of the training set.

We consider a set of the patterns T_t covered by the tth min-max hyperbox HB_t. The HB_t is determined using two vertexes, the max vertex $\overline{\omega_t} = [\omega_{t1}\omega_{t2}\dots\omega_{tn}]$ and the min vertex $\bar{v}_t = [v_{t1}v_{t2}\dots v_{tn}]$, where $\omega_{tj} = \max{(x_{ij}|\bar{x}_i \in T_t)}$ and $v_{tj} = \min{(x_{ij}|\bar{x}_i \in T_t)}$. If T_t consists of the patterns associated with the cluster labeling m only, then the HB_t will be considered as a pure hyperbox labeling m, and denoted $pHB_t^{(m)}$. An HB can be considered as a crisp frame on which different types of membership functions (MFs) can be adapted. Here, the original Simpson's MF is adopted, in which the slope outside the HB is established by the value of the fuzziness parameter γ.

$$\mu_{pHB_t^{(m)}}(\bar{x}_i) = \frac{1}{n}\sum_{j=1}^{n}[1 - f(x_{ij} - \omega_{tj}, \gamma) - f(v_{tj} - x_{ij}, \gamma)]$$

$$f(x, y) = \begin{cases} 1 & if \ xy > 1 \\ xy & if \ 0 \leq xy \leq 1 \\ 0 & if \ xy < 0 \end{cases} \tag{2}$$

where $t = R_m$; R_m is the number of pure hyperboxes labeling m. Several pHB can be associated with the same cluster labeling m, thus the overall input MF, $\mu_{B_t^{(m)}}(\bar{x}_i)$, is calculated as follows:

$$\mu_{B_t^{(m)}}(\bar{x}_i) = \max{\{\mu_{pHB_1^{(m)}}(\bar{x}_i), \dots, \mu_{pHB_{R_m}^{(m)}}(\bar{x}_i)\}} \tag{3}$$

There are several approaches for min-max classification such as Simpson's min–max models, ARC in [21] and CSHL in [17]. Here, the CSHL algorithm is applied for this work.

The combination of the hyperplane clustering followed by the min-max classification on the input space can be determined of the ANFIS network for a given number of rules. A previously proposed structure of the conjunct space clustering-based ANFIS in [17] is depicted as *Figure 1*.

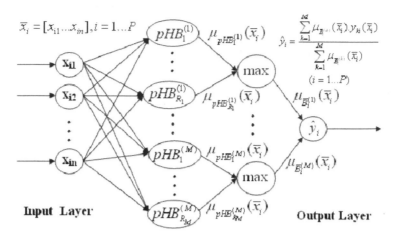

Fig. 1. Structure of the Conjunct Space Cluster-based ANFIS

The process of the conjunct space clustering-based ANFIS can be summarized as follows:

Step 0. Initialization: Let M_{min} and M_{max} be a minimal number and maximal number of fuzzy rules used for the initial training set Γ including P patterns, respectively.

-    $k = M_{min} - 1$;

Step 1. Training dataset clustering

-    $k = k + 1$;
-    Use the k-hyperplane clustering algorithm to split the initial training terns Γ into k clusters;
-    Use the CSHL [17] to build set L_p of pure hyperboxes pHB covering all the patterns belonging to the initial training set Γ;

Step 2. Establish the fuzzy-neural network:

-    Calculate values of MFs based on equations (2) and (3);
-    The output of the neuro-fuzzy network is calculated as following equations:

$$\hat{y}_i = \frac{\Sigma_{i=1}^{k} \mu_{\overline{B_i^{(m)}}}(\overline{x}_i) * y_{mi}(\overline{x}_i)}{\Sigma_{i=1}^{k} \mu_{\overline{B_i^{(m)}}}(\overline{x}_i)} \tag{4}$$

$$y_{mi} = \Sigma_{j=1}^{n} a_j^{(k)} x_{mj} + a_0^{(k)}, k = 1..M \tag{5}$$

-    Calculate the Mean Squared Error

$$E = \frac{1}{P} \sum_{i=1}^{P} (y_i - \hat{y}_i)^2$$

Step 3. Check for stopping condition
- If $k < M_{max}$, goto **Step 1**;
- If $k \geq M_{max}$, goto **Step 4**;

Step 4. Choose the optimal neuro-fuzzy network based on priorities: $E \leq \varepsilon$ and k is enough small, the structure of the ANFIS is chosen;

4   Experimental Result

4.1   Dataset

According to literature, the ENSO events have impacted to the different characteristics in typhoon's activity in Western North Pacific, South China Sea and in Vietnam. It causes the changes in the origin of typhoon formation, frequency, intensity, track, and in other characteristics of acted typhoons in these regions. We collect factors related to the formation and activity of the storms in the study area. In particular, the indices of El Niño–Southern Oscillation (ENSO) including warming phase (El Niño), cooling phase (La Niña) and neutral phase relates to the activity of the tropical storms and tropical depressions in the Vietnamese coast. In addition to ENSO, other global climate factors (such as the stratospheric Quasi-Biennial Oscillation, Pacific decadal oscillation (PDO), North Atlantic Oscillation, Arctic Oscillation, Antarctic Oscillation, the northern hemisphere oscillation, long-wave radiation equatorial Pacific, etc.). The local factors (such as sea surface temperature, monsoon intensity and rainfall, sea level pressures, tropospheric vertical shear, precipitable water, low-level relative vorticity, and vertical wind shear) can also help modulate tropical cyclone variability. The dataset including 30 independent factors and a dependent factor (the annual number of tropical storms) is available in the site (http://co-intelligence.vn/ENSO).

The annual number of tropical depressions in the Vietnamese coast from 1952 to 2011 is shown as *Table* 1.

Table 1. The number of tropical depressions in the Vietnamese coast (1951-2011)

Year	Number of landfall TC	Year	Number of landfall TC	Year	Number of landfall TC
1951	2	1972	6	1993	5
1952	8	1973	10	1994	6
1953	1	1974	7	1995	7
1954	3	1975	2	1996	8
1955	2	1976	2	1997	4
1956	3	1977	3	1998	6
1957	1	1978	6	1999	2
1958	0	1979	3	2000	2
1959	1	1980	4	2001	2
1960	3	1981	3	2002	0

Table 1. *(Continued)*

1961	2	1982	4	2003	2
1962	4	1983	6	2004	2
1963	2	1984	4	2005	5
1964	10	1985	4	2006	4
1965	3	1986	4	2007	5
1966	2	1987	4	2008	5
1967	1	1988	1	2009	5
1968	5	1989	8	2010	4
1969	2	1990	8	2011	4
1970	3	1991	3		
1971	7	1992	5		

We leant that the frequency of landfall TC in Vietnam is concentrated in the summer months, from July to November. Therefore, we divide the dataset into two group including first six months and last six months. For this, we have 61 samples corresponding to years, from 1951 to 2011. Each sample consists of 60 independent factors of input and output is the number of landfall TC.

4.2   Result

We used the samples from 1951 to 2000 to train the ANFIS network. The test samples are from 2001 to 2011.

The prediction error for the ith factor is calculated as follows:

$$error_i = \frac{factor_i - factor_i^{NF}}{factor_i} * 100(\%)$$

Figure 2 and 3 present the prediction error of 60 factors affecting to TC in the years, from 2001 to 2010.

Figure 4 shows the predicted number of TC versus reality in the years, from 2001 to 2011. The discrepancy between the number of predicted TC and reality in the years, from 2001 to 2011 is presented as *Table* 2 and *Figure* 5.

The difference between the actual number of TCs and the number of TCs that was predicted for a given period (year) is measured as follows:

$$err_i = y_i^{data} - y_i^{NF}$$

where, y_i^{data} and y_i^{NF} are the actual and predicted number of TCs in year i, respectively.

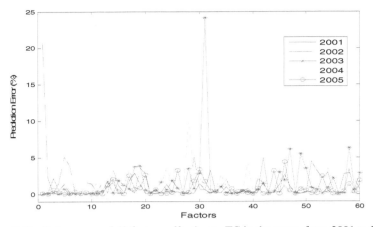

Fig. 2. Prediction error of 60 factors affecting to TC in the years, from 2001 to 2005

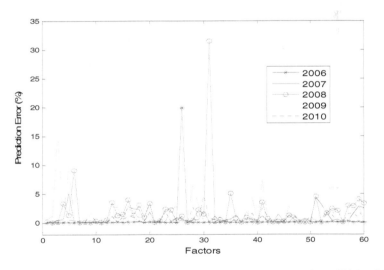

Fig. 3. Prediction error of 60 factors affecting to TC in the years, from 2006 to 2010

The mean absolute deviation:

$$err_{mean} = \frac{\sum_{i=2001}^{2011} |err_i|}{11\ (year)} = 0.099$$

The discrepancy between the actual number and the predicted number of TCs indicated that the forecast is very slightly lower than reality (see *Figure* 5 and *Table* 2). The mean error in the set of forecasts is only 0.099. From *Figure* 4, we can say that the accuracy of prediction is significantly high.

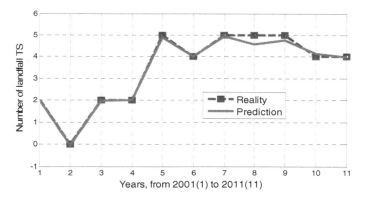

Fig. 4. The number of predicted TC and reality in the years from 2001 to 2011

Table 2. Discrepancy between the number of predicted TC and reality in the years, from 2001 to 2011

Year	Number of TC in reality, y_i^{data}	Error prediction for period i, $err_i = y_i^{data} - y_i^{NF}$
2001	2	0.018
2002	0	0.054
2003	2	0.015
2004	2	0.002
2005	5	0.111
2006	4	-0.004
2007	5	0.052
2008	5	0.421
2009	5	0.249
2010	4	-0.139
2011	4	0.024

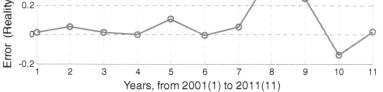

Fig. 5. Discrepancy between the number of predicted TC and reality in the years, from 2001 to 2011

5   Conclusion

This paper presented a conjunct space cluster-based ANFIS for the seasonal forecasting of tropical cyclones making landfall along the Vietnam coast. The experimental result indicated that the conjunct space clustering-based ANFIS is an effective approach with high accuracy for the seasonal forecasting of tropical cyclones.

References

1. Azizi, A., Ali, Y.A., Ping, L.W.: Model Development and Comparative Study of Bayesian and ANFIS Inferences for Uncertain Variables of Production Line in Tile Industry. Wseas Transactions on Systems 11(1), 22–37 (2012)
2. Chan, J.C.L.: Tropical cyclone activity in the Northwest Pacific in relation to the El Nino/Southern Oscillation phenomenon. Mon. Wea. Rev. 113, 599–606 (1985)
3. Chan, J.C.L.: Interannual variability of tropical cyclones making landfall over China. ESCAP/WMO Typhoon Committee Annual Review Rev., 85–94 (1993)
4. Chu, P.-S.: Bayesian Forecasting of Seasonal Typhoon Activity: A Track-Pattern-Oriented Categorization Approach. Journal of Climate 23, 6654–6668 (2010)
5. Chu, P.-S., Zhao, X.: Bayesian change-point analysis of tropical cyclone activity: The central North Pacific case. J. Climate 17, 4893–4901 (2004)
6. Chu, P.-S., Zhao, X.: A Bayesian regression approach for predicting seasonal tropical cyclone activity over the central North Pacific. J. Climate 15, 4002–4013 (2007)
7. Chu, P.S., et al.: Climate Prediction of Tropical Cyclone Activity in the Vicinity of Taiwan Using the Multivariate Least Absolute Deviation Regression Method. Terr. Atmos. Ocean. Sci. 18(4), 805–825 (2007)
8. Dong, K.: El Nino and tropical frequency in the Australian region and the northwest Pacific. Australian Meteor. Mag. 36, 219–225 (1988)
9. Fan, K., Wang, H.: A New Approach to Forecasting Typhoon Frequency over the Western North Pacific (2009)
10. Goh, A.Z.C., Chan, J.C.L.: Variations and prediction of the annual number of tropical cyclones affecting Korea and Japan. Int. J. Climatology (2010a)
11. Goh, A.Z.C., Chan, J.C.L.: Interannual and interdecadal variations of tropical cyclone activity in the South China Sea. International Journal of Climatology 30, 827–843 (2010b)
12. Goh, A.Z.C., Chan, J.C.L.: An Improved Statistical Scheme for the Prediction of Tropical Cyclones Making Landfall in South China (2010c)
13. Gray, W.M., et al.: Atlantic seasonal hurricane frequency: Part I: El Nino and 30mb quasi-biennial oscillation influences. Mon. Wea. Rev. 112, 1649–1668 (1984)
14. Gray, W.M., et al.: Predicting Atlantic seasonal Hurricane activity 6-11 Months in Advance. Weather and forecasting 7, 440–455 (1992)
15. Kim, H.S., et al.: Seasonal prediction of summertime tropical cyclone activity over the East China Sea using the least absolute deviation regression and the Poisson regression. Int. J. Climatology. 30, 210–219 (2010)
16. Landsea, C.W.: El Niño-Southern Oscillation and the seasonal predictability of tropical cyclones. In: Diaz, H.F., Markgraf, V. (eds.) El Niño: Impacts of Multiscale Variability on Natural Ecosystems and Society (2000) (in press)
17. Nguyen, S.D., Ngo, K.N.: An adaptive input data space parting solution to the synthesis of neuro-fuzzy models. Int. J. Control Autom. Syst. 6, 928–938 (2008)

18. Nicholls, N.: A possible method for predicting seasonal tropical cyclone activity in the Australian region. Mon. Wea. Rev. 107, 1221–1224 (1979)
19. Nicholls, N.: Recent performance of a method for forecasting Australian seasonal tropical cyclone activity. Aust. Mete. Mag. 40, 105–110 (1992)
20. Panella, M., Rizzi, A., Frattale Mascioli, F.M., Martinelli, G.: ANFIS synthesis by hyperplane clustering. In: Proc. of IFSA/NAFIPS, Vancouver, Canada, vol. 1, pp. 340–345 (2001)
21. Panella, M., Gallo, A.S.: An input-output clustering approach to the synthesis of ANFIS networks. IEEE T. Fuzzy Systems 13(1), 69–81 (2005, 2012)
22. Shapiro, L.J.: The relationship of the quasi-biennial oscillation to Atlantic tropical storm activity. Mon. Wea. Rev. 117, 1545–1552 (1989)

An Application of Computational Collective Intelligence to Governance and Policy Modelling

Tien Van Do[1], Tamas Krejczinger[1], Michal Laclavik[4], Nguyen-Thinh Le[3], Marcin Maleszka[2], Giang Nguyen[4], and Ngoc Thanh Nguyen[2]

[1] Department of Networked Systems and Services,
Budapest University of Technology and Economics, Hungary
[2] Institute of Informatics, Wroclaw University of Technology, Poland
[3] Department of Informatics, Clausthal University of Technology, Germany
[4] Institute of Informatics, Slovak Academy of Sciences, Slovakia

Abstract. The spread of social media provides a great opportunity to enhance the transparency, participation and collaboration in modern democracies. Since nothing is perfect, a best practice engineering approach can be used to continuously monitor processes and operations that are applied in increasing the participation of people and improving public service provision. This paper outlines some basic concept of our initiative called "Knowledge Management Tools for Quality of Experience Evaluation and Policy Modeling-KNOWN" that aims at the modeling and analysis of data collected from social media and other online sources. The purpose is to provide quantitative information and feedbacks regarding the quality of open government and public service provision.

1 Motivation and Main Goal

To organize and run a modern, democratic and well-developed society, huge costs are needed. The tax payers' (people and companies) money are used to organize public services, build infrastructure, stimulate economy, etc. Therefore, it is the joint and long-term interest of the society for a better future that more and more people are involved in a political process. However, people are less interested in common issues in new democratic and developed democratic countries as well, which are due to various reasons (e.g., the complex political processes, the lack of transparency in decision-making, the political inactivity, etc.). Furthermore, the quality and the efficiency of public service (to go to the local government to arrange some things –e.g., to obtain a new passport, identity card, etc.) provision are often questionable.

Therefore, processes, measures and best of practices are needed to increase the participation of people and to improve public service provision, which principles are laid by the Open Government Partnership (www.opengovpartnership.org) initiated in 2011. Since the establishment of the initiative, a number of countries have agreed to join forces in the Open Government Partnership to increase transparent and participatory government.

N.T. Nguyen, T. Van Do, and H.A. Le Thi (Eds.): *ICCSAMA 2013*, SCI 479, pp. 237–249.
DOI: 10.1007/978-3-319-00293-4_18   © Springer International Publishing Switzerland 2013

The spread of social media (Facebook, MySpace, Twitter, Youtube, Flickr, Foursquare, Wiki and Google Doc) can provide a great opportunity to implement and to enhance the three principles (transparency, participation and collaboration) of open government [15]. Indeed, there are various examples that local governments and public organizations have a presence on Facebook, Twitter, and Youtube [6][7].

Furthermore, the research on monitoring open government based on data from social media is still in its infancy. The information extraction and the analysis of social media (that are built using Information and Communication Technology--ICT) can provide an efficient opportunity to quantitatively monitor and to evaluate public service provision and open government as a consequence of human activities and decisions, which provides a motivation for this work.

To efficiently support the efforts by the governments and agencies of the Open Government Partnership, we propose a joint effort and initiative called "Knowledge Management Tools for QoE Evaluation and Policy Modeling-KNOWN". The KNOWN project applies computational collective intelligence to model and analyze "open government-related" data collected from social media. The purpose of the modeling and analysis is to provide quantitative information and feedbacks regarding the quality of open government and public service provision. To our best knowledge, this proposal is a pioneering initiative in the field. To raise awareness to the problems, this paper presents a concept and proposed methods that will be applied in our initiative.

The rest of this paper is organized as follows. Section 2 covers the challenges of the initiative, describes the applied methodology. Section 3 presents the overview of the software tools. Finally, Section 4 provides general conclusions of the initiative.

2   Concept, Approach and Methodology

2.1   Challenges

We are convinced that intangible aspects related to human activities, society, ways people (politicians and citizens), society and economic environments interact with each other, should be taken into account (see Fig. 1.). This leads to the following major challenges:

- What are intangible aspects that are to be considered?
- What quantitative measures can be proposed to describe and to monitor the performance of public service provision?
- How computable models can be constructed that are able to characterize intangible aspects and processes?
- How can we determine the parameters of computable models?
- How accurate are predictions based on computable models (how can we validate models based on scientific evidence and to obtain scientific evidence)?

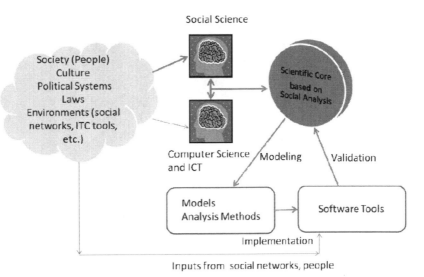

Fig. 1. The context and the environment, and the concept sketch of the KNOWN project

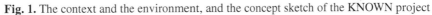

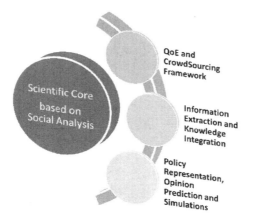

Fig. 2. The scientific methods

2.2   Concept

To achieve our goals, a systematic approach (Fig. 1) is performed as follows.

- Understanding the details of problems/challenges through application scenarios: To really serve the real needs we should understand problems in depth. We have to identify exactly scenarios and opportunities to identify the needs of society and end-users with the aim to improve the life of citizens, the participation in policy decisions, etc.
- Research directions are to be identified and ICT toolsets developed to solve problems that emerge in the scenarios. That is, the KNOWN project proposes and develops algorithms and methods that will be implemented in the

KNOWN software framework. Then, the methods and procedures developed will be subjected to extensive empirical validation. The test and validation of models and methods will be performed through comparing results obtained by mining and analysis of data collected from social media and results collected by the classical opinion polling. The comparison will give feedbacks to improve models and methods.

The scientific objectives to be achieved are categorized into three main frameworks as follows (Fig. 2):

- A Quality of Experience (QoE) framework allows us and the users of our software to define the mapping of opinions into quantitative measures

  o The survey and classification of a set of QoE (Quality of Experience) measures to evaluate policy decisions and public service provision,
  o The specification and investigation of methods to get qualitative solutions from crowd-sourcing techniques that are used in public service provision,

- a framework that is capable to perform large scale data extraction, processing and knowledge integration

  o The development of methods and procedures for the large scale extraction of information from various online sources,
  o The proposal of procedures to build semantic knowledge graph and integrate knowledge,

- a modelling, prediction and simulation framework that incorporates scientific techniques to evaluate and predict QoE measures and opinions based on extracted and processed data.

  o The definition of methodologies (argumentation graphs, conflict resolution) for representing policy and mining opinions, the use of Sentiment Analysis (SA) and Multidimension Scaling (MDS) for visualizing opinions and exploring similarities or dissimilarities in opinions,
  o The development of methods for evaluating and predicting the QoE measures and opinions based on mining data,
  o The development of stochastic models that are able to characterize the dynamic and stochastic behaviour aspects of participants (voters and politicians) in Governance and Policy based on mined information and data from various sources. We will validate proposed stochastic models and provide a tool to compute parameters using real data collected from social networks, crowd-sourcing and collaborative feedbacks in response to policy decisions.

Functions incorporating scientific methods and framework are proposed to be implemented in a scalable and extendable software architecture using advanced software engineering [13] and service oriented software architecture [1].

2.3   Scientific Methods

To achieve our aims, we plan to apply and to develop advanced computational collective intelligence methods. In what follows we outline some key methods and framework.

QoE framework

Lee and Kwak [7] have proposed an open government maturity model to access and guide open government initiatives after field studies regarding U.S. federal healthcare administration agencies. They closed their study by concluding:

> "*Furthermore, open government metrics are currently under-developed. Future research needs to develop reliable and valid metrics to measure and demonstrate the return on investment in open government.*" [7]

Therefore, we will do research on defining a set of metrics we term as the collection of Quality of Experience (QoE) measures that can be used to characterize aspects and public opinions related to open government. Therefore, QoE measures/metrics can be interpreted as the quantitative measures of public opinions in the general sense. Furthermore, QoE measures should reflect the true feelings of people from their perspective when they watch policy processes, participate in some aspects of open government and use public services. We are aware this is the complex issue since mapping involves subjective judgements as well. Therefore, we will work out a QoE framework that is configurable by the users of the KNOWN software framework using the XML schema.

Information Extraction and Knowledge Integration

Methods suitable to be used for this purpose include: A/B testing, association rule learning, classification, cluster analysis, crowdsourcing, data fusion and integration, ensemble learning, genetic algorithms, machine learning, natural language processing, neural networks, pattern recognition, predictive modelling, regression, sentiment analysis, signal processing, supervised and unsupervised learning, simulation, time series analysis and visualisation.

Based on the current state of the art in the field of information retrieval [8], the Semantic Web [2], processing of information networks and graphs [10] and unstructured text [9], as well as our previous experience in these areas, we would like to develop new methods which can integrate, process, search and visualize structured and unstructured sources and deliver them as knowledge bases. Text is still the most used medium for information sharing and communication, available ubiquitously on the Web, emails or new social media as well in organizational digital assets. This textual data often points to graph/network data through Web links, communication links, transactions or social links and tags, but describes a large part of their knowledge in unstructured textual form as well. In the area of natural language processing and extraction of information, we shall develop methods for transforming text to structured semantic data, such as tags, annotations, semantic trees and graphs.

We have to model knowledge graphs from text and network data. This research will improve and generalize our existing approach applied successfully on business data such as emails and documents. We will model knowledge graphs from textual interconnected documents. The novel approach of exploitation view and semantic search will be also developed with the aim to improve the quality of collected data and gained structures.

The process of creating a knowledge graph or an ontology based on different data sources requires attention to problems known in the knowledge integration area. In particular, the following issues must be addressed:

- Eliminating data inconsistency between sources.
- Integratiing different sources into a single knowledge graph.
- Eliminating knowledge inconsistencies in output graph.

All issues may be addressed in a single integration algorithm, but it is preferable to approach them independently to improve the overall quality of the result.

Eliminating inconsistency between different data sources is a process well known in the integration theory area, but generally not considered for more complex structures, like graphs. An example of such inconsistency would be if the same user states pro-A in one source and against-A in another (note that at data level this conflict is direct; on knowledge level such conflict may be hidden and require additional inference to determine). For simple structures such as conjunctive or multi-value structures, we have to solve the inconsistency [11]. The solution may be a single statement of opinion, a statement of uncertainty or eliminating the issue from this user's representation (note that on the knowledge level other solutions are possible). This method may be used, among other things, to eliminate uncertain or fake users.

The second issue, integration of different sources into a single knowledge graph, presents another kind of challenges. The most important one is the size of the dataset under processing. In a related field of ontology matching, the challenge of large scale ontology matching is still an open research question. The largest previously processed graph structures, where mapping related elements between different sources was done, were cross-lingual cases in [3] consisting of tens of thousands of entities. In this project, large social networks may be considered, containing up to hundreds of thousands of users. As each user may possess multiple opinions for mining, the overall graph may contain number millions of entities.

The third issue is related to the first one, but operating on knowledge level requires another set of tools for eliminating inconsistencies. If some inconsistencies are not detected on data level, then more complex analysis on a knowledge level allows finding those conflicts. This step will be responsible for finding fake profiles, which are not consistent with others. An example of such inconsistency would be if the same user states A in some issue and states B in some related issue but these two opinions A and B are mutually exclusive (note that in the first step it was not possible to find this without analysis on the semantic level). Another problem appears when we get empty opinions during the integration process. Based on collected data we can infer missing opinions. Similarly like in the first step, we need multiple methods of solving inconsistencies in ontologies and resolving conflicts which may be used to eliminate fake users, empty opinions and other uncertainties.

Graph-Based Policy Representation and Policy Simulation

A policy can be considered as a principle or a rule which serves to achieve a rational impact. In order to simulate the impact of a policy, it is required to represent a policy. Since a policy is released on the basis of several supporting arguments. We need to support a policy maker (e.g., a politician) to sketch her argumentation process. One approach of representing the argumentation process is using an argumentation graph which may contain elements such as: facts, contra arguments, pro arguments, and hypotheses. There exist various computer-supported tools for argumentation [12]. The benefits of using a computer-supported argumentation tools include: 1) an incorrect argumentation process (e.g., a cyclic argumentation) will be detected automatically by the system, and 2) the graph-based representation of a policy can be used directly for simulation. We will investigate which tool is appropriate to represent policies and whether the tool can be integrated into the tools framework of this project.

After collecting data and transforming them into representations which capture semantics, we use these data for simulating reaction of people on a policy. The classification of opinions can be more complex. We will develop an algorithm to calculate the pro-/contra-scale for each individual opinion with respect to a policy. For this purpose, Bayesian networks can be deployed. The goal is to visualize how many people agree with the policy to which extent. Alternatively, methods [11] for solving inconsistencies in multi-attribute and multi-value data may be modified to achieve similar results. The advantage of this alternative method is that it groups the users without the additional processing step for calculating the scale. On the other hand, compared to the Bayesian network approach, it is significantly harder to influence the number of positions on the scale (if more or less is required).

According to the social impact theory, social influence is one of the pervasive forces that operate in groups and societies [5]. That is, each individual person can influence the opinion of her neighbors. This theory suggests that the amount of impact other people have on an individual depends on three parameters: 1) the number of people influencing or being influenced, 2) the strength of these people, and 3) their immediacy to each other. The strength of each individual represents the ability to influence other's opinions. One challenging problem is detecting the strength of each individual in a social network. For example, we can assume that a Facebook user who has more friends can be considered more influential (i.e., she has more strength) than a user with less friends. This issue deserves investigation in this project. The third parameter, the immediacy, is relevant in the social impact theory, because people interact most often and are mostly influenced by those who are close to them (such as family members, friends, and colleagues) and their neighbors (i.e., those who live close to them in physical space). The distance between the two individuals in social space can be mapped to their immediacy. Whether in virtual social networks from which we intend to collect data, the immediacy can also be determined in a similar way as in physical space, needs further investigation. Once we have determined values for these three parameters, the number of people in a social network, the strength of these people, and their immediacy, we are able to model and to simulate the change of opinions of people until their opinions reach a constant state. In order to model the opinions of people in a social network Stocker et al. [14] proposed three modeling approaches: random network structures, hierarchical network structures, and scale-free network structures. These structures are represented as graphs. Each graph consists of a number of nodes which represent an individual of a network and are connected by edges which represent communication for exchanging information. Social science researchers usually have used random network structures for representing a social network in order to provide useful insights into dynamic and structural patterns. In this project, we intend to apply random network structures to model virtual social networks because they are more appropriate than hierarchical and network structures. Hierarchical network structures are well-suited for representing an organization, e.g., a company. Scale-free network structures have the property of having a few highly connected nodes and many with fewer connections and in a virtual social network such a pattern is rare.

Additionally, in this project we intend to use integration tools to determine group opinions in social networks. Consensus is an especially useful tool in this area. In the axiomatic approach to consensus theory [11], one of the postulates requires that the solution is the closest to all inputs. This is called optimality, as the task to solve is to determine the minimum of the sum of distances to all inputs. Depending on the distance measure used, different things may be determined. In this project this may be either the consensus opinion of the group, or the most influential member of the group. This presents an alternative, previously not researched, approach to social networks.

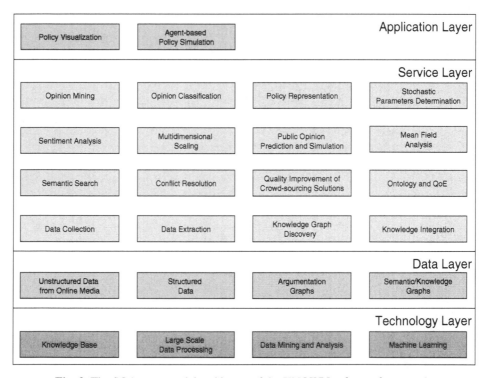

Fig. 3. The SOA-meta model and layers of the KNOWN software framework

3   Design of Software Framework

3.1   Software Architecture

Based on Service Oriented Software Architecture, we propose a SOA-meta model (see Fig. 3) for the KNOWN software framework that consists of the following layers:

- *Technology Layer*: The technologies which will be utilized by the system implementation.
- *Data Layer*: This layer contains some data structures and representations of information introduced and utilized by the project. These data will be stored in our cluster.
- *Service Layer*: The collection of algorithms and processes, which implements data building, manipulation and analysis tasks, and can be utilized as services by the applications developed in the project. Note that the results (methods, algorithms) of the KNOWN research will be implemented as services.
- *Application Layer*: This layer contains the applications developed by the project. These applications will use the services, and their operation can be defined using BPM [12] flowchart.

The relationship of the scientific and technical topics (to be presented in what follows) pursued in the KNOWN project is depicted in Fig.

The results (methods, algorithms) of the KNOWN research will be realized as services with well-defined interfaces. These services will be implemented as web services, using a proper software framework (for example: Java and the Spring Framework[1]).

For large scale data processing, MapReduce[2] developed by Google, with an open source implementation Apache Hadoop[3], or Spark[4] will be used in combination. Distributed databases, like Apache HBase[5], Apache Cassandra[6] (developed by Facebook), Voldemort[7] (developed by LinkedIn) or Dynamo[8] (developed by Amazon) will be considered for large scale data storage. For distributed processing and analysis of data stored in relational databases, there are several tools and frameworks like Pig[9] (developed by Yahoo!), Hive[10] (developed by Facebook) or Stratosphere[11], which is a generalization of the MapReduce framework optimized for processing big relational data. A distributed alternative to standalone statistical or machine learning frameworks is Apache Mahout[12], which can be used on Apache Hadoop distributed architecture.

---

[1]   http://www.springsource.org/
[2]   http://research.google.com/archive/mapreduce.html
[3]   http://hadoop.apache.org/
[4]   http://spark-project.org/
[5]   http://hbase.apache.org/
[6]   http://cassandra.apache.org/
[7]   http://www.project-voldemort.com/voldemort/
[8]   http://www.allthingsdistributed.com/2007/10/amazons_dynamo.html
[9]   http://pig.apache.org/
[10]   http://hive.apache.org/
[11]   https://www.stratosphere.eu/
[12] http://mahout.apache.org/

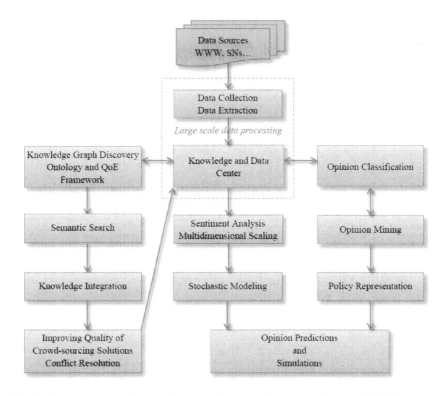

Fig. 4. The relationship of scientific and technical topics pursued in the KNOWN project

3.2   Applications

The applications of the KNOW Software Framework are interpreted as the orchestration of services that communicate through the defined interfaces and the jBPM integration mechanism[13]. The specific orchestration of services results in a process flows to satisfy a specific user need (e.g., to monitor the media appearances of policies issued by politicians/legislators and public opinions).

Information from focused data sources such as social networks, news portals and blogospheres will be exhaustively collected and extracted as raw data, which will be available in the Knowledge and Data Center. The center will be developed using advanced technology such as Hadoop and MapReduce to handle large information sources and keep up the system's performance. Further, raw data will be transformed by Knowledge Graph Discovery and Knowledge Integration (cf. Section 2.3) into structured data. In this process, the data will also be cleaned from inconsistencies by Conflict Resolution [11]. An exploitation view with interactions into data structures will be provided by Semantic Search. Ultimately, the structured and cleaned data will become available in the Knowledge and Data Center.

---

[13] http://docs.jboss.org/jbpm/

For the opinion predictions and simulations purpose, structured data will be processed further by the QoE and Crowd-Sourcing Framework, which defines a set of QoE metrics and quantitative measures of public opinions in general. The technique to explore distributional properties of citizens' opinions from the collected data and the method to map popular responses to policies, events and political processes will be done through Sentiment Analysis and Multidimensional Scaling techniques. Stochastic Modelling will simulate large-sized population projections, while traditional public opinion polling data will be collected on selected issues to validate the results obtained on a much more massive and cost-efficient way with our new tool. The output will be well visualized through the above mentioned opinion prediction and simulation.

Structured data can either be the sources for Opinion Mining and Classification, which is used as the input for Policy Simulation. The Policy Simulation uses graph-based policy representation to simulate reactions of citizens to policies based mined and analysed data. The mined and analysed opinions can also be used to predict and to simulate possible changes of citizens' opinions. These two simultaneous approaches of policy and opinion predictions provide complement results to each other because each technique can be used for different cases depending on the number of citizens.

In what follows, we provide an example for the use of the KNOWN framework to solve problems. Two scenarios are planned:

The purposes of scenario 1 are two-fold: 1) monitor new policies which will be issued by politicians and legislators and 2) trace public opinion and reaction to these new policy issues. For these purposes, we intend to develop a tool to visualize policies being issued and a tool for simulating public opinions (Fig. 5). The former tool can be developed through the tasks: 1) collection and extraction of unstructured data, 2) converting raw data into semantic and knowledge graphs, 3) transforming knowledge graphs into argumentation graphs which represent policies and as a result, policies can be visualized using a computer-supported argumentation tool. The development of the latter tool will require, in addition to the first two tasks, that semantic data in the form of knowledge/semantic graphs are analysed in order to mine and classify public opinion with respect to the policy being issued, and build agent-based simulation models. Deploying the developed simulation models we are able to simulate public responses to policies.

The second scenario aims at predicting public opinion with respect to a policy to be issued. As a new policy decision is introduced, it will be published in different media in relatively short time: news articles, blog posts, and social media commentary will appear online. We select and manually code in terms of objective characteristics (such as subject matter and the political actors involved) relevant content in a wide range of news media that cover the entire political spectrum of a country. Next, we trace the social media commentary (e.g. Facebook likes) that these news media products receive. Our new tools will be used to automatically collect and analyse the feedback of people to existing topics, providing input for policy simulation and opinion prediction (see Fig. 5). This is done automatically through scheduling tasks that run regularly and periodically. In particular, we will obtain measures of the intensity, valence, over-time dynamics of public responses to a story, and can break these down according to the political sympathies and media preferences of social media users that can be determined from their publicly available postings in social media. This classification will be automated via our tool after a human-assisted learning period.

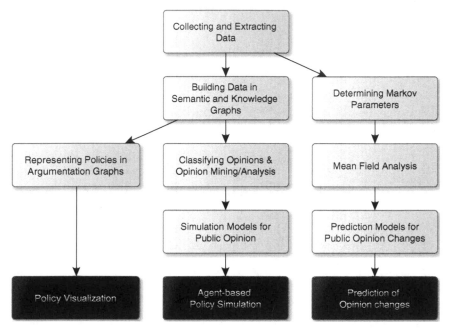

Fig. 5. Process-oriented description of the applications

With these data we also intend to develop a tool for predicting public opinion and possible changes (Fig. 5) because in a social network opinion leaders influence other participants in relatively predictable ways [4]. The development of this tool will be carried out through the following steps. First, we determine parameters for Markov chains and second, we use the mean field analysis and Multidimensional Scaling technique to map public opinion. In addition, we define a set of metrics which are referred to as Quality of Experience (QoE) measures that can be used to characterize the feelings and opinions of the public with respect to a policy issue.

4   Conclusions

This paper outlines the S&T challenges and methodologies used in the KNOWN initiative. At present, we are in the stage of the preparation of the project. We hope that our software tools can be implemented, which then can be used to monitor many measures related to public service provision and open government.

References

1. Bell, M.: Introduction to Service-Oriented Modelling. In: Service-Oriented Modelling: Service Analysis, Design, and Architecture. Wiley & Sons (2008)
2. Berners-Lee, T., et al.: The Semantic Web, pp. 29–37. Scientific American (2001)
3. Caracciolo, C., et al.: Results of the ontology alignment evaluation initiative 2008. In: Proc. of the 3rd International Workshop on Ontology Matching, held at the International Semantic Web Conference, pp. 73–119 (2008)
4. Huckfeldt, R., Johnson, P.E., Sprague, J.: Political Disagreement. The Survival of Diverse Opinions within Communication Networks. Cambridge University Press (2004)
5. Latane, B.: The psychology of social impact. Am. Psychol. 36, 343 (1981)
6. Lathrop, D., Ruma, L.: Open government: Collaboration, transparency, and participation in practice. O'Reilly Media (2010)
7. Lee, G., Kwak, Y.H.: An Open Government Maturity Model for social media-based public engagement. Government Information Quarterly 29, 492–503 (2012)
8. Manning, C.D., et al.: Introduction to Information Retrieval. Cambridge University Press (2008)
9. Mihalcea, R.F., Radev, D.R.: Graph-Based Natural Language Processing and Information Retrieval. Cambridge University Press, New York (2011)
10. Newman, M.: Networks: An Introduction. Oxford University Press, Inc., New York (2010)
11. Nguyen, N.T.: A Method for Ontology Conflict Resolution and Integration on Relation Level. Cybernetics and Systems 38(8), 781–797 (2007)
12. Scheuer, O., et al.: Computer-Supported Argumentation: A Review of the State-of-the-Art. Int. Journal of CSCL (2010)
13. Sommerville, I.: Software Engineering, 9/e edn. Addison-Wiley (2011)
14. Stocker, R., et al.: Network structures and agreement in social network simulations. Journal of Artificial societies and social simulation 5(4) (2002)
15. The White House. Memorandum for the heads of executive departments and agencies: Transparency and open government (2009)

A Framework for Building Intelligent Tutoring Systems

Adrianna Kozierkiewicz-Hetmańska and Ngoc Thanh Nguyen

Institute of Informatics, Wroclaw University of Technology, Poland
adrianna.kozierkiewicz@pwr.wroc.pl, ngoc-thanh.nguyen@pwr.edu.pl

Abstract. Intelligent tutoring systems provide customized instruction or feedback to learners, without intervention from a human teacher. This feature causes that intelligent tutoring systems attract attention because they allow learning everywhere, every time and the cost of courses is cheaper than traditional in-class learning. In this work we propose a formal framework for building intelligent tutoring systems. The particular elements of those systems such as: learner profile, domain model, methods for determination and modification of a learning scenario and for computer adaptive tests are presented. Additionally, we describe an application of rough classification in e-learning systems. The conducted experiments and analysis demonstrate that the personalization has a significant influence on a learning process and the probability of passing all lessons from the learning scenario is greater if the opening learning scenario is selected using a worked-out methods than chosen in a random way. The obtained results proof the correctness of our assumptions and have significant implications for development of intelligent tutoring systems.

1   Introduction

The popularity of intelligent tutoring systems (called also in this work e-learning systems, systems for distance education) is increasing from one year to another. However, there are no researches dedicated to universal e-learning systems. The systems are created for a certain domain (e.g. mathematics, foreign languages, programming languages etc.) and only simply learning platforms are available for free. It is difficult to find the real systems which offer the personalization and recommendation on each step of the learning process. In none of the researches the learner profile contains all significant data such as: demographic data, learner style, interests, abilities, history interaction with system. Only a few systems offer a computer adaptive testing, however in none of the works a learner profile is considered for selecting next test's items. The problem of determination of a learning path to student's characteristic and current knowledge level were solved only by some researches.

The mentioned conclusions incline to conduct a researches for developing an intelligent tutoring systems. The aim of this work is to present a formal framework for building intelligent tutoring systems which offer the personalized learning environment.

The idea of designing the system is the assumption that similar student will learn in the same or a very similar way [14]. Therefore, in the first step the system collects

N.T. Nguyen, T. Van Do, and H.A. Le Thi (Eds.): *ICCSAMA 2013*, SCI 479, pp. 251–265.
DOI: 10.1007/978-3-319-00293-4_19   © Springer International Publishing Switzerland 2013

information about a student and stores them in a learner profile. Next, a user is classified to a group of similar students. The criterion of classification is a subset of the user's attributes selected by an expert. This criterion could be change during the functioning of the system and based on collected data. For this task the rough classification is applied. User can start the learning process according to an opening learning scenario determined by the intelligent tutoring system. The opening learning scenario is chosen based on final, successful scenarios of students who belongs to the same class as the new one. A student reads and learns the subsequent lessons from the learning scenario and after each lesson tries to pass a computer adaptive test. If he does not achieve the assumed test score (he fails more than 50% of questions) it is a signal for the system that the learning scenario is not suitable for a student and a user needs a change. The system has to offer a repetition of the learning material and modify the opening learning scenario. All changes are stored by the system. The course is graduated if tests related to all lessons from the learning scenario are passed.

As it was mentioned above, the created intelligent tutoring systems have some limitations. The first intelligent tutoring system was created in about 1995-1996 and has been called ELM-ART [1]. In this system many techniques which allow the personalization of the learning process are applied. Authors proposed adaptive annotation of links, individual curriculum sequencing and a simple adaptive test method. The knowledge is represented by means of a multilayered overlay model that supports adaptation in the system. The learner profile is a quite poor and does not contain information about learner's styles. Described recommendation methods are simple and not very efficient.

System EDUCE [4,5] has a similar functionality as ELM-ART but for personalization process also student's learner styles are taken into consideration. The naïve Bayes algorithms is applied for recommendation of learning materials. EDUCE has no computer adaptive tests.

In [2] Bayesian networks are applied to model a learner and a knowledge, to plan learning paths and for suggesting pedagogical actions. The content of a learner profile proposed by authors is poor. Moreover, the system is able to offer only traditional tests.

The interesting features is proposed in DEPTH [3]. Despite the fact that DEPTH provides the learning material adapted only to the student's performance and cognitive capacity, this system is one of the few systems which offers the modification of the learning path after changing user's knowledge level. In DEPTH the method for dynamical selection of test's items based on student's characteristic is implemented.

In Learning Assistant [15] neural networks are used to infer using metadata describing pupils and didactic materials. The learning path for a new user is generated based on the individual information about this pupil: learner styles, skills and abilities among others. Authors assumed that after each test the plan of course is modified which is the unnecessary task. More efficient is the modification of learning path in case of test failure.

Some intelligent tutoring systems have implemented more advanced methods and techniques for the learning personalization. However, the most popular are still systems where only adaptive navigation and presentations are applied, traditional tests are offered to users, systems collect only some information about their students and the personalization process is limited: KBS Hyperbook, InterBook, PAT, Inter-Book, CALAT, VC Prolog Tutor, ELM-ART-II, AST, ADI, ART-Web, ACE and ILESA [1].

In the next section, designed elements of an intelligent tutoring system such as: the content and structure of a learner profile, knowledge structures, methods for a learning scenario determination and modification, algorithm for computer adaptive test are described. The section 3 contains the results of experiments and mathematical analysis. Finally, conclusions and future works are presented.

2   Model of Intelligent Tutoring System

2.1   Learner and Domain Model

The appropriate content and structure of learner model/profile is the key to good rec-ommendation. If we have knowledge about student's needs, preferences, learning styles, activities, interests, knowledge levels we are able to offer the best suitable learning material. In our work we assume that a learner profile is represented as a tuple of values defined in the following way:

$$t : A \rightarrow V$$

where A is a finite set of the profile attributes and V- the domain of all attributes from A, $V = \bigcup_{a \in A} V_a$, $\forall_{a \in A} (t(a) \in V_a)$.

The learner profile stores two types of data: the user data which refers to user's personal characteristic and the usage data which is related to a history of interaction and user's behaviour. We propose a user model which consists of the following attributes [11]: demographic data (such as login, name, sex, educational level, IQ), learning style (related to perception, receiving, processing, and understanding of in-formation by a student), abilities (verbal comprehension, word fluency, computational ability, spatial visualization, associative memory, perceptual speed, reasoning), per-sonal character traits (such as concentration, motivation, ambition, self-esteem, level of anxiety, locus of control, open mindedness, impetuosity, perfectionism), and interests (humanistic science, formal science, natural science, economics and law, technical science, business and administration, sports and tourism, artistic science, management and organization, education). The data is provided directly or achieved by analyzing psychological tests and questionnaires filled in during a registration process. Additionally, during user's interaction with the system information is col-lected and stored such as: current and finished scenario, test's scores, number of failed tests, time of learning etc with reference to lessons. The approach to personali-zation process of selected attributes were analyzed.

Education process requires the appropriate learning material. It has been reported that students prefer learning materials divided into smaller pieces. In our work we propose two knowledge structures [6],[9],[10]. In the first of them, learning material is divided into units and lessons. Each lesson exists in one of the following forms:

textual, graphical, interactive or combination of the mentioned forms. Between units and lessons linear orders occurred. This means that all lessons from all units need to be learned, but some of them should be learned before others.

Let $C = \{c_0,...,c_q\}$ be a finite set of units. By P_i we denote the set of lessons corresponding to unit c_i where $i \in \{0,...,q\}$, $P = \bigcup_{i=0,...,q} P_i$. By p_{ij} we denote j-th lessons belonging to unit c_i . Each lesson p_{ij} is related to their version v_{ijk} for $k \in \{1,...,m\}$.

We assumed that V_{ij} is a set of versions of lesson p_{ij} for unit c_i , $V_i = \bigcup_{j=1}^{r_i} V_{ij}$,

$V = \bigcup_{i=0}^{q} V_i$ for $r_i = card(P_i)$. Rc is called a set of linear relations between units and $R_P(P_i)$ is called a set of partial linear relations between lessons where $Rp = \bigcup_{i=1,...,q} Rp(P_i)$, $i \in \{1,...,q\}$. A binary relation γ is called linear if the relation is reflexive, transitive, antisymmetric and total.

Definition 1. *The relational knowledge structure is defined by the following tuple*

$$(C, P, V, R_C, R_P)$$

Definition 2. *By a learning scenario s based on orders $\alpha \in R_C$ and $\beta_i \in R_P(P_i)$ for each $i \in \{0,...,q\}$ we call a sequence:*

$$s = < w_0, w_1,..., w_q >$$

A learning scenario fulfils the following conditions:

1. $w_i \neq w_{i+1}$ *for each $i \in \{0,...,q\}$* ,

2. w_0 *is sequence of versions of lesson belongs to set V_o,*

3. $\alpha \in R_C$ *, therefore $(c_i,c_z) \in \alpha \Leftrightarrow c_z = \alpha(c_i)$ for each $i \in \{0,...,q\}$, $z \in \{1,...,q\}$, w_{ρ_z} is sequence of versions of lesson belongs to set V_z, ρ_z -the position of unit c_z in scenario s, $\rho_z \in \{0,...,q\}$* ,

4. *the order of versions of lessons in sequence w_{ρ_z} should correspond to the order of lessons $(p_{zj}, p_{zx}) \in \beta_z$, $\beta_{\rho_z} \in R_P(P_z)$ for $\rho_z \in \{0,...,q\}$, $j,x \in \{1,...,r_z\}$* .

5. *If sequence w_i contains version of lesson $v_{ijk,}$ $i \in \{0,...,q\}$, $j \in \{1,...,r_i\}$, $k \in \{1,...,m\}$, sequence w_i does not contain v_{ihy}, where $h = j$, $y = k$, $h \in \{1,...,r_i\}$, $y \in \{1,...,m\}$* .

The second knowledge structure [7,9,10] is similar to previous one but consists only of lessons, relations between them and their versions. The main advantage of this knowledge structure is the additional information stored on edges' labels. The intelligent tutoring system during his functioning collects and stores information such as the average score for each lesson, the average time of learning of each lesson and the difficulty degree of each lesson, which is measured by the number of failed tests. This data is stored separately in two-dimensional arrays for different student classes and different lesson orders. Therefore, by $W = [w_{iz}]_{i=0,...,q,}^{z=1,...,q}$ we denote weight matrix where

w_{iz} could equal the average score, the average time of learning or the difficulty degree of lesson p_z which was learnt after lesson p_i, for $i \in \{0,...,q\}$, $z \in \{1,...,q\}$. Let us assume that P is a finite set of lessons. Each lesson $p_i \in P$, $i \in \{0,...,q\}$ is a set of different versions $v_k^{(i)} \in p_i, k \in \{1,...,m\}$, $V = \bigcup_{i=0,...,q} p_i$, $i \in \{0,...,q\}$. R_C are called linear orders on P. The graph-based knowledge structure is defined in the following way:

Definition 3. *The graph-based knowledge structure is the labeled and directed graph:*

$$Gr = (P, E, \mu)$$

where: P is the set of nodes, E is the set of edges, $\mu : E \rightarrow L$ is function assigning labels to edges, , $L = \bigcup_{f=1}^{card(R_C)} L(\alpha_f)$ is the set of labels, where $L(\alpha_f) = (W, \alpha_f)$, $f \in \{1,..., card(R_C)\}$, $\alpha \in R_C$.

Definition 4. *By the Hamiltonian path based on order $\alpha \in R_C$ in the graph Gr we mean the sequence of nodes:*

$$hp = < p_0,..., p_q >$$

Where:
(a) for each $i \in \{0,..., q\}$ $p_i \neq p_{i+1}$
(b) for each $e \in E$ $\mu(e) \in L(\alpha_f)$, where $f \in \{1,..., card(R_C)\}$

Definition 5. *By the learning scenario s we mean a Hamiltonian path hp based on an order $\alpha \in R_C$ in which exactly one element from each node p_i , $i \in \{0,..., q\}$ occurs:*

$$s = < v_k^{(0)},..., v_n^{(q)} >$$

where: $v_k^{(0)} \in p_0,..., v_n^{(q)} \in p_q$ for $k, n \in \{1,...m\}$

2.2   Determination of an Opening Learning Scenario

The main goal of the intelligent tutoring system is to provide the learning material suitable for student's needs and preferences. The opening learning scenario is the first learning scenario offered to a student. After the registration process the system has a poor knowledge about students, therefore for choosing an opening learning scenario the system used only information collected by questionnaires, psychological tests and information about final successful scenario. After registering in the system and providing information about himself, a new learner is classified to a group of similar students. The classification criterion is a set of the user's attributes selected by an expert. The opening learning scenario is chosen from final scenarios of students who belong to the same class as the new student. For this task we propose two algorithms ADOLS I and ADOLS II based on consensus theory [6,10,11] . The problem of determination of an opening learning scenario is defined in the following way: *For given learning scenarios* $s^{(1)}, s^{(2)},..., s^{(n)}$ *one should determine a scenario* $s* \in S_C$ *such that the condition is satisfied* $\sum_{i=1}^{n} d(s*, s^{(i)}) = \min_{s} \sum_{i=1}^{n} d(s, s^{(i)})$, *where:* S_C *is the set of* all possible scenarios following the defined knowledge structure.

In the problem solved by the consensus theory the key issue is the distance function. The distance between scenario $s^{(1)}$ and $s^{(2)}$ for the relational knowledge structure is defined in the following way:

Definition 6. *By* $d: S_C \times S_C \to [0,1]$ *we call a distance function between scenarios* $s^{(1)}$ *and* $s^{(2)}$ *which is calculated as:*

$$d(s^{(1)}, s^{(2)}) = \lambda_1 \sigma(\alpha^{(1)}, \alpha^{(2)}) + \lambda_2 \frac{\sum_{i=0}^{q} \sigma(w^{(1)}{}_{g_i}, w^{(2)}{}_{h_i})}{q+1} + \lambda_3 \frac{\sum_{i=0}^{q} \delta(w^{(1)}{}_{g_i}, w^{(2)}{}_{h_i})}{q+1}$$

where : $\lambda_1, \lambda_2, \lambda_3 \in [0,1]$, $\lambda_1 + \lambda_2 + \lambda_3 = 1$, g_i, h_i - *the position of module* c_i *in scenario* $s^{(1)}$ *and* $s^{(2)}$, $i \in \{0,...,q\}$, *scenarios* $s^{(1)}$ *and* $s^{(2)}$ *are based on order* $\alpha^{(1)}$ *and* $\alpha^{(2)}$, *respectively.*

The values of distance functions are calculated in three steps:

1) $\sigma(\alpha^{(1)}, \alpha^{(2)}) = \dfrac{\sum_{i=0}^{q} \dfrac{|k^{(1)} - k^{(2)}|}{q+1}}{q+1}$ *where:* $k^{(1)}$ *and* $k^{(2)}$ *are the position of unit c in* $\alpha^{(1)}$ *and* $\alpha^{(2)}$ *respectively.*

2) $\sigma(w^{(1)}{}_{g_i}, w^{(2)}{}_{h_i}) = \dfrac{\sum_{i=1}^{r_i} (\dfrac{|k^{(1)} - k^{(2)}|}{r_i})}{r_i}$ *where: lesson p occurs in sequence* $w^{(1)}{}_{g_i}$ *on position and in sequence* $w^{(2)}{}_{h_i}$ *on position* $k^{(2)}$, $r_i = card(P_i)$.

3) $\delta(w^{(1)}{}_{g_i}, w^{(2)}{}_{h_i}) = \dfrac{\sum_{j=1}^{r_i} \theta(v^{(1)}{}_{ijk}, v^{(2)}{}_{ijy})}{r_i}$ *where:*

$$\theta(v^{(1)}{}_{ijk}, v^{(2)}{}_{ijy}) = \begin{cases} 1, if\ v^{(1)}{}_{ijk} \neq v^{(2)}{}_{ijy} \\ 0\ w\ p.p. \end{cases}, for\ k, y \in \{1,...,m\}$$

Algorithm ADOLS I is dedicated for relational knowledge structure. The opening learning scenario is conducted in four steps: the first one depends on a proper order of units, in the next a proper orders of lessons are determined and in the third one the system chooses suitable versions of lessons. Finally, the selected opening learning scenario is checked if it belongs to S_C.

In the first step for each lesson unit c_i a multiset $I_c(c_i)$ and an index $J_c(c_i)$, $i \in \{0,...,q\}$ are calculated where:

$$I_c(c_i) = \{ jm : if\ there\ exists\ a\ scenario\ that\ c_i\ occurs\ on\ its$$

$$jmth\ position\ \};$$

$$J_c(c_i) = \sum_{jm \in I_c(c_i)} jm$$

Based on increasing value $J_c(c_i)$ the units order is set in relation to $\alpha *$.

For each lesson p_{ij} a multiset $I_p(p_{ij})$ and an index $J_p(p_{ij})$, $i \in \{0,...,q\}$, $j \in \{1,...,r_i\}$ are calculated where:

$$I_c(p_{ij}) = \{ jl : if\ there\ exists\ a\ scenario\ that\ p_{ij}\ occurs\ on\ its$$

$$jlth\ position\ \};$$

$$J_p(p_{ij}) = \sum_{jl \in I_p(p_{ij})} jl$$

Based on the increasing value of $J_p(p_{ij})$ the lesson orders are set in relation to $\beta_i *$ for each unit c_i, $i \in \{0,...,q\}$.

Next, for each lesson p_{ij} the number of appearances of versions of lesson v_{ijk} in sequences $w_{g_i}^{(1)}, w_{h_i}^{(2)},..., w_{d_i}^{(n)}$ are estimated. For each lesson p_{ij} the version of lesson v_{ijt} such that $f(v_{ijt}) = \max_{k \in \{1,...,m\}} f(v_{ijk})$, $i \in \{0,...,q\}$, $j \in \{1,...,r_i\}$, $t \in \{1,...,m\}$. Finally, it is determined whether the opening learning scenario belongs to S_C. If not, from set $s^{(1)}, s^{(2)},..., s^{(n)}$ the opening learning scenario is chosen that satisfies the following condition: $\sum_{i=1}^{n} d(s*, s^{(i)}) = \min_{s} \sum_{i=1}^{n} d(s, s^{(i)})$.

Algorithm ADOLS II is modification of ADOLS I and is adapted to the graph-based knowledge structure. The distance function for this problem is defined in the following way:

Definition 7. By $d : S_C \times S_C \to [0,1]$ we call a distance function between scenarios $s^{(1)}$ and $s^{(2)}$ which is calculated as:

$$d(s^{(1)}, s^{(2)}) = \lambda_1 \sigma(\alpha^{(1)}, \alpha^{(2)}) + \lambda_2 \delta(\alpha^{(1)}, \alpha^{(2)})$$

where : $\lambda_1, \lambda_2 \in [0,1]$, $\lambda_1 + \lambda_2 = 1$

The values of distance functions are calculated in two steps:

1) $\sigma(\alpha^{(1)}, \alpha^{(2)}) = \dfrac{\sum\limits_{i=0}^{q} \dfrac{|k^{(1)} - k^{(2)}|}{q+1}}{q+1}$ where: $k^{(1)}$ and $k^{(2)}$ are the position of lesson in $\alpha^{(1)}$ and $\alpha^{(2)}$, respectively.

3) $\delta(\alpha^{(1)}, \alpha^{(2)}) = \dfrac{\sum\limits_{i=0}^{q} \theta(v_k^{(1)(i)}, v_y^{(2)(i)})}{q+1}$ where:

$\theta(v_k^{(1)(i)}, v_y^{(2)(i)}) = \begin{cases} 1, & if\ (v_k^{(1)(i)} \neq v_y^{(2)(i)} \\ 0 & w\ p.p. \end{cases}$, for $k, y \in \{1,...,m\}$

Algorithm ADOLS II is conducted in 3 three steps: selection of the lesson order, selection of version of lessons and checking the correctness of obtained results. Similarly as in previous algorithm, the selection of order lesson is based on a multiset $I_p(p_i)$ and increasing value of an index $J_p(p_i)$ which determine the relation $\alpha*$.

The version of lesson is chosen depending on the maximum number of appearances in scenarios: $s^{(1)}, s^{(2)},..., s^{(n)}$.

The ADOLS I and ADOLS II allows adapting a learning scenario to student's preferences. Both methods are simple and have a relatively low computational complexity equal to $O(n^2)$.

2.3   Modification of a Learning Scenario during Learning Process

The system collects information about a student during the whole time. It means that the system can learn more about users. Additionally, user's needs could change during interaction with the system. If the opening learning scenario turned out not sufficient for a learner, system should be able to modify the learning scenario during the learning process. For this task two algorithms AMAM and BAM were proposed. Both methods require the graph-based knowledge structure and information stored in a learner profile.

Algorithm AMAM [7] depends on the identification of student's mistakes and elimination of them. Authors distinguish three probable reasons of mistakes:

1. Lack of concentration, motivation, the bad condition;
2. The wrong order of lessons;
3. The wrong student's classification.

Authors take into consideration all of the mentioned above problems with passing test and proposed method which consists of three steps. If a student has a problem with passing a test for the first time he is offered a repetition of the same lesson but in a different version. It is possible that a student has worse day, was tired and his problem with a test was not caused by the wrong learning scenario. After the next failure the system changes lessons' order based on data of students who belongs to the same class. The system assumed that material from one lesson could help understand another one. The third step is the final chance. A student could provide false information about himself so it might have happened that he was classified to the improper group. If this possibility is taken into account in the last step a user is offered the modification of lessons' order based on all collected data.

For choosing the another version of lesson in the first step we applied the Naïve Bayesian Classification method. The second and the third step can be brought down to finding the shortest Hamiltonian path in the graph Gr. In the second step the learning path which was learnt the most effective by students (students achieved the best score) is found. In the third step the easiest learning path (students failed the tests the rarest) is found. The shortest Hamiltonian path problem is NP complete problem and it could be solved by using a brute force method or heuristic algorithms such as the nearest neighbourhood method or genetic algorithm.

BAM algorithm [9,10,11] is based on Bayesian network. Let us assume that a student u was learning according to learning scenario $s = < p_0, ..., p_i, p_z, ..., p_q >$, based on $\alpha_f \in R_C$, $f \in \{1, ..., card(R_C)\}$ and passes a test for lesson p_i but fails a test for lesson p_z, $i \in \{0, ..., q\}$, $z \in \{1, ..., q\}$. In that case the system builds a Bayesian network that is used to modify the learning scenario. The procedure of creation of a Bayesian model consists of two steps: construction of a graph representing the qualitative influences of the modeled situation, and assignment of probability tables to each node in the graph. In our model, the following variables are considered:

- *time* - the time of learning lesson p_z which was learnt after lesson p_i, the set of values: *average, less , more* ;
- *number* - the number of tests taken for lesson p_z, the set of values equals $1,2,3, > 3$;
- *score* - the last of test scores in percent for lesson p_z, which was learnt after lesson p_i, the set of values: $\{1,2,3,4,5\}$;
- *version* - the number of versions of the lesson which occurs in the opening learning scenario s most often before learning lesson p_z; the set of values equals $1,2,...,m$;
- *lesson*: pi, the number of failed lesson and their version v_k^z, $k \in \{1,...,m\}$;
- *lessons* $p_z, ..., p_q$

If a student fails lesson p_z which was learnt after the lesson p_i, $i \in \{0, ..., q\}$ the system dynamically creates a Bayesian network based on collected data. In the first step the graph Gr' is created by elimination from the graph Gr nodes $p_0, ..., p_i$ and all edges connected with those nodes. For the new defined knowledge structure the all Hamiltonian path hp are determined. In Bayesian network we have Hamiltonian path based on orders, therefore if $(p_i, p_{i+1}) \in \alpha_f$ then $p_{i+1} = \alpha_f(p_i)$ for $f \in \{1, ..., card(R_C)\}$. The second step in creating the Bayesian network is the assignment of probability tables to each node in the graph. The probabilities are estimated based on observing the student's interaction with the system. It is necessary to assess the local distributions:

- $p(time = wt)$ for each $wt \in \{l, a, m\}$,

- $p(number = wn)$ for each $wn \in \{1, 2, 3, > 3\}$,

- $p(score = ws)$ for each $ws \in \{1, 2, 3, 4, 5\}$,

- $p(lesson = v_k^{(i)})$ for each $k \in \{1, ..., m\}$, $i \in \{0, ..., q\}$,

- $p(version = k)$ for each $k \in \{1, ..., m\}$.

and conditional distributions:

- $p(v_l^{(i+1)} \in \alpha_f(p_i) \mid time = wt, number = wn, score = ws, lesson = v_n^{(i)}, version = k)$

- $p(v_k^{(i+2)} \in \alpha_f(\alpha_f(p_i)) \mid v_l^{(i+1)} \in \alpha_f(p_i))$

$$\vdots$$

- $p(v_n^{(i+d+1)} \in \alpha_f(...(\alpha_f(\alpha_f(p_i)))) \mid v_k^{(i+d)} \in \alpha_f(...(\alpha_f(p_i))))$

For each $wt \in \{l, a, m\}$, $wn \in \{1, 2, 3, > 3\}$, $ws \in \{1, 2, 3, 4, 5\}$, $k, n, l \in \{1, ..., m\}$, $i \in \{0, ..., q\}$, $z \in \{1, ..., q\}$ $f \in \{1, ..., card(R_C)\}$, $d \in \{1, ..., (q-i)\}$.

In the first step the lesson order is determined. The choice is based on time of learning the lesson p_z, test score for the lesson p_z, version of lesson $v_k^{(z)}$, the number of test failure for p_z and the most popular version of the lesson in a scenario s before learning the lesson p_z. The algorithm of modification of a learning scenario is based on the highest posterior probability, and therefore we choose a lesson and a version of the lesson for which the following condition is satisfied:

$$\arg\max_{k,f}\{p(v_k^{(i+1)} \in \alpha_f(p_z), time, number, score, lesson, version)\}$$

The choice of subsequent lessons depends only on the previous one . Let us assume that in the previous step the order α_{mi} has been chosen, where $mi \in \{1, ..., card(R_C)\}$. In subsequent steps versions of lessons should be chosen which satisfy the following condition:

$$\arg\max_{k}\{p(v_k^{(i+d+1)}, v_n^{(i+d)})\}$$

where: $d \in \{1, ..., (q-i)\}$, $k, n \in \{1, ..., m\}$, $i \in \{0, ..., q\}$.

The computational complexity of BAM algorithm is equal to $O(n^2)$.

2.4   Computer Adaptive Testing

The evaluation process is associated with the student's stress that his knowledge has not been assessed properly. For avoiding the mentioned problem the method for computer adaptive testing P-CAT was worked out. The presentation of each test's item in computer adaptive tests is adapted to the student's proficiency level [8]. It is assumed that the item pool is calibrated and contains multi-choice questions. Each question is described by the level of difficulty of question b_j, $j \in (1,...,J)$ where J is the cardinality of the item pool. The P-CAT is the iterative algorithm consists of the following procedures: the first item selection method, the next item selection method and the knowledge level estimation. The mentioned steps are repeated until stop criterion is met.

The first step in P-CAT is the selection of the test's item. We do not have any information about student's proficiency level therefore for choosing a first item we analyze a student's learner profile. Selection of the first item is based on the index FI which is equal to 0 at the beginning. If the student is older, global, perfectionist, or characterized by a high level of concentration, motivation, self-esteem, ambition and a low level of anxiety, then index FI is increased by 1 for the value of each of the above-mentioned attributes that appear in the learner's model. The index FI decreases by 1 for each attribute value that appears in the learner model as follows: younger, sequential understanding, impetuous, sensitive perception, low level of concentration, low level of self-esteem, low level of ambition, and high level of anxiety. The difficulty of the first item depends on the estimated index FI. Belonging the index FI to the following intervals: [-8,-6], [-4,-2], 0, [2,4], [6,8] implicates that a student starts a test with a random very easy, easy, medium, hard and very hard items, respectively.

After presentation of test's item student's level of proficiency is estimated by using the following formula:

$$\hat{\theta}_{s+1} = \hat{\theta}_s + \frac{\sum_{i=1}^{n}(u_i - P_i(\hat{\theta}_s))}{\sum_{i=1}^{n}P_i(\hat{\theta}_s)(1 - P_i(\hat{\theta}_s))}$$

where: u_i- the answer to the item, $u_j=1$ for a correct response and $u_j=0$ for an incorrect response, $j \in (1,...,J)$, J- the cardinality of item pool, s- number of the iteration.

For selecting the subsequent question the item response theory (Rasch model) is applied. Firstly, the probability that a person with a given proficiency level θ will answer each question correctly is estimated:

$$P_i(X_i = correct \mid \theta) = \frac{e^{(\theta-b_i)}}{1+e^{(\theta-b_i)}}$$

where: θ is the student's knowledge level; X_i is the answer to question j, $i \in (1,...,J)$.

Next, the student is offered a question that maximizes the item information function $I_i(\theta)$, for each $i \in (1,...,J)$ and for the actual proficiency level θ:

$$I_i(\theta) = P_i(\theta)[1 - P_i(\theta)] = \frac{1}{1+e^{-(\theta-b_i)}}$$

The test is finished after an assumed number of questions (20) or if the standard deviation of the distribution of student's knowledge is smaller than a fixed threshold (0.3), where:

$$SE(\theta) = \frac{1}{\sqrt{\sum_{i=1}^{M} I_i(\theta)}}$$

2.5   A Method for Determining the Proper Attributes for User Classification

The problem for determining the set of demographic attributes which is the basis of the user classification process, is very important. One would like to have it minimal (containing of as little attributes as possible) and the most proper. The criterion "most proper" means that the partition of user set generated by these attributes should be very close to the partition arising as the result of clustering the set of scenarios with which the users passed the course offered by the system. It turned out that the same system acting in different environments (different sets of users) can have different sets of these attributes.

In our framework we consider the following problem: *For a given classification Q of set U which is generated by set A of attributes, one should determine such minimal set B of attributes from A that the distance between the classification P generated by attributes from B and partition Q is minimal.* In this problem Q represents the partition generated be the clustering process for the set of scenarios with which the users passed the course offered by the system, and classification P represents the partition represented by the set of demographic attributes.

This approach has been originally presented in work [16] and developed in [17]. In this paper we give only a brief description of its solution and applications for e-learning systems and user interface management in recommendation processes.

3   The Results of Experiments and Analysis

The quality evaluation of the designed intelligent tutoring system is based on the assessment of students' learning results which they have achieved during their learning process by using the proposed system. In our works we implemented the prototype of an e-learning system. The intelligent tutoring system was prepared based on the content management system Joomla and its plug-in VirtueMart (with custom modifications). The e-lessons were created using HTML and CSS and tests were created using Hot Potatoes. The results of the experiment were stored in MySQL database. Additionally, 11 learning scenarios were worked out. Learning scenarios differ from each other in the lessons order and presentations methods. The educational material focuses on intersections, road signs related to intersections and right-of-way laws. The idea of experiments was based on dividing users into two groups: a control group and an experimental group. Both groups should have the similar distribution of women, man, young people, old people, middle-age people, educated people, uneducated people, with and without driving license. If a student is assigned to the first group he is offered the personalized learning scenario suited to his learning styles. If a student is classified to a control group he is offered a universal learning scenario the same as the other students in that group, which is a combination of dedicated learning scenarios and contains some pictures, text, films, tasks etc.

297 users took part in our experiment. This group contained 123 persons who did not have driving licenses for whom we assumed that they have learned the presented learning material for the first time. It turned out that in both groups (with and without having driving license) the most frequently the interactive-graphical-sequential version of the learning scenario was assigned. It means that people who took part in the experiment were visualizers, active processing of information and learning from details to whole idea.

The detailed analysis of results of experiments allow to infer general conclusion that the personalization of the learning scenario increases the effectiveness of the learning process but only in case when the learning material is presented for the first time. The mean test score of experimental group is greater by more than 7.182% and less than 7.976% than the mean score of the control group. The personalization have the statistically significant influence on improvement of the learning result in case that users are underage, man and uneducated people [12,13]. All analysis were made for significance level equal to 0.05.

Similar conclusions were obtained by mathematical analysis. We conduct an analytical proof that the probability of graduating a course (passing the tests for all lessons from the learning scenario) is greater if the opening learning scenario is determined using method worked out by us (algorithm ADOLS II) than the opening learning scenario is chosen in a random way from the set of all possible scenarios. The conception of proof depends on pointing out the probability of graduating the whole course based on collected and stored data in system and by applying Bayes theorem [11].

4   Conclusion and Further Work

This paper contains framework for building intelligent tutoring systems. In this work we describe the following parts of the proposed system:

-   learner profile - the content and structure were presented;
-   knowledge representations (relational and graphical);
-   methods for determination of an opening learning scenario (suitable for both knowledge structures);
-   methods for modification of a learning scenario during the learning process (based on Bayesian Network and identification of failure's reasons);
-   method for computer adaptive testing based on Item Response Theory and analysis of content of learner profile;
-   methods for rough classification.

The worked out algorithms are simple and have a low computational complexity. We examined our assumptions and approaches in a specially implemented environment and by analytical analysis. The researches show us that personalization considerably improve the learning results in case that learning materials were presented for the first time.

In the further work the ontology will be considered as a tool for knowledge representation. Methods for knowledge integration which have been worked out could be useful for determination an opening learning scenario. Additionally, we want to elaborate on the technique for collaborative learning in the intelligent tutoring system. Our last target is to apply the presented model in a real intelligent tutoring system.

References

1. Brusilovsky, P., Schwarz, E., Weber, G.: ELM-ART: An intelligent tutoring system on World Wide Web. In: Lesgold, A.M., Frasson, C., Gauthier, G. (eds.) ITS 1996. LNCS, vol. 1086, pp. 261–269. Springer, Heidelberg (1996)
2. Gamboa, H., Fred, A.: Designing intelligent tutoring systems: A Bayesian approach. In: Proceedings of ICEIS 2001, Setubal, Portugal, July 7–10 (2001)
3. Jeremic, Z., Jovanovic, J., Gasevic, D.: Evaluating an intelligent tutoring system for design patterns: The DEPTHS experience. Journal of Educational Technology & Society 12(2), 111–130 (2009)
4. Kelly, D., Tangney, B.: Incorporating learning characteristics into an intelligent tutor. In: Cerri, S.A., Gouardéres, G., Paraguaçu, F. (eds.) ITS 2002. LNCS, vol. 2363, pp. 729–739. Springer, Heidelberg (2002)
5. Kelly, D., Tangney, B.: Predicting learning characteristics in a multiple intelligence based tutoring system. In: Lester, J.C., Vicari, R.M., Paraguaçu, F. (eds.) ITS 2004. LNCS, vol. 3220, pp. 678–688. Springer, Heidelberg (2004)
6. Kozierkiewicz, A.: Determination of opening learning scenarios in intelligent tutoring systems. In: Siemiński, A., Zgrzywa, A., Choroś, K. (eds.) New trend in Multimedia and Network Information Systems. IOS Press, Amsterdam (2008)
7. Kozierkiewicz-Hetmańska, A.: A conception for modification of learning scenario in an intelligent E-learning system. In: Nguyen, N.T., Kowalczyk, R., Chen, S.-M. (eds.) ICCCI 2009. LNCS (LNAI), vol. 5796, pp. 87–96. Springer, Heidelberg (2009)
8. Kozierkiewicz-Hetmańska, A., Nguyen, N.T.: A computer adaptive testing method for intelligent tutoring systems. In: Setchi, R., Jordanov, I., Howlett, R.J., Jain, L.C. (eds.) KES 2010, Part I. LNCS, vol. 6276, pp. 281–289. Springer, Heidelberg (2010)
9. Kozierkiewicz-Hetmańska, A., Nguyen, N.T.: A method for scenario modification in intelligent E-learning systems using graph-based structure of knowledge. In: Nguyen, N.T., Katarzyniak, R., Chen, S.-M. (eds.) Advances in Intelligent Information and Database Systems. SCI, vol. 283, pp. 169–179. Springer, Heidelberg (2010)
10. Kozierkiewicz-Hetmanska, A., Nguyen, N.T.: A method for learning scenario determination and modification in intelligent tutoring system. International Journal Applied of Mathematics and Computer Science 21(1) (2011)
11. Kozierkiewicz-Hetmanska, A.: A Method for Scenario Recommendation in Intelligent E- learning System. Cybernetics and Systems 42, 82–99 (2011)
12. Kozierkiewicz-Hetmanska, A.: Evaluation of an intelligent tutoring system incorporating learning profile to determine learning scenario. In: Jezic, G., Kusek, M., Nguyen, N.-T., Howlett, R.J., Jain, L.C. (eds.) KES-AMSTA 2012. LNCS, vol. 7327, pp. 44–53. Springer, Heidelberg (2012)

13. Kozierkiewicz-Hetmańska, A.: Evaluating the effectiveness of intelligent tutoring system offering personalized learning scenario. In: Pan, J.-S., Chen, S.-M., Nguyen, N.T. (eds.) ACIIDS 2012, Part I. LNCS, vol. 7196, pp. 310–319. Springer, Heidelberg (2012)
14. Kukla, E.E., Nguyen, N.T., Danilowicz, C., Sobecki, J., Lenar, M.: A model conception for optimal scenario determination in an intelligent learning system. ITSE—International Journal of Interactive Technology and Smart Education 1(3), 171–184 (2004)
15. Kwasnicka, H., Szul, D., Markowska-Kaczmar, U., Myszkowski, P.: Learning Assistant – Personalizing learning paths in elearning environments. In: 7th Computer Information Systems and Industrial Management Applications. IEEE (2008)
16. Nguyen, N.T.: A Method for Ontology Conflict Resolution and Integration on Relation Level. Cybernetics and Systems 38(8), 781–797 (2007)
17. Nguyen, N.T.: Advanced Methods for Inconsistent Knowledge Management. Springer, London (2008)

A Review of AI-Supported Tutoring Approaches for Learning Programming

Nguyen-Thinh Le, Sven Strickroth, Sebastian Gross, and Niels Pinkwart

Clausthal University of Technology,
Department of Informatics,
Julius-Albert-Str. 4, 38678 Clausthal-Zellerfeld, Germany
{nguyen-thinh.le,sven.strickroth,sebastian.gross,
niels.pinkwart}@tu-clausthal.de

Abstract. In this paper, we review tutoring approaches of computer-supported systems for learning programming. From the survey we have learned three lessons. First, various AI-supported tutoring approaches have been developed and most existing systems use a feedback-based tutoring approach for supporting students. Second, the AI techniques deployed to support feedback-based tutoring approaches are able to identify the student's intention, i.e. the solution strategy implemented in the student solution. Third, most reviewed tutoring approaches only support individual learning. In order to fill this research gap, we propose an approach to pair learning which supports two students who solve a programming problem face-to-face.

Keywords: Computer-supported collaborative learning, Intelligent Tutoring Systems, programming, pair learning.

1 Introduction

For programming as a Computer Science subject, researchers have found that learning cannot only be achieved through memorizing facts or programming concepts, rather students need to apply the learned concepts to solve programming problems [1]. That is, researchers suggest focusing on constructivistic learning/teaching approaches for Computer Science courses, because the goal is to develop and to master programming skills. For this purpose, a variety of AI-supported tutoring approaches have been developed in order to help beginner students learn programming.

How can AI techniques be deployed to engage students in the process of learning, and can students play an active role in these settings? That is the motivation of this paper. For this purpose, we have reviewed several AI-supported tutoring approaches for the domain of programming: example-based, simulation-based, collaboration-based, dialogue-based, program analysis-based, and feedback-based approaches. All the tutoring approaches under review in this paper have been evaluated in empirical studies with students.

N.T. Nguyen, T. Van Do, and H.A. Le Thi (Eds.): *ICCSAMA 2013*, SCI 479, pp. 267–279.
DOI: 10.1007/978-3-319-00293-4_20   © Springer International Publishing Switzerland 2013

2   AI-Supported Tutoring Approaches

2.1   Example-Based Approaches

One popular approach to teaching programming is explaining example problems and solutions and then asking students to solve similar new problems. This way, students transfer learned ways of solving problems to new problems of the same type. Yudelson and Brusilovsky [2] applied this tutoring approach for building learning systems for programming. NavEx [2] is a web-based tool for exploring programming examples. Each example is annotated with textual explanations for each important line. Thus, students can open the explanation line-by-line, or go straight to the lines which are difficult to understand, or display explanations of several lines at the same time. This system has several benefits compared to normal textbooks: 1) the code of each example can be shown as a block, instead of being dissected by comments as usually found in textbooks; 2) explanations can be accessed in different ways in accordance with the student's need; and 3) students can learn with examples in an exploratory manner, instead of reading examples passively. NavEx provides adaptive navigation support to access a large set of programming examples, that is, for each individual student only examples which correspond to the knowledge level of that student are recommended. This is the feature which requires the contribution of AI techniques. The adaptive navigation support is based on a *concept-based mechanism*. Applying this technique, first, a list of programming concepts is identified in each example and is separated into *pre-requisite concepts* and *outcome concepts*. The process of separation is supported by the structure of a specific course, which is defined by having the teacher assign examples to the ordered sequence of lectures. This process assumes that for the first lecture, all associated examples have only outcome concepts. The algorithm advances through lectures sequentially until all example concepts are effectively covered. The pre-requisites and the outcome concepts of each example and the current state of the individual user model determine which example should be recommended next to the student. The student's degree of engagement within an example is measured by counting the number of clicks on annotated lines of code.

Another type of example-based learning is the completion learning strategy. Learning tools present a programming problem and a solution template to be filled in. Gegg-Harrison [3] developed the tutoring system ADAPT for Prolog applying this tutoring approach. ADAPT exploits the existence of Prolog schemata which represent solutions for a class of problems. The tutor first introduces students with several solution examples and shows significant components of the schema underlying these solutions. After that, the tutor asks students to solve a similar problem given a solution template with slots to be filled. If the student is stuck and needs help, then the tutor breaks the template down into its recursive functional components and explains the role of each component. After explaining each of these components, the student has a chance to complete them or she can ask for further help. If the student still needs help, the tutor provides her with a more specific template, i.e., a template which is already filled with a partial solution. If the student is still unable to solve the problem, then the complete solution is given and explained. The AI technique which supports this tutoring approach is using a library of Prolog schemata. The purpose is giving feedback to student solutions. A Prolog schema represents the common structure for a class of programs. Prolog schemata are used in ADAPT in order to diagnose errors in students' solutions. The process of diagnosis consists of two phases. First, ADAPT tries to recognize the algorithm underlying the student solution. It uses the normal form program for each schema in the schema library to generate a class of representative implementations and then transforms the most appropriate implementation into a structure that best matches the student solution. In the second phase, ADAPT tries to explain mismatches between the transformed normal form program produced in the first phase and the student solution. For this purpose, ADAPT uses a hierarchical bug library to classify bugs found in incorrect programs. If there are no mismatches, then the student solution is correct.

Chang and colleagues [4] followed a similar completion strategy. In addition to the fill-in-the-blank approach proposed in ADAPT, they also proposed modification and extension tasks. Modification tasks are to rewrite a program to improve it, to simplify it, or to use another program to perform the same function. Extension tasks include adding new criteria or new commands to original questions (for each question, several exercises are proposed), and students are asked to rewrite a program by adding or removing some aspects. In this system, the AI technique has been deployed to give feedback to a student solution. Here, program templates are used. For each problem (from which several completion, extension, or modification tasks are generated), a model program is defined by several templates which represent required semantic components of that model program. Each template is defined as a representation of a basic programming concept. The basic programming concepts are those which the system wants the students to learn working with the problem. In order to understand the student solution, pre-specified templates for each problem are expanded into source code and thus, the model program is generated. The generated model program serves to create exercises (completion, modification, and extension tasks) and to evaluate the student solution. Mismatches between the model program and the student solution are considered as errors and feedback is returned to students.

In this subsection we have introduced three AI techniques. While the concept-based mechanism has been used for providing adaptive navigation support, schemata-based and template-based techniques have been developed to provide feedback to student's solutions. The schemata-based technique is really useful if a schema can represent a broad class of programs which can be considered solutions for a class of programming problems. This is fulfilled for Prolog, a declarative programming language. However, for imperative programming languages, schemata can hardly be determined. Instead, templates which represent a basic programming concept can be used as proposed by Chang and colleagues [4].

2.2   Simulation-Based Approaches

The simulation-based teaching approach deals with the problem that a program has a dynamic nature (i.e., the state of variables is changing at run time) and students can not realize how a piece of program code works. Researchers have suggested visualizing the process underlying an algorithm. One way to lower the abstraction of programs is using micro worlds which are represented by concrete entities, e.g., turtles using NetLogo[1]. The student can program the movements of entities and control their behavior. Several environments have been developed following this approach, e.g. Alice[2] and Scratch[3]. The goal is to make program flows more visible. Through such simulation-based environments, students can develop their basic programming skills in order to later use them within more complex programming environments.

Recently, other simulation-based approaches have been developed to make visualizations more helpful: engaging visualization, explanatory visualization and adaptive visualization. Engaging visualization emphasizes the active role of student in learning, instead of just observing a teacher prepared animations. Explanatory visualization augments visual representations with natural language explanations that can help students understand simulation. Adaptive visualization adapts the details of visual representations to the difficulties of underlying concepts and to the knowledge level of each student. Several simulation tools have been developed applying the first two approaches in order to engage students and provide them with explanatory feedback: SICAS [5], OOP-AMIN [6], and PESEN [7]. SICAS and OOP-AMIN were developed to support learning of basic procedural programming concepts (e.g., selection and repetition) and object-oriented concepts (class, object or inheritance), respectively. PESEN is another simulation system which is intended to support weak students. Here, the goal is learning basic concepts of sequential, conditional and repetition execution which are frequently used in computer programs. These systems focus on the design and implementation of algorithms. Algorithms are designed by using an iconic environment where the student builds a flowchart that represents a solution. After a solution to a problem has been created, the algorithm is simulated and the student observes if the algorithm works as expected. The simulator can be used to detect and correct errors. While the tasks for students using SICAS and OOP-AMIN include designing and implementing an algorithm, PESEN just asks the student to program a very simple algorithm which consists of movements of elementary geometrical shapes using simple commands. The AI technique supporting these systems for detecting and correcting errors is relatively simple. Each programming problem is composed of a statement that describes it, one or more algorithms that solve it, and a solution with some test cases. The existence of several algorithms for the same problem serves to provide different ways of reasoning about the solution strategy chosen by the student. The test cases allow a student to verify if the designed algorithm really solves

---

[1] http://ccl.northwestern.edu/netlogo
[2] http://www.alice.org
[3] http://scratch.mit.edu

the problem. Both the existence of alternative algorithms and test cases allow the simulation system to diagnose errors in the student solution.

Brusilovsky and Spring [8] developed WADEin to allow students to explore the process of expression evaluation. The goal of WADEin is to provide adaptive and engaging visualization. The student can work with the system in two different modes. In the exploration mode, the user can type in expressions or select them from a menu of expressions. Then, the system starts visualizing the execution of each operation. The student can observe the process of evaluating an expression in C progamming language step-by-step. The evaluation mode provides another way to engage students and to evaluate their knowledge. The student works with the new expression by indicating the order of execution of the operators in the expression. After that, the system corrects errors, shows the correct order of execution, and starts evaluating the expression. Here, the contribution of AI is very little. What the system has to do is storing the operators of an expression and possible reformulations a priori. Using this information, the order of operators in an expression submitted by the student is evaluated. In addition, the value of expressions can be evaluated "on-the-fly".

The AI techniques which have have been deployed to support the simulation-based tutoring approaches introduced in this subsection have been used for detecting errors in student solutions in order to provide feedback.

2.3   Dialogue-Based Approaches

One way of coaching students is to communicate with them in form of mixed-initiative dialogues. That is, both the tutor and the student are able to initiate a question. Lane [9] developed PROPL, a tutor which helps students build a natural-language style pseudocode solution to a given problem. The system initiates four types of questions: 1) identifying a programming goal, 2) describing a schema for attaining this goal, 3) suggesting pseudocode steps that achieve the goal, and 4) placing the steps within the pseudocode. Through conversations, the system tries to remediate student's errors and misconceptions. If the student's answer is not ideal (i.e., it can not be understood or interpreted as correct by the system), sub-dialogues are initiated with the goal of soliciting a better answer. The sub-dialogues will, for example, try to refine vague answers, ask students to complete incomplete answers, or to redirect to concepts of greater relevance. The AI contribution of PROPL is the ability of understanding student's answers and communicating with students using natural language. For this purpose, PROPL has a knowledge source which is a library of Knowledge Construction Dialogues (KDCs) representing directed lines of tutorial reasoning. They consist of a sequence of tutorial goals, each realized as a question, and sets of expected answers to those questions. The KCD author is responsible for creating both the content of questions and the forms of utterances in the expected answer lists. Each answer is either associated with another KCD that performs remediation or is classified as a correct response. KCDs therefore have a hierarchical structure and follow a recursive, finite-state based approach to dialogue management.

2.4   Program Analysis-Based Approaches

In contrast to other tutoring approaches which request students to synthesize a program or an algorithm, the program analysis-based approach intends to support students learn by analyzing a program.

Kumar [10] developed intelligent tutoring systems to help students learn the C++ programming language by analyzing and debugging C++ code segments. These systems focus on tutoring programming constructs rather than on the entire programming. Semantic and run-time errors in C++ programs are the objectives of tutoring. The problem-solving tasks addressed by these systems include: evaluating expressions step-by-step, predicting the result of a program line by line, and identifying buggy code in a program. For instance, for a debugging task, a tutor asks the student to identify a bug and to explain why the code is erroneous. If the student does not solve the problem correctly, the tutor provides an explanation of the step-by-step execution of the program. The AI technique which has been deployed in these systems is the model-based reasoning approach. In model-based reasoning, a model of a C++ domain consists of the structure (the objects in the language) and behavior of the domain (the mechanisms of interaction). This model is used to simulate the correct behavior of an artifact in the domain, e.g., the expected behavior of C++ pointers. The correct behavior generated from the model is compared with the behavior predicted by the student for that artifact. The discrepancies between these two behaviors are used to hypothesize structural discrepancies in the mental model of the student which are used to generate feedback for the student.

2.5   Feedback-Based Approaches

Feedback on student solutions can be used in a specific tutoring approach, e.g., ADAPT [3] is able to provide feedback during example-based tutoring. There exists a variety of learning systems for programming which make use of feedback on student solutions as the only means for tutoring. Systems of this class assume that feedback would be useful to help students solve programming problems, and thus programming skills can be improved. To be able to analyze student solutions and give feedback, learning systems for programming require a domain model and appropriate error diagnosis techniques. Here, we discuss the AI techniques applied in learning systems which have shown to be effective in helping students: model-tracing, machine learning, libraries of plans and bugs, and a weighted constraint-based model.

The core of a *model-tracing* tutoring system is the *cognitive model* which consists of "ideal rules" which represent correct problem-solving steps of an expert and "buggy rules" which represent typical erroneous behaviors of students. When the student inputs a solution, the system monitors her action and generates a set of correct and buggy solution paths. Whenever a student's action can be recognized as belonging to a correct path, the student is allowed to go on. If the student's action deviates from the correct solution paths, the system generates instructions to guide the student towards a correct solution. Applying this approach, Anderson and Reiser [11] developed an intelligent tutoring system for LISP. This tutor presents to the student a problem description containing highlighted identifiers for functions and parameters which have to be used in the implementation. To solve a programming problem, the student is provided with a structured editor which guides the student through a sequence of templates to be filled in. Applying the same approach, Sykes [12] developed JITS, an intelligent tutoring system for Java. JITS is intended to "intelligently" examine the student's submitted code and determines appropriate feedback with respect to the grammar of Java. For this purpose, production rules have been used to model the grammar of Java and to correct compiler errors. Since this system does not check semantic requirements which are specific for each programming problem, it does not require representations of problem-specific information.

Since the task of modeling "buggy rules" is laborious and time-consuming, Suarez and Sison applied a multi-strategy machine learning approach to automatically construct a library of Java errors which novice programmers often make [13]. Using this bug library, the authors developed a system called JavaBugs which is able to examine a small Java program. The process of identifying errors in the student solution takes place in two phases. First, the system uses Beam search to compare the student solution to a set of reference programs and to identify the most similar correct program to the student's solution. The solution strategy intended by the student to solve a programming problem is considered the student's intention. After the student's intention has been identified, differences between the reference program and the student solution are extracted. In the second phase, misconception definitions are learned based on the discrepancies identified in the first phase. These misconceptions are used to update the error library and to return appropriate feedback to the student.

Another class of systems uses a library of programming plans or bugs for modeling domain knowledge and diagnosing errors. Representatives for this class include, for example, PROUST [14], ELM-ART [15], and APROPOS2 [16]. The process of error diagnosis of these systems builds on the same principles: 1) modeling the domain knowledge using programming plans or algorithms and buggy rules, 2) identifying the student's intention, and 3) detecting errors using buggy rules. In PROUST, a tutoring system for Pascal, each programming problem is represented internally by a set of programming goals and data objects. A programming goal represents the requirements which must be satisfied, while data objects are manipulated by the program. A programming goal can be realized by alternative programming plans. A programming plan represents a way to implement a corresponding programming goal. In contrast to PROUST, APROPOS2 (a tutoring system for Prolog) uses the concept of algorithms to represent different ways for solving a Prolog problem instead of programming plans, because Prolog does not have keywords (the only keyword is ":-" which separates the head and the body of a Prolog clause) to define programming plans and to anchor program analysis like in the case of Pascal. In addition to modeling domain knowledge using programming plans or algorithms, common bugs are collected from empirical studies and represented as *buggy rules* or *buggy clauses*. While

PROUST contains only buggy rules, ELM-ART adds two more types: good and sub-optimal rules which are used to comment on good programs and less efficient programs (with respect to computing resources or time), respectively. After the domain model has been specified by means of programming plans (or algorithms) and buggy rules, systems of this class perform error diagnosis in two steps: identifying the student's intention and detecting errors. The process of identifying student's intention starts by using pre-specified programming plans (or algorithms) to generate a variety of different ways of implementations for a programming goal. Then, it continues to derive hypotheses about the programming plan (or algorithm) the student may have used to satisfy each goal. If the hypothesized programming plan (or algorithm) matches the student program, the goal is implemented correctly. Otherwise, the system looks up the database of buggy rules to explain the discrepancies between the student solution and the hypothesized programming plan (or algorithm).

Instead of using programmings/algorithms and buggy rules for modeling a domain, Le and Menzel [17] used a semantic table and a set of weighted constraints to develop a weighted constraint-based model for a tutoring system (INCOM) for Prolog. The semantic table is used to model alternative solution strategies and to represent generalized components required for each solution strategy. Constraints are used to check the semantic correctness of a student solution with respect to the requirements specified in the semantic table and to examine general well-formedness conditions for a solution. Constraints' weight values are primarily used to control the process of error diagnosis. In their approach, the process of diagnosing errors in a student solution consists of two steps which take place on two levels: hypotheses generation and hypotheses evaluation. First, on the *strategy level*, the system generates hypotheses about the student's intention by iteratively matching the student solution against multiple solution strategies specified in a semantic table. After each solution strategy has been matched, on the *implementation variant level* the process generates hypotheses about the student's implementation variant by matching components of the student solution against corresponding components of the selected solution strategy. The generated hypotheses are evaluated with respect to their plausibility by aggregating the weight value of violated constraints. As a consequence, the most plausible solution strategy intended by the student is determined.

All the AI techniques presented in this subsection share one common point: they are able to identify the student's intention first, then detect errors in student solutions. The model tracing technique diagnoses the student's intention by relating each step of the student's problem-solving with generated correct and erroneous solution paths. The approach in JavaBugs uses several reference programs to hypothesize the student's implemented strategy. Other systems use a library of programming plans/algorithms to recognize the student's intention. The weighted constraint-based model uses a semantic table, which is specified with different solution strategies, to hypothesize a solution strategy underlying a student solution.

2.6 Collaboration-Based Approaches

In comparison to other tutoring approaches (presented in previous subsections) which support individual learning, a collaboration-based tutoring approach assumes that learning can be supported through collaboration between several students.

HABIPRO [18] is a collaborative environment that supports students to develop good programming habits. The system provides four types of exercises: 1) finding a mistake in a program, 2) putting a program in the correct order, 3) predicting the result, and 4) completing a program. When a student group proposes an incorrect solution, the system shows four types of help: 1) giving ideas to the student how they can solve the problem, 2) showing and explaining the solution, 3) showing a similar example of the problem and its solution, and 4) displaying only the solution. The system has a simulated agent that is able to intervene in students' learning, to avoid off-topic situations, and to avoid passive students. This system exploits several AI techniques. First, it is able to guide students in conversations using natural language. Second, it uses knowledge representation techniques to model a group of students and an interaction model. The interaction model defines a set of patterns which represent possible characteristics of group interaction (e.g., the group prefers to look at the solution without seeing an explanation). During the collaboration, the simulated agent uses the group model to compare the current state of interaction to these patterns, and proposes actions such as withholding solutions until the students have tried to solve the problem. Third, the system exploits information from a student model which includes features that enable the system to reason about the student's collaborative behavior: e.g., frequency of interaction, type of interaction, level of knowledge, personal beliefs, and mistakes (individual misconceptions). The simulated agent makes use of information from both the student models and the group model to decide when and how to intervene in the collaboration.

3 Discussion

We summarize the tutoring approaches and AI techniques presented in the previous section in Table 1. From this table we can learn three things. First, we have found that most learning systems have been developed applying the feedback-based tutoring approach in comparison to other categories of tutoring. This does not mean that the feedback-based tutoring approach is more effective than others. Rather, the feedback-based tutoring approach can be regarded as the first attempt to building intelligent learning systems. Indeed, we can deploy any AI technique for building feedback-based tutoring systems to support other categories of tutoring as shown in case of ADAPT. That is, feedback on student solutions can be used as a means within different tutoring scenarios. Second, from the last column of the table we can identify that AI techniques, which have been deployed in different tutoring approaches, serve three purposes: to support adaptive navigation, to analyze student solutions, and to enable a conversation with students. In addition, we can also recognize that AI techniques developed for feedback-based tutoring approaches are able to identify the student's intention underlying the student solution. This is necessary to generate feedback on student solutions, because there are often different ways for solving a programming problem. If feedback returned to the student does not correspond to her intention, feedback might be misleading. Hence, the first step of diagnosing errors in a student solution is identifying the student's intention. The third lesson we have learned from the survey above is that most systems support individual learning. HABIPRO is the only system (to our best knowledge) which has been designed to support collaborative programming activities.

Table 1. AI-supported Tutoring Approaches

| Type | System | AI Support | |
		Techniques	Function
Example	NavEx	concept-based mechanism	adaptive navigation
	ADAPT	Prolog schemata	solution analysis
	Chang's	template-based	solution analysis
Simulation	SICAS	test cases	solution analysis
	OOP-AMIN	test cases	solution analysis
	PESEN	test cases	solution analysis
	WADEin	expression evaluation	solution analysis
Dialogue	PROPL	natural language processing	conversation
Program Analysis	Kumar's	model-based reasoning	solution analysis
Feedback	LISP-Tutor	model-tracing	solution analysis
	JITS	model-tracing	solution analysis
	JavaBugs	machine learning	solution analysis
	PROUST	programming plans & buggy rules	solution analysis
	APROPOS2	algorithms & buggy rules	solution analysis
	ELM-ART	algorithms & buggy, good, sub-optimal rules	solution analysis
	INCOM	semantic table & weighted constraints	solution analysis
Collaboration	HABIPRO	natural language processing	conversation

4 Intelligent Peer Learning Support

One of the lessons we have learned from the survey above is that most tutoring approaches have been developed to support individual learning. Yet, in the context of solving programming problems, collaboration and peer learning have been shown to be beneficial [19], [20], [21].

From a course "Foundations of programming" in which primarily the programming language Java is used, we have collected thousands of student solutions over three winter semesters (2009-2012). During the semester, every two weeks students were given an exercise sheet. For each task, students could achieve maximal 10 points. Students were asked to submit their solutions in pairs or alone using the GATE submission system [22]. By comparing the score of pair submissions with the score of single submissions we have learned that the paired students achieved higher scores than single submissions: there were 2485 submissions in total (1522 single, 963 pair submissions); the average score of single submissions and of pair submissions were 6.59 and 6.91 points, respectively (p=0.008). This suggests that pair submissions correlate with higher achievement. This can be explained by the reason that solving programming problems in pairs, students could contribute complementarily, and thus a solution for a given problem could be found easier by a pair of students than by individual learners.

Identifying the need of supporting solve programming problems in pairs, in this paper, we propose an approach for intelligent pair learning support which can take place when two students work together. Researchers agreed that the programming process consists of the following phases: 1) problem understanding, 2) planning, 3) design, and 4) coding (implementation in a specific programming language). After reviewing literature of learning systems for programming, to our best knowledge, only SOLVEIT [23] is able to support all four phases. However, SOLVEIT does not provide feedback to the student. We propose to support pair learning during these four phases. In the first phase, the system poses questions to ask a pair of students to analyze a problem statement, e.g., "Can you identify given information in the problem statement?" A list of keywords which represent given information in a problem statement can be associated with each problem. Using this list, the system can give hints to the students. In the planning phase, the system asks a pair of students to plan a solution strategy, e.g., "Which strategy can be used to solve this problem?" or "What kind of loop can be used to solve this problem?" Similarly to the first phase, to each problem, a list of alternative solution strategies can be stored which are used to propose to the student. PROPL [9] is able to support this phase applying the dialogue-based tutoring approach. In the design phase, the system asks the pair of students to transform the chosen solution strategy into an algorithm which can be represented, e.g., in activity diagrams. Required components of an activity diagram can be modeled using one of the AI techniques presented in the previous section in order to check the solution of the pair of students. In the fourth phase, the system can ask the students to transform the designed algorithm into code of a specific programming language, e.g., "How can we code the input action which reads the value of a variable?" Again, we can apply one of the AI techniques introduced above to examine the semantic correctness of the student solution.

We need to identify and to combine strengths and weaknesses of individual learners in order to strengthen the collaboration between students and to help learners overcome weaknesses. For this purpose, models for individual learners need to be developed. User models can be used to parametrize group learning [24] as well as to build learning groups intelligently [25].

In order to build learner models, a system which is intended to support pair learning needs to distinguish contributions between different learners. Contrary to recognizing individual learners using unique identifiers (e.g., user name and password), pair learning systems should be able to distinguish learners (who are working in collaboration) by identifying contributions of each peer in order to associate individual progress of a learner. Modeling characteristics of learners and groups could then be used to adapt a pair learning supporting system to individual and group needs. For example, recognizing learners' inactivity could lead to a stimulation by the system. How can we realize a pair learning supporting system technically? We might apply visual recognition techniques in order to distinguish different learners or exploit speech recognition techniques in order to recognize activity and to assign contributions to learners.

5   Conclusion and Future Work

We have reviewed various tutoring approaches which are supported by AI techniques for the domain of programming. We have learned that most existing learning systems aim at analyzing student's solutions and providing feedback which serve as an important means for improving programming skills. We identified that AI techniques, which have been deployed in different tutoring approaches, serve three purposes: to support adaptive navigation, to analyze student solutions, and to enable a conversation with students. Especially, AI techniques which have been applied for feedback-based tutoring approaches, are able to identify the student's intention before detecting errors. The third lesson we have learned is that most systems support individual learning. This is a research gap, because collaborative learning is beneficial for solving programming problems. Hence, we propose in this paper an approach for pair learning which is intended to coach students along typical phases of programming and is supported by a conversational approach. In the future, we plan to implement this approach using a conversational agent.

References

1. Radosevic, D., Lovrencic, A., Orehovacki, T.: New Approaches and Tools in Teaching Programming. In: Central European Conference on Information and Intelligent Systems (2009)
2. Yudelson, M., Brusilovsky, P.: NavEx: Providing Navigation Support for Adaptive Browsing of Annotated Code Examples. In: The 12th International Conference on AI in Education, pp. 710–717. IOS Press (2005)
3. Gegg-Harrison, T.S.: Exploiting Program Schemata in a Prolog Tutoring System. Phd Thesis, Duke University, Durham, North Carolina 27708-0129 (1993)
4. Chang, K.E., Chiao, B.C., Chen, S.W., Hsiao, R.S.: A Programming Learning System for Beginners - A Completion Strategy Approach. Journal IEEE Transactions on Education 43(2), 211–220 (1997)
5. Gomes, A., Mendes, A.J.: SICAS: Interactive system for algorithm development and simulation. In: Computers and Education - Towards an Interconnected Society, pp. 159–166 (2001)
6. Esteves, M., Mendes, A.: A simulation tool to help learning of object oriented programming basics. In: FIE 34th Annual on Frontiers in Edu., vol. 2, pp. F4C7-12 (2004)

7. Mendes, A., Jordanova, N., Marcelino, M.: PENSEN - A visual programming environment to support initial programming learning. In: International Conference on Computer Systems and Technologies - CompSysTech 2005 (2005)

8. Brusilovsky, P., Spring, M.: Adaptive, Engaging, and Explanatory Visualization in a C Programming Course. In: World Conference on Educational Multimedia, Hypermedia and Telecommunications (ED-MEDIA), pp. 21–26 (2004)

9. Lane, H.C., Vanlehn, K.: Teaching the tacit knowledge of programming to novices with natural language tutoring. J. Computer Science Education 15, 183–201 (2005)

10. Kumar, A.N.: Explanation of step-by-step execution as feedback for problems on program analysis, and its generation in model-based problem-solving tutors. Journal Technology, Instruction, Cognition and Learning 4 (2006)

11. Anderson, J.R., Reiser, B.: The Lisp Tutor. Journal Byte 10, 159–175 (1985)

12. Sykes, E.R.: Qualitative Evaluation of the Java Intelligent Tutoring System. Journal of Systemics, Cyber (2006)

13. Suarez, M., Sison, R.C.: Automatic construction of a bug library for object-oriented novice java programmer errors. In: Woolf, B.P., Aïmeur, E., Nkambou, R., Lajoie, S. (eds.) ITS 2008. LNCS, vol. 5091, pp. 184–193. Springer, Heidelberg (2008)

14. Johnson, W.L.: Understanding and debugging novice programs. Journal Artificial Intelligence 42, 51–97 (1990)

15. Brusilovsky, P., Schwarz, E.W., Weber, G.: ELM-ART: An Intelligent Tutoring System on World Wide Web. In: Lesgold, A.M., Frasson, C., Gauthier, G. (eds.) ITS 1996. LNCS, vol. 1086, pp. 261–269. Springer, Heidelberg (1996)

16. Looi, C.-K.: Automatic Debugging of Prolog Programs in a Prolog Intelligent Tutoring System. Journal Instructional Science 20, 215–263 (1991)

17. Le, N.T., Menzel, W.: Using Weighted Constraints to Diagnose Errors in Logic Programming - The case of an ill-defined domain. Journal of AI in Education 19, 381–400 (2009)

18. Vizcaíno, A.: A Simulated Student Can Improve Collaborative Learning. Journal AI in Education 15, 3–40 (2005)

19. Van Gorp, M.J., Grissom, S.: An Empirical Evaluation of Using Constructive Classroom Activities to Teach Introductory Programming. Journal of Computer Science Education 11(3), 247–260 (2001)

20. Nagappan, N., Williams, L., Ferzli, M., Wiebe, E., Yang, K., Miller, C., Balik, S.: Improving the CS1 experience with pair programming. In: The SIGCSE Technical Symposium on CS Education, pp. 359–362. ACM (2003)

21. Wills, C.E., Deremer, D., McCauley, R.A., Null, L.: Studying the use of peer learning in the introductory computer science curriculum. Journal of Computer Science Education 9(2), 71–88 (1999)

22. Strickroth, S., Olivier, H., Pinkwart, N.: Das GATE-System: Qualitätssteigerung durch Selbsttests für Studenten bei der Onlineabgabe von Übungsaufgaben? In: GI LNI, vol. P-188, pp. 115–126. Köllen Verlag (2010)

23. Deek, F.P., Ho, K.-W., Ramadhan, H.: A critical analysis and evaluation of web-based environments for program development. Journal of Internet and Higher Education 3, 223–269 (2000)

24. Hoppe, H.U.: The use of multiple student modeling to parameterize group learning. In: Int. Conference on AIED (1995)

25. Muehlenbrock, M.: Formation of Learning Groups by using Learner Profiles and Context Information. In: Int. Conference on AI in Education, pp. 507–514 (2005)

Supporting Career Counseling with User Modeling and Job Matching[*]

Cuong Duc Nguyen, Khoi Duy Vo, and Dung Tien Nguyen

International University – VNU-HCM
{ndcuong,vdkhoi,ntdung}@hcmiu.edu.vn

Abstract. After graduation, an engineer expects to have a desired position of predesigned jobs, e.g. a young IT engineer with a great programming experience usually prefers to a software developer or a tester than a teaching job. For each selected job position, a student has made a good preparation during the study period, such as, right selection of a major or a minor, choosing related elective courses and topics for graduation thesis, taking supportive professional certificates, etc. Therefore, a career counselor is very helpful for students to select a suitable job position to which they would prefer after graduation. This paper presents an ontology-based job matching to facilitate finding the most appropriate job position for a student based on the job's demand and the student's capabilities. Ontologies are employed to model industrial job positions and student's capabilities. Matching functions are proposed to match between the job modeling and the student modeling for job recommendation. The evaluation of job matching result shows a high correlation between the proposed system's recommendation and the recruiter's choice.

Keywords: Job matching, user model, job model, career counseling.

1 Introduction

At 18 years old, everyone has to make one of the most important decision in his/her own life: he/she has to choose which career to pursuit and which program to study to take up that selected career. Career counseling has been very effectively used in high schools to support pupil's choice a suitable career for their own life [1][2].

Even after getting into a university, students still need the support of career advising. With broadly educational objectives, after finishing a program, graduates can work in various roles with different duties in industry. For instances, a graduate with a Computer Science degree can be a programmer, a developer, a tester or an IT consultant. To fulfill that diversification, training programs always consists of a compulsory part and a flexible part. The compulsory component contains fixed courses that provide background knowledge and common skills. In the flexible component, students can select electives, project and graduating thesis topics that are suitable for their

---

[*] This paper is based in part upon work supported by the VNU Science Foundation under Grant Numbers B2012-28-12.

N.T. Nguyen, T. Van Do, and H.A. Le Thi (Eds.): *ICCSAMA 2013*, SCI 479, pp. 281–292.
DOI: 10.1007/978-3-319-00293-4_21   © Springer International Publishing Switzerland 2013

future career. Therefore, with the consulting with Career Advisor or Director of Study, they can choice an appropriate future job title suitable with their capabilities and then studies appropriate knowledge and skills related to their selected job.

To provide a good counseling to students, student advisors have to well manage the following three folds. Firstly, they have to have a good understanding about their students: attitudes, abilities, interests, ambitions, etc. [2]. Secondly, they know about typical jobs with essential and desired skills of those vocations. Lastly, they are able to compare the potential capabilities of a student and the demand of a job to find good matches.

There are several methods to help a student counselor know about his students. The most traditional and popular way having been used in universities is face-to-face interaction between advisors and students. With the rapid development of information management systems in universities, richer sources of information about students can be provided to advisors. In addition, user modeling techniques can formulate student models to support advisors to efficiently manage and reason on students' information.

Job Matching techniques can support student advisors to find good matches between student models and aimed vocations. Such techniques are often applied on job seeker profiles and job descriptions. However, they can be adjusted to match the current capabilities of students with the requirements of aimed job description for newly graduates. The good matches between a student model and an aimed job can show a high appropriateness of that job for the student. With such a support from the job matching techniques, student advisors are able to show potential vocations for students.

There are two main approaches in job matching. In the macro approach, economists focus on constructing economical models for the labor market of the whole nation or continent. These models can be used to predict the unemployment rate [22], specify the insurance parameters [23], etc. To construct these models, only general information, such as degree, wage, years of experience, of job candidates and profiles is used in matching. Such information is often collected in national census databases.

On another side, in the micro approach, matching models are built to support job seekers find better opportunities or recruiters find more suitable candidates. Specific information about candidate's attitudes, fields of study, previous positions or profile's demand about particular technologies, languages is used to improve the matching quality. Therefore, in this approach, matching requires more research efforts to combine various kinds of information about candidates and jobs to improve its quality. This kind of matching is more suitable student counseling, so it is used in this paper.

Several ontologies, such as HR-Ontology (Human Resource Ontology) [11][12], HR-ontology [24] (similar to HR-Ontology of Mochol's), SkillOnt [9], DOGMA (Developing Ontology-Grounded Methods and Applications) [21], Competency Ontology [8][14] [18], have been developed to capture main features of candidates and job descriptions. Based on constructed ontologies, semantic methods have been used to improve the matching process. RDF queries is applied on the ontology to find suitable job seekers for a job advertisement [11][12][18]. Description logics is used to seek jobs or applicants [9]. Ontology commitments are executed to match the model of school actors with the model of job actors [21].

Beside the ontology usage, profiles are used to keep few important features of job seekers and descriptions. In iCoper project [15], profiles follow draft standard IEEE Reusable Competency Definitions (RCD) [25]. An expert system of Drigas [17] for evaluation of the unemployed uses a simple user profile with 6 simple fields (Age, Education, Addition Education, Previous Employment, Foreign Language, Computer Knowledge), each field only take a few values. In project TEN Competence, the profiles of job seekers are used to store information as in their resume. In these such projects, a profile cannot keep much data and often has a simple hierarchical structure, because it has to able to be edited and managed by its owner. Relationship between data inside profiles is very limited. The reasoning on profiles is also simple.

Structured Relevance Models (SRM) [20], semi-structured documents, were developed to store employee's resumes and job descriptions in text forms. Seeking suitable resumes or jobs on SRM is based on the matching of text fields using the searching mechanism of database servers.

In these reviewed researches, the user and job models are often represented in the qualitative manner based on clear facts such as an educational degree, a language certificate... These representation schemes are only suitable for job seekers. During the studying period, students are not yet achieved any degree or certificate so that there is not any clear facts about their skills and existing methods of user modeling cannot be used to represent student models.

In this paper, an ontology-based user model is used to represent job models in Section 2. Matching functions are developed to match these models in Section 3. In Section 4 of the paper, the recommended job of the system is evaluated with the recruiters' selection.

2   Student and Job Modeling

A detailed ontology-based student model has been constructed for Information Technology (IT) students [26]. In this model, 214 popular IT skills are selected from the IT curriculum and then organized in the tree-like ontology. A part of this ontology is displayed in Fig. 1.

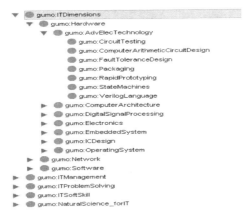

Fig. 1. A part of User Model Ontology

Each node in the ontology has a skill-meter that represents for the level of know-ledge and experience of as student on that skill. A ten-degree scale is used to describe for those levels, from unknown to the expert levels. This scale is enough to represent the ability of a particular skill of a student, from unknown level at the beginning of his/her studying program to very experienced levels which he/she can reach to in a few years after graduation. The description of this ten-degree scale is described in Table 1.

Table 1. Scales of knowledge and experience on a skill

Scale	Level of Knowledge & Experience	Indicator
0	No knowledge	No information
1	A little	Taken the course teaching the skill but failed (mark < 50)
2	Limited	Taken the course teaching the skill but just passed (50 ≤ mark < 60)
3	Understanding fundamentals	Medium theoretical knowledge. Taken the course teaching the skill but just passed Taken the course teaching the skill and got good mark (60 ≤ mark < 70)
4	Applicable	Good theoretical knowledge. Taken the course teaching the skill and got excellent mark (70 ≤ mark < 85)
5	Mastery of fundamental knowledge	Excellent theoretical knowledge Taken the course teaching the skill and got excellent mark (85 ≤ mark, good feedbacks from the lecturer of the course)
6	Some practical experience	Have some practical experience, such as "doing thesis with the skill as the main tool" or "have internship experience with the skill as the main tool" or "have 6 months or 1 year of experience in industry".
7	Experienced	Good experience, have 1.5 – 3 years of experience in industry. Master degree.
8	Very experienced	Long period of using the skill in several projects in a leading role, "have more than 3 years of experience in industry" or "use the skill at the job and play a team leader role"
9	Expert	Have more than 5 years of experience in industry or be a technical manager
		Note: Grading of a course is based on the 100 scale. 50 is the pass mark.

The studying result of a student is used to fill the leaf nodes of the student model of that student. In the curriculum design, each course of the program trains a set of skills for students in a supportive or a highly supportive modes. In another side, a skill can be trained in different levels and aspects by several courses that are organized in different years of the program. In general, the skill-meter of a skill is determined by the training mode and the academic level of courses that train the skill.

The studying result only helps to fill a part of student models. The fewer number of courses the student studied, the lower number of leaf nodes in the student model the system can be filled. Thus, there are several skills in student models having a empty skill-meter. In practice, IT skills are often related to each others. It means that when a student are good on a skill, he/she has some knowledge on related skills of that skill. Therefore, skills with a empty skill-meter can be induced from known skills. There

Table 2. Three induction rules on the skill ontology

Sample Tree
 • A (V, D)
      ○ $A_1(V_1, D_1)$
      ○ $A_2(V_2, D_2)$
      ○   ...
      ○ $A_i(V_i, D_i)$
 In the sample tree, A is name of skill, V is value of skill meter, and D is value of dissimilarity. $A_1, A_2, ... A_i$ are children nodes of A.	

Sibling induction rule	*Sibling induction rule* is used to estimate the skill-meter value of a node from its sibling skills. Select parameter k=2. "k-max of a set" means the sub-set k of biggest members from that set Suppose that A_1, A_2, A_3 are the known skills with $V_1 > V_2 > V_3$. The skill-meter value of remaining skills $A_4, A_5, ..., A_i$ are: $V_4 = V_5 = ... = V_i = $ Median of k-max of $\{V1, V2, V3\} - f_d(D)$;
Upward induction rule	*Upward induction rule* is used to estimate the skill-meter a skill from its children skills. Suppose that $A_1, A_2, ..., A_i$ are the known skills. Skill-meter value of skill A is: $V = $ average $(V_1, V_2, ..., V_i)$;
Downward induction rule	*Downward induction rule* is used to estimate the skill-meter of children skills from the known skill. Suppose that A is the known skills. Skill-meter value of skill A is: $V_1 = V_2 = ... = V_i = V$
Function $f_d(x)$ is used to calculate dissimilarity points from a dissimilarity value x. $f_d(x) = \begin{cases} 1, & x=1 \\ 2, & x=2 \\ 4, & x=3 \end{cases}$	

are three cases of skill induction. Firstly, in the sibling induction, a skill on a node can be induced from its sibling skills. Secondly, in the upward induction, a skill can be induced from its known children. Thirdly, in the downward induction, children skills of a filled skill can be induced from their super-skill. These three kinds of induction are differently developed in SoNITS (see Table 2).

The sibling induction depends on the similarity between the sibling skills. The more similar they are, the smaller the difference between the induced value and known skill-meters is. For instance, high-level programming languages are very similar in their manner, so that if an engineer is very skillful about "C++ programming", he can have a good programming skill on similar languages, such as "C", "Java" or "C#". In other wise, system administration skill is much different in various operating systems, so that a very experienced user of Windows systems may not know much about Unix or Linux systems. Therefore, a dissimilarity value is introduced in each middle node of the skill ontology to support the sibling induction.

In upward induction, a more general skill can be induced from their children skills. The "Programming" skill of a student can concluded from his skills on all particular programming languages. In otherwise, in downward induction, a unknown skill can be induced from its super-skill. For example, if a student is good in "System Administration" but there is no information about his skills on a particular operating system, it can be assumed that he is good in "System Administration on Unix systems" or "System Administration on Windows systems".

These three skill induction methods are constructed based on the human reasoning in practice. When only able collecting a part of information about a person, the other information of that person can be estimated based on the organization and similarity of information. This process is popularly used when screening curriculum vitas or interviewing candidates for job vacancies. The order of applying induction methods is sibling, upward and then downward.

Another part of system used in this research is job. In this research, job is referred to Standard Occupational Classification (SOC) website [7]. This website is constructed by the Department of Labor, the United States, to classify recognized industrial job positions. For each job position job, a short description and related duties is described. The 8 following occupations, which are typically common in Vietnam IT industry, are selected from the version 2010 of SOC, as the aimed jobs for IT students:

- 15-1121 Computer Systems Analysts
- 15-1132 Software Developers, Applications
- 15-1134 Web Developers
- 15-1141 Database Administrators
- 15-1142 Network and Computer Systems Administrators
- 17-2061 Computer Hardware Engineers
- 15-1133 Software Developers, Systems Software
- 15-1131 Computer Programmer

Based on the description of selected occupations, the job models of each job title has been constructed in the exactly same way as constructing student models. They are called "job models" in this paper. To construct job models, using mapping method, some skill-meters of required skills are directly filled from job specifications. Other skills which are not clearly specified in SOC are induced. This induction is used to make job models and student models are corresponding.

3   Matching Functions

Job Matching technique is core of Career Counseling System. There are some matching functions are implemented and evaluated to find out the most possible function. All of them require student models and job models to return matching rates and show defective skills of student to get his/her desire job. As mentioned above, these both models actually are instances of a structure of user model but contain different skill-meter values. By the definition, the matching functions are to simply measure the distance between two instances of that structure which actually represents by skills. The distance mentioned here is Euclidean method.

The models inputting to matching functions can be student models without skill induction or with skill induction. Four following matching functions between a student model S and a job model J are used in the paper:

- $f_1(S, J)$ measures on the leaf nodes of the induced models.
- $f_2(S, J)$ measures on the intermediate nodes of the induced models.
- $f_3(S, J)$ measures on all nodes of the induced models.
- $f_4(S, J)$ measures on the intermediate nodes of input models.

Other combinations are collapsed and return same result with one of above functions. The functions are also employed to identify which skills are defected and need to be studied or improved to fulfill job demands. Above functions require some parameters and operators as follows:

- U: user model structure (represented by ontology).
- S: student model structure (represented by ontology).
- J: job model structure (represented by ontology).
- $applyRule(V)$: applying induction rules to model V follows induction rules [26].
- $inter(V)$: set of intermediate skill node of model V.
- $leaf(V)$: set of skill leaf node of model V.
- $isDirect(i)$: check if skill-meter values are explicitly set by mapping functions.
- $value(i, V)$: return skill-meter values of skill i in model V.

Pseudo codes of the functions are expressed in Table 3.

Table 3. Matching functions' pseudo code

Function No.	Pseudo code
$f_1(S, J)$	$applyRule(S);$ $applyRule(J);$ $For\ each\ i\ in\ \textbf{leaf}(U)$ $Begin$ $distance\ =\ distance\ +\ (value(i,J)-\ value\ (i,S))^2;$ End $return\ \sqrt{distance};$
$f_2(S, J)$	$applyRule(S);$ $applyRule(J);$ $For\ each\ i\ in\ \textbf{inter}(U)$ $Begin$ $distance\ =\ distance\ +\ (value(i,J)-\ value\ (i,S))^2;$ End $return\ \sqrt{distance};$
$f_3(S, J)$	$applyRule(S);$ $applyRule(J);$ $For\ each\ i\ in\ \textbf{U}$ $Begin$ $distance\ =\ distance\ +\ (value(i,J)-\ value\ (i,S))^2;$ End $return\ \sqrt{distance};$
$f_4(S, J)$	$For\ each\ i\ in\ inter(U)$ $Begin$ $if\ isDirect(i)$ $distance\ =\ distance\ +\ (value(i,J)-\ value\ (i,S))^2;$ End $return\ \sqrt{distance};$

4   Result and Evaluation

4.1   Results

A Job Matching service is constructed in SoNITS [6] to help IT students to find out their strength and follow it. The compatible percentages are represented as matching rates between a student model and the designed jobs and displayed as in Fig. 2.

Career Counseling portlet developed by User Modeling group - International University, Vietnam National University Hochiminh city

Fig. 2. Career Orientation Pane with Distances of a Student to 8 SOC's jobs

The system also depicts how a progress of a student is during 4 years in university. Advisors can monitor matching results every year and give significant advices to counsel student to pick up suitable courses for their selected career. The figures bellow (Fig. 3) show the distances between a student capability and job demands during years. In different perspective, we can see progress growth of student during studying lifetime. Moreover, besides progress visualization, the system can also show potential companies and list of courses need to be taken to fulfill demands of a specific job as in Fig. 4.

Fig. 3. A student and job model visualization in 4 years

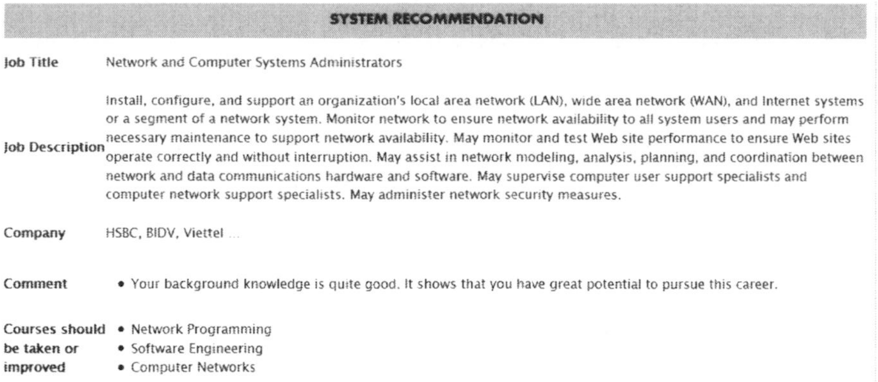

| | SYSTEM RECOMMENDATION | |

Job Title   Network and Computer Systems Administrators

Job Description Install, configure, and support an organization's local area network (LAN), wide area network (WAN), and Internet systems or a segment of a network system. Monitor network to ensure network availability to all system users and may perform necessary maintenance to support network availability. May monitor and test Web site performance to ensure Web sites operate correctly and without interruption. May assist in network modeling, analysis, planning, and coordination between network and data communications hardware and software. May supervise computer user support specialists and computer network support specialists. May administer network security measures.

Company   HSBC, BIDV, Viettel ...

Comment   • Your background knowledge is quite good. It shows that you have great potential to pursue this career.

Courses should • Network Programming
be taken or   • Software Engineering
improved   • Computer Networks

Fig. 4. System Recommendation Pane

4.2   Evaluation

The recruiter selection after reviewing the studying result of a student will be used as the most suitable job for that student based on his capabilities. The feedbacks from several recruiters can be different so the voting method is used to find the expected job. Several transcripts of senior students are sent to recruiters to get their feedback only 28 feedbacks are collected.

The matching results of 4 mentioned functions, the Collaborative Filtering method (f_5) and the student own choice (f_6) is compared with the recruiter selection in Table 4. In Table 4, function f_4 achieve the highest result with 81.2% correct cases. This is a reasonable result in career counseling.

The result of Collaborative Filtering (f_5) is quite low, only 17.3%, due to its binary measurement. In this method, when a student learn course C that trains skill S, that student is considered to have skill S without any information about the studying performance of course C. Therefore, it cannot correctly measure the capability of a student and the requirement of a job, so its matching result is not the same as the expected result.

Table 4. Compare matching results with the recruiter selection

	Number of correct matching (on 28 cases)	Correct Percentage
f_1	3	10.7%
f_2	19	67.8%
f_3	21	75%
f_4	23	82.1%
f_5	4	17.3%
f_6	4	17.3%

The student choice is also much different with the recruiter feedback as in the result of function f_6 in Table 4. To make that choice, students read the job specification of 8 designed jobs and choose the most interested job. In Table 4, only 4 student choices are matched with the recruiter selection. Some reasons for this low matching are the misunderstanding of students about a job, the mismatch between the student's capability and the job's requirement or the wrong selection of learning majors.

5   Conclusion

In this paper, a job matching system are introduced and used to support IT students to develop their own capability to reach their desire job position. Based on ontology-based student and job models, job matching functions are used to recommend suitable job positions to students.

This study also point out that this system is quite necessary for student because it can help student recognize the difference between their capability and their desire. Finally, this system can help student adjust themselves by changing their desire job or adjusting studying strategy.

References

1. Whiston, S.C., Oliver, L.W.: Career Counseling Process and Outcome. In: Walsh, W.B., Savickas, M.L. (eds.) Handbook of Vocational Psychology - Theory, Research, and Practice, 3rd edn., pp. 155–194. Lawrence Erlbaum Associates, Inc. (2005)
2. Brown, D.: Associates: Career choice and development, 4th edn. John Wiley & Sons, Inc. (2002)
3. Brusilovsky, P., Millán, E.: User models for adaptive hypermedia and adaptive educational systems. In: Brusilovsky, P., Kobsa, A., Nejdl, W. (eds.) Adaptive Web 2007. LNCS, vol. 4321, pp. 3–53. Springer, Heidelberg (2007)
4. Bull, S.: Supporting Learning with Open Learner Models. In: 4th Hellenic Conference with International Participation: Information and Communication Technologies in Education. Keynote, Athens (2004)
5. Kelly, D., Tangney, B.: Adapting to Intelligence Profile in an Adaptive Educational System. Journal of Interacting with Computers 18(3), 385–409 (2006)
6. Nguyen, C.D., Bui, B.D., Vo, D.K., Nguyen, V.S.: Educational Social Network. Journal of Science and Technology, Vietnamese Academy of Science and Technology, 379–382 (2011)
7. Standard Occupational Classification System (SOC), http://www.bls.gov/soc
8. Colucci, S., Di Noia, T., Di Sciascio, E., Donini, F.M., Mongiello, M., Mottola, M.: A Formal Approach to Ontology-Based Semantic Match of Skills Descriptions. Journal of Universal Computer Science 9(12), 1437–1454 (2003)
9. Zarandi, M.F., Fox, M.S.: Semantic Matchmaking for Job Recruitment: An Ontology-Based Hybrid Approach. In: Proceedings of the 3rd International Workshop on Service Matchmaking and Resource Retrieval, Washington, DC (2009)
10. Bizer, C., Heese, R., Mochol, M., Oldakowski, R., Tolksdorf, R., Eckstein, R.: The Impact of Semantic Web Technologies on Job Recruitment Processes. In: 7th International Conference Wirtschaftsinformatik, pp. 1367–1381 (2005)

11. Mochol, M., Wache, H., Nixon, L.: Improving the Accuracy of Job Search with Semantic Techniques. In: Abramowicz, W. (ed.) BIS 2007. LNCS, vol. 4439, pp. 301–313. Springer, Heidelberg (2007)
12. Mochol, M., Wache, H., Nixon, L.: Improving the recruitment process through ontology-based querying. In: Procedings of the 1st International Workshop on Applications and Business Aspects of the Semantic Web (SEBIZ 2006), collocated with the 5th International Semantic Web Conference (ISWC 2006), Athens, Georgia, USA (2006)
13. De Coi, J.L., Herder, E., Koesling, A., Lofi, C., Olmedilla, D., Papapetrou, O., Siberski, W.: A Model for Competence Gap Analysis. In: Proceedings of the Third International Conference on Web Information Systems and Technologies: Internet Technology, Web Interface and Applications (WEBIST 2007), Barcelona, Spain (2007)
14. Paquette, G.: An Ontology and a Software Framework for Competency Modeling and Management. Educational Technology & Society 10(3), 1–21 (2007)
15. Najjar, J., Simon, B.: Learning Outcome Based Higher Education: iCoper Use Cases. In: Proceedings of Ninth IEEE International Conference on Advanced Learning Technologies, pp. 718–719 (2009)
16. Schickel-Zuber, V., Faltings, B.: OSS: A Semantic Similarity Function based on Hierarchical Ontologies. In: Proceedings of the 20th International Joint Conference on Artificial Intelligence (IJCAI 2007), pp. 551–556 (2007)
17. Drigas, A., Kouremenos, S., Vrettos, S., Vrettaros, J., Kouremenos, D.: An expert system for job matching of the unemployed. Expert Systems with Applications 26, 217–224 (2006)
18. Draganidis, F., Chamopoulou, P., Mentzas, G.: An Ontology Based Tool for Competency Management and Learning Paths. In: Proceedings of I-KNOW 2006, pp. 1–10 (2006)
19. Herder, E., Kärger, P., Kawase, R.: Competence Matching Tool, Competence Gap Matching, Editors for Competences and Job Profiles. Internal Project Deliverable Report, TEN Competence (2009)
20. Yi, X., Allan, J., Croft, W.B.: Matching Resumes and Jobs Based on Relevance Models. In: Proceedings of SIGIR 2007, pp. 809–810 (2007)
21. De Leenheer, P., Christiaens, S.: Mind the Gap! Transcending the Tunnel View on Ontology Engineering. In: Proceedings of the 2nd International Conference on Pragmatic Web (ICPW 2007), pp. 75–82 (2007)
22. Silva, J.I., Toledo, M.: The Unemployment Volatility Puzzle: The Role of Matching Costs Revisited Economic Inquiry. Western Economic Association International 51(1), 836–843 (2011)
23. Centeno, M., Corrêa, M.: Job matching, unexpected obligations and retirement decisions. Pensions: An International Journal 13(3), 159–166 (2006)
24. Maniu, I., Maniu, G.: A Human Resource Ontology for Recruitment Process. Review of General Management 10(2), 12–18 (2009)
25. Reusable Competency Definitions Version 0.4,
    http://ltsc.ieee.org/wg20/files/RCD_0_4.pdf
26. Nguyen, C.D., Vo, D.K., Bui, B.D., Nguyen, T.D.: An Ontology-based IT Student Model in an Educational Social Network. In: Proceedings of the 13th International Conference on Information Integration and Web-based Applications and Services, pp. 379–382 (2011)

Part IV

Knowledge Engineering with Cloud and Grid Computing

An Abstraction Model for High-Level Application Programming on the Cloud

Binh Minh Nguyen, Viet Tran, and Ladislav Hluchy

Institute of Informatics, Slovak Academy of Sciences,
Dubravska cesta 9, 845 07 Bratislava, Slovakia
{minh.ui,viet.ui,hluchy.ui}@savba.sk

Abstract. One of the biggest obstacles in the widespread take-up of cloud computing is the difficulties that users meet in developing and deploying applications into clouds. Although PaaS cloud type offers advanced features such as available platform, automatic scaling, it ties users/developers into certain technologies provided by vendors. Meanwhile, due to the gap of a suitable programming environment for IaaS cloud type, the development and deployment applications into IaaS clouds become a complex task. In this paper, we present a novel abstraction model for programming and deploying applications on the cloud. The approach also enables service migration and interoperability among different clouds.

Keywords: cloud computing, distributed application, abstraction, object-oriented programming, interoperability.

1 Introduction

Cloud computing gathers key features like high availability, flexibility and elasticity, that intends to reduce total cost and decrease risk for both users and providers. Today, consumers can buy computation resources, platforms or applications over cloud infrastructures. In the language of this market, the commodities are usually referred to X as a service (XaaS) paradigm[1], which mainly includes Infrastructure as a Service (IaaS), Platform as a Service (PaaS) and Software as a Service (SaaS). In principle, cloud-based appliances or services often provide higher availability and lower cost than the traditional IT operations. For this reason, there is now a strong trend of deploying and migrating appliances to cloud. This process could be realized by using PaaS or IaaS. On the one hand, the PaaS clouds provide platform (programming language, databases and messaging) for implementing services and environments for hosting them. The platform also manages the execution of these appliances and optionally offers some advanced features like automatic scaling. Thus, this model allows developers to simply create their cloud services without the need to manually configure and deploy virtual machines (VMs). On the other hand, the IaaS clouds provide raw resources (VMs, storages) where the users have full access to the resources and manipulate with them directly in order to create their own platform and deploy services on this.

---

[1] XaaS is commonly used to express "anything/something" as a service.

N.T. Nguyen, T. Van Do, and H.A. Le Thi (Eds.): *ICCSAMA 2013*, SCI 479, pp. 295–306.
DOI: 10.1007/978-3-319-00293-4_22   © Springer International Publishing Switzerland 2013

It can be realized easily that, while PaaS binds developers into certain existing platforms, building cloud services in IaaS will be their choice to meet specific requirements. However, the use of IaaS is perceived as difficult with the developers/users, requiring advance computer skills for the installation and usage. Otherwise, since several commercial cloud systems have been already marketed, the problem of interoperability between these systems also arises, i.e. it would be possible and feasible to move appliances from a cloud provider to another, or to deploy an existing appliance on resources provided by several different cloud providers. Such possibility would have very large impact on the competitiveness of providers, as it would require them to offer better quality services at lower prices without forcing customer to rely on their resources through vendor lock-in.

Although several standardizations and solutions in this area have emerged, they have not yet brought any comprehensive solution for the service development and deployment issue on IaaS clouds. Therefore, from the view of general cloud users, they need to have an instrument, which can solve the problem. The work presented in this document is dedicated to innovative research and development of an elastic instrument (called high-level Cloud Abstraction Layer - CAL) allowing easy development and deployment of services on resources of multiple infrastructure-as-a-service (IaaS) clouds simultaneously. The CAL provides a total novel approach with emphasis on abstraction, inheritance and code reuse. Then, cloud-based services can be developed easily by extending available abstractions classes provided by the CAL or other developers. The interoperability between different clouds is solved by the basic abstraction classes of the CAL and all services are inherited and benefited from the advantage. The work does not only stop on theoretical work but also continues on applying CAL to build cloud services in order to deal with real problems.

2   Related Works

Open Virtualization Format (OVF) [1] is one of the numerous projects of Distributed Management Task Force (DMTF) [2]. Conceptually, it aims at creating a VM standard that ensures a flexible, secure, portable and efficient way to distribute VMs between different clouds. The VM standard is called virtual appliances [3]. Users can package a VM in OVF and distribute it on a hypervisor. The OVF file is an XML file that describes a VM and its configuration. Application-specific configurations can be packaged and optimized for cloud deployment, as multiple VMs, and maintained as a single entity in OVF format. Although appliances/services can move among various clouds along with the OVF image, creating cloud-based services still require many efforts from developers, who must carry out complicated steps such as preparing VM and platform, deploying services and management. Additionally, all of the operations are performed without any utility functionalities or any supports from cloud providers. This causes time-consuming and increased cost. Otherwise, the incompatible APIs issue also brings about the service operation are not guaranteed when they migrate from a cloud to another.

Open Cloud Computing Interface (OCCI) [4] is a recommendation of the Open Grid Forum [5]. This project is identified by the provisioning and management research direction in cloud computing. The OCCI is a RESTful Protocol [6] and API for all kinds of management tasks. OCCI was originally initiated to create a standard management API for existing IaaS clouds, allowing users to manage and use various clouds under single unified interface that intends to enable cloud interoperability. Although OCCI enables cloud users to manage resources from different clouds at the same time, unfortunately, the users still have to directly connect to VMs in order to create their own platform for service development and deployment. In this way, the service migration is almost impossible.

The weakness of the cloud standardizations presented above is that they force cloud providers to accept and support their products. Such scenario would have a very large impact on the competitiveness of the cloud vendors, as it would require them to offer better quality services at lower prices without locking customers to rely on their resources.

The API abstractions (e.g. Simple Cloud API [7], Deltacloud [8], SAGA API [9]) have been created for managing resources from various clouds at the same time. These APIs support a set of functionalities that are a common denominator among different providers and hide technological differences by their interfaces. Therefore, the APIs do not require any support from the cloud vendors. For example, Deltacloud defines a REST-based API for managing and manipulating cloud resources in a manner that isolates the API client as much as possible from the particulars of specific cloud API's. To simplify operation of the REST interface, the Deltacloud project provides a CLI (command-line interface) tool, as well as client libraries in Ruby, Java, C, and Python. This makes Deltacloud differ from other solutions including Simple Cloud API, Apache Libcloud [10], etc. in which all these libraries are tied to a specific programming language. SAGA API is another API for deploying applications to distributed computing. The API defines an abstraction layer for scheduling jobs to grid or cloud infrastructures, including execution, management functions of files or data. This abstraction scheduler is essentially a resource broker (which is about negotiating between those who want to consume resources and providers). Applications are thus represented as jobs. However, since it is built based on gird computing, there are drawbacks when applying it to cloud. In real time, the SAGA API applications are not services. The life cycle of the applications consists of three stages: submit to computational resources, run or calculate, return result or output. After finishing, the used resources will be terminated. Consequently, the API is only suitable for computational applications (e.g. calculation, simulation), not for development and use of cloud services.

As pointed out before, the advantage of the API abstractions is independent of cloud vendors. However, like OCCI, these abstractions do not help developers develop and deploy services more easily than the traditional way. The developers still have to prepare a platform and develop the services by connecting directly to VMs. Although there are API abstractions, which offer support for service deployment via scheduling job mechanism like Simple Cloud API SAGA API, due to limitations of the mechanism, they cannot be used for developing services. At present, no APIs have provided a comprehensive solution for both development and deployment tasks.

Clearly, the problem of service development and deployment is a big gap of cloud computing today. There are three reasons for this:

- PaaS cloud type limits developers/users to concrete platforms and APIs.
- Lack of suitable programming model for service development on IaaS.
- Lack of interoperability between different IaaS clouds.

3   General Approach

Currently, services in IaaS clouds are usually developed and released in form of pre-defined images, which are very cumbersome, difficult to modify, and generally non-transferable between different cloud providers. For example, marketplace of AWS appliances [11], VMware Virtual Appliance [12] StratusLab [13], etc. provide a large number of services in that way. Though OVF can solve in part of image move problem, it almost cannot be applied to all commercial clouds that have held the largest market share of cloud computing uses (see section 2).

For this purpose, **the approach releases cloud services as installation and configuration packages that will be installed on VMs** already deployed in cloud resources and containing only the base OS. Then, users just choose an OS provided by clouds (e.g. Ubuntu 12.04 LTS) and deploy the configuration packages to create their own services. The advantages of this service approach are as follows:

- Most cloud infrastructures support images with base OS, so the services are easily transferrable.
- The base OS images provided by the infrastructures are usually kept up-to-date, so they are secure.
- VMs are always started correctly with cloud middlewares, unlike image delivery of the existing approaches like OVF and the marketplaces presented above, which must have acceptance and support from providers.
- Services can work on unknown infrastructures without changing codes implemented before.
- The approach allows taking full advantages of many existing applications (especially legacy ones) that already have had install/setup tools or via package managers of base OS (e.g. *apt, rpm* and *git* package of Debian/Ubuntu).
- Allowing automatic deployment of developed services with near zero VM tuning.

One of the techniques support perfectly for the abstraction approach, namely **object-oriented programming** (OOP). Because the encapsulation and polymorphism mechanism of OOP allow hiding state details and standardizing different interaction types that are the beginnings of the abstraction. Through these mechanisms, OOP bring the ability of defining a common interface over different implementations and behaviors of various systems. In addition, the OOP inheritance mechanism also allows creating easily new abstraction layers over an initial abstraction layer.

Nowadays, many appliances are already provided in a package which specifies the configuration interface (e.g. Android, iOS appliances for example) allowing developers to hide appliance details and thus optimize them in the manner the developers want. Since services and their functions are defined in form of OOP objects, they can be extended and customized in order to create new services based on the existing codes. In this way, **developers can reuse service codes via the inheritance mechanism without learning implementation details of the origin**. To achieve this, CAL first provides basic functionalities of VMs with base OS. Developers will inherit the code and modify/add functions related to the services. CAL then acts as foundation for development and deployment of cloud services on IaaS clouds. The advantages of the code reuse are as follows:

- Service developers do not have to use any middleware functionality directly. As a consequence, the codes of services are portable between different clouds.
- Developers just focus on service aspects, not on the clouds; in this way, the approach reduces the efforts to learn about cloud middlewares.
- Enabling developers themselves to create PaaS, SaaS based on IaaS.

4   Design of High-Level Cloud Abstraction Layer

4.1   Design of CAL

As mentioned before, CAL is designed with number of programming functions in order to manage and interact with VMs that belong to different clouds. These functions are divided into functionality groups:

- **Setting Cloud:** enables developers to set which cloud will be used. This group has only setCloud() function.
- **Provisioning:** consists of start() and stop() function to create and terminate the VMs.
- **Monitoring:** getting actual information of the machines (cloud provider, IP address, ID instance and so on) by status() function.
- **Execution:** running commands on VMs by execute() function.
- **Transfer** includes two functions: put_data() to upload and get_data() to download data to/from the VMs.
- **VM Snaphost**: creates/restores snapshot of VM into an image. The group involves create_snapshot() and restore_snaphsot() function.

To interact with various clouds, for each of them, we design a driver, which uses its specific API to manage VMs. Note that, the drivers do not need to implement all API functionalities that are provided by cloud vendors. Each driver only uses necessary API actions to shape CAL functions, including `start()`, `stop()`, `status()`, `create_snapshot ()` and `restore_snaphsot()`. The `setCloud()` function is call to set a driver (cloud) to use.

Otherwise, Execution and Transfer do not use any APIs because no APIs provide functionalities to carry out those operations. Thus, CAL abstracts the connection and realization process and hides implementation details by the `execute()`, `put_data()` and `get_data()` functions.

Through abstract functions above, detailed interactions between CAL and clouds are hidden. Therefore, developers can manipulate multiple clouds at the same time under the unified interface without caring about how each cloud works.

4.2   Inheritance and Software Layering of CAL

The developers can easily create new cloud service by using the abstraction layer. In this way, the developers only have to inherit the functions of the abstraction layer for developing their service functions. The functionalities of the service include:

- **Initialization:** the developer just call the `setDefault()` function of the abstraction layer to choose a cloud in order to deploy their service, then use `start()` function to create a virtual machine on the cloud. The developer can add commands to install software packages on newly created machine by the `execute()` function. He or she also can terminate the virtual machine that contains the service by the `stop()` function.
- **Backup/Restoration:** to create backup/restoration functions for the service, the developer will inherit the `backup()` and `restore()` function of the abstraction layer.
- **Service functionalities:** the developer can create several functionalities for their service. For examples, for database service, they can add a number of functions to import database, make query, and so on by inheriting the `execute()`, `get_data()` and `put_data()` function.

Users (distinguish from the developers) thus will use the service via its interface. The users would not require worrying about how the service deploys. Meanwhile, the developer can choose simply the target cloud to deploy their service without having to care about incompatible API because during the development and deployment, the developer does not use directly any middleware APIs. They only reuse the functions that are provided by the abstraction layer.

In addition, a software layer, which has created by a developer, also can be used and extended further by other developers in the same way. Thus, the first developer defines the first software layer with new functionalities on demand from his or her users. Similarly, the later developers can also define functionalities for second software over the first layer according to their users' needs by inheriting the first layer functionalities. As the result, each layer is practically a platform-as-a-service by itself, because the users can use the service via a clear interface provided by the previous developers. For example: a developer can create a LAMP (Linux-Apache-PHP-MySQL) layer that is equivalent to web-hosting platforms for his or her users. Another developer can use this LAMP layer to provide web applications (e.g. wiki, forum) without manipulating with the cloud infrastructures.

5   Case Studies

5.1   Experimental Setup

Our current implementation of CAL prototype bases on the installations of three middlewares: OpenStack Folsom release, Eucalyptus 2.0.3 and OpenNebula 3.6 release. The purpose of this setup is to provision VMs that belong to OpenStack, Eucalyptus and OpenNebula middlewares at the same time. Tests are successful if and only if all VMs have SSH access. Each of them consists of a controller node, a management network (switch) and at least two compute nodes. For controller nodes, each server blade is equipped with processor Xeon including 16 cores (2.93 GHz), 24GB of RAM and 1TB hard drive, meanwhile for compute nodes, each server blade is equipped with processor Xeon with 24 cores (2.93 GHz), 48GB of RAM and 2TB hard drive. Linux is installed for all servers as OS. KVM hypervisor is used for all three systems. An Ubuntu 12.04 images are created and deployed on the clouds. While OpenStack, Eucalyptus are configured with Glance [14] and Walrus [15] respectively as internal image storage services, OpenNebula uses non-shared file systems [16] with transferring image via SSH for test purpose.

5.2   Development and Deployment of Cloud Services

The realistic services are the best way to show and demonstrate the CAL effects. In the direction, this section presents the development and deployment of cloud monitoring service for distributed systems based on CAL. Additionally, the layering feature of the services is also demonstrated in the section.

As Python language [17], the high-level abstraction layer is represented as a class (CAL) which provides the basic functions of VM. For each cloud infrastructure, we respectively define separate classes: Eucalyptus, OpenNebula and Openstack, which are the drivers of these clouds. Each the class uses the middleware-specific API and utility mechanisms (execution, transfer functionality) for the implementation. Structurally, the functions of the CAL class are derived from the separate classes. If we need to extend our implementation for a new cloud infrastructure, the only thing we have to do is to create the new derived class (driver) for its specific middleware functions.

The case study carried out process of development and deployment of a concrete service using CAL. Inspired by the fact that most users need a cloud-based hosting for their applications, a webhosting was built. Purposes of the service are:

- Providing a platform for development and deployment of web applications (e.g. websites or WordPress [18] blog).
- The platform can be deployed into well-known clouds of CAL without any changes in its code.
- Limiting interaction with VMs for the service user.

The service is developed based on Apache web server (LAMP). It is an open source platform that uses Linux as OS, Apache as web server, MySQL as the relational database management systems and PHP as the OOP language. The service class is programmed as follows:

```
class webhosting(CAL):

  def setCloud(self, cloud):
    CAL.setCloud(self, cloud)

  def config(mysql_password):
    ...

  def start(self, image, VM_type)
      CAL.start(self, image, VM_type)
  CAL.execute(self, 'install_LAMP_command')
```

The backup and restoration functionality also are written shortly inside the webhosting class:

```
  def create_snapshot(self):
     CAL.create_snapshot(self)

  def restore_snapshot(self):
     CAL.restore_snapshot(self)

  def data_backup(self):
    CAL.get_data(self, ' ')

  def data_restore(self):
    CAL.put_data(self, ' ')
```

Besides fundamental functions such as setCloud(), config(), start(), stop(), create_snapshot() and restore_snapshot(), the webhosting provides specific function for web application developers, including:

- data_backup() and data_restore() backs up/restores only user data (e.g. web pages) on the server to local.
- upload() uploads files or web packages into the hosting server. The function will return URL of the websites.
- run() runs web configuration commands on the server (e.g. changing directory name of the web package, copying files)
- mysql_command() runs MySQL statements (e.g. creating database, username, password or setting access privilege for a database).

```
  def upload(self, dir_package):
    CAL.put_data(self, dir_package)
    CAL.execute(self, ' ')

  def run(self, user_command):
     CAL.execute(self, user_command)
```

```
def mysql_command(self, MYSQL_statement):
  CAL.execute(self, MySQL_statement)
```

After development, the service can be deployed simply by running commands:

```
service = webhosting()
service.setCloud('cloud_name')
service.config('mypass')
service.start()
```

At the time, users can use the hosting service through its existing functions. There are two types of web applications:

- Flat sites are simple web pages. They are called *flat* because they do not use any database on the server.
- Dynamic sites are built based on database. Examples of this type are e-commerce websites, forums (content management system) and blogs.

The webhosting service supports both types. For flat sites, web designers just use upload() function to upload their site packages into the server. Then, the webhosting service will return URL for these sites. For example, a flat site with name *"my-site"* is hosted on the server by:

```
service.upload('mysite')
```

For dynamic sites, developers can program new software layers by inheriting webhosting service functions. Each the new software layer serves a specific purpose of a dynamic site. In order to demonstrate expansibility of the webhosting service as well as CAL, we continue to develop a blog WordPress service dealing with the layer of webhosting. This service is inherited all existing functions and deployed on the server provided by the webhosting. The following presents some codes of it:

```
class wordpress(webhosting):

  def config(self, database_pass):
    webhosting.mysql_command(self, ' ')
  def start(self):
    webhosting.run(self, 'install wordpress')
    ...

  def stop(self):
    ...
```

Blog users only have to enter password for their database on the server when creating their blogs. A blog is created simply by commands:

```
blog = wordpress()
blog.config('mysecret')
blog.start()
```

In the studies, hosting service is the first software layer. According to the user requirements, we can go further with providing many other hosting functions such as FTP transfer or phpMyAdmin [19] interface for MySQL databases. Meanwhile the WordPress service acts as second software layer (equivalent to Software as a Service - SaaS) over the first layer. We also can to deploy other SaaS services (wiki pages, forum for instance) based on the webhosting. The most important thing is the codes of both software layers can be deployed on any cloud infrastructures by changing the name of target cloud in the `setCloud()` function. Therefore, CAL enables the ability to deliver services among different clouds without any obstacle. Furthermore, users of the services do not need to worry about the VM management because the process of service deployment is automatic.

5.3   Experimental Results

To evaluate operation of webhosting service, WordPress service and CAL, the process of service deployment is tested on three existing cloud installations with various VM types. The experimental measurement is repeated 20 times for each of the VM type of each cloud. The average values of deployment time that are summarized in Table 1. The duration time is calculated in seconds.

Table 1. Deployment Time of cloud services

	VM Type	Webhosting	WordPress
OpenNebula	*small*	655.718	113.014
	medium	666.898	112.128
	large	674.093	110.768
Eucalyptus	*small*	365.66	84.66
	medium	373.248	79.66
	large	388.702	78.652
OpenStack	*small*	215.7	48.495
	medium	225.132	46.215
	large	236.491	45.985

The results also are illustrated by diagrams in Fig. 1 and Fig. 2. There are some observations that can be made from inspecting the results. First, CAL operates well with the well-known clouds. Second, the webhosting and WordPress service can be deployed on all these clouds. Third, in the case of comparison between clouds, deploying the webhosting service on OpenStack is faster than on Eucalyptus (approx. 41%) and OpenNebula (approx. 67%). The reason is that OpenNebula installation uses non shared file system and image is transferred between nodes via SSH. Additionally, while OpenStack is kept up-to-date with consecutive versions, Eucalyptus only supports open source with an old version. This is importance factor that can explain why OpenStack achieves higher performance than Eucalyptus cloud. Four, in the same cloud, since attributes of VM types are different. Therefore, deployment of the webhosting service into medium or large type is faster than small type. However, using medium and large VM, the process of service deployment is still slower because the VM startup needs more time. Finally, deploying WordPress on webhosting server of

OpenStack requires less time than Eucalyptus (approx. 43%) and OpenNebula (approx. 57%). Although the process of deploying services takes a long time. However, since the process is realized automatically, the time for deployment is always less than manipulation of the traditional approaches.

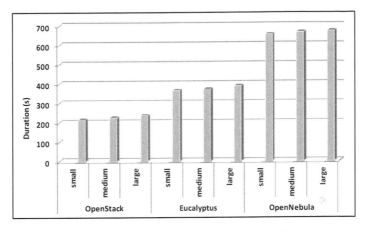

Fig. 1. Deployment time of webhosting service

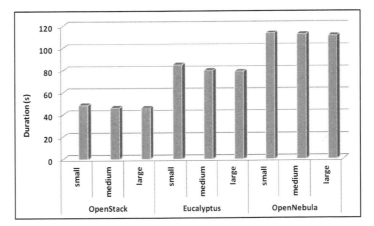

Fig. 2. Deployment time of WordPress service

6   Conclusion

In this paper, we presented the novel approach for developing interoperable cloud-based services that are treated as objects with strongly defined interfaces. The foundation of the approach is a high-level abstraction layer that provides basic functionalities of VM for each known cloud middleware. Based on the layer, process of service development and deployment is easier: developers will build their services by inheriting the existing functionalities of the abstraction layer without using any middleware

APIs as well as directly connecting to the VMs. Thus, developed services are independent of infrastructures and they can be deployed on the diverse clouds. In this way, our approach enables the service interoperability, which is one of the invaluable features for cloud computing.

Since CAL services are independent from underlying infrastructures, they can be published in a marketplace that allows other service developers or pure users to download and use them without coding. Consequently, in the near future, we will continue to build the marketplace for CAL services.

Acknowledgment. This work is supported by projects SMART II ITMS: 26240120029, CRISIS ITMS: 26240220060, VEGA No. 2/0054/12, CLAN No. APVV 0809-11.

References

1. Open Virtualization Format, http://www.dmtf.org/standards/ovf
2. Distributed Management Task Force, http://dmtf.org
3. Kecskemeti, G., Terstyanszky, G., Kacsu, P., Neméth, Z.: An Approach for Virtual Appliance Distribution for Service Deployment. Future Generation Computer Systems 27(3), 280–289 (2011)
4. Open Cloud Computing Interface, Open Grid Forum, http://occi-wg.org
5. Open Grid Form, http://www.gridforum.org/
6. Edmonds, A., Metsch, T., Papaspyrou, A., Richardson, A.: Toward an Open Cloud Standard. Internet Computing 16(4), 15–25 (2012)
7. Simple Cloud API, http://www.simplecloud.org
8. Apache Deltacloud, http://deltacloud.apache.org
9. Kawai, Y., Iwai, G., Sasaki, T., Watase, Y.: SAGA-based File Access Application over Multi-filesystem Middleware. In: 19th International Symposium on High Performance Distributed Computing, Chicago (2010)
10. Apache Libcloud, http://libcloud.apache.org
11. Amazon Web Service Marketplace, AWS, https://aws.amazon.com/marketplace
12. VMware Virtual Appliance Marketplace, VMware, http://www.vmware.com/appliances
13. StratusLab Marketplace, StratusLab, http://stratuslab.eu
14. von Laszewski, G., Diaz, J., Wang, F., Fox, G.C.: Comparison of Multiple Cloud Frameworks. In: Proceeding of IEEE Cloud 2012, pp. 734–741 (2012)
15. Lonea, A.M.: A survey of management interfaces for eucalyptus cloud. In: Proceeding of 7th IEEE International Symposium on Applied Computational Intelligence and Informatics (SACI), pp. 261–266 (2012)
16. Wen, X., Gu, G., Li, Q., Gao, Y., Zhang, X.: Comparison of open-source cloud management platforms: OpenStack and OpenNebula. In: Wen, X., Gu, G., Li, Q., Gao, Y., Zhang, X. (eds.) Proceeding of IEEE 9th International Conference on Fuzzy Systems and Knowledge Discovery, pp. 2457–2461 (2012)
17. Python programming language, http://python.org
18. WordPress blog source, http://wordpress.org
19. phpMyAdmin, http://phpmyadmin.net

Agent Based Modeling of Ecodistricts
with Smart Grid

Murat Ahat, Soufian Ben Amor, and Alain Bui

Laboratory of PRiSM
University of Versailles St-Quentin-en-Yvelines
45 avenue des Etats-Unis, 78035 Versailles, France
murat.ahat@prism.uvsq.fr,
{soufian.benamor,alain.bui}@uvsq.fr

Abstract. Studies have been carried out on the different aspects of the smart grid using models and simulations, however little is done to combine those efforts to study the smart grid and ecodistrict in the same context. In this paper we discuss the role of smart grid technologies in an ecodistrict context and the mathematical tools such as complex networks theory and game theory for modeling smart grid and ecodistrict. Further more, from a modeling point of view, we choose and discuss the multiagent based modeling approach due to the complex and unpredictable characteristics of the to be modeled system. With the help of multiagent modeling guideline ASPECS, an example of ecodistrict model is discussed, as well. The example can be extended to become a real world model to study different problems in smart grid or ecodistrict designing.

Keywords: ecodistrict, smart grid, agent based modeling, game theory.

1 Introduction

Now we are living in an era that majority of world population lives in cities, and those urban areas are expecting even more population growth. People are facing with global problems like climate change, ever increasing demand for energy resources and pollution, etc. The pressing nature of those problems are urging people to manage resources more efficiently and decrease the ecological harm. Ecodistricts try to provide solution to those problems at a neighborhood level [29].

Ecodistricts are being experimented in different countries under diverse environments with dissimilar approaches, and have achieved various economical, ecological and social successes. For instance, Portland's ecodistricts and Stockholm's ecodistricts are creating a new generation of integrated district-scale community investment strategies at a scale large enough to create significant social and environmental benefits, but small enough to support quick innovation cycles in public policy, governance, technology development and consumer behavior [10,4]. Smart grid technologies play a key role in ecodistrict development by facilitating the pursuit of environmental and energy related goals of ecodistricts [27,6,7].

N.T. Nguyen, T. Van Do, and H.A. Le Thi (Eds.): *ICCSAMA 2013*, SCI 479, pp. 307–318.
DOI: 10.1007/978-3-319-00293-4_23   © Springer International Publishing Switzerland 2013

Our goal in this paper is to layout the fundamentals for modeling smart grid and ecodistricts, in order to help decision making through computer simulation in ecodistrict and smart grid designing and development. Apart from addressing problems in modeling ecodistrict and smart grid, the example model is can be easily extended to a complete model to study real world or academically problems.

The paper is organized as the following: in section 2, we introduce basic concepts of ecodistrict and smart grid; in section 3, we discuss theoretical approaches for smart grid and ecodistrict modeling; in section 4, we propose a conceptual ecodistrict model; and finally in section 5, we give the conclusion and future directions.

2   Basic Concepts

In the previous section, we discussed shallowly the concepts of ecodistrict, smart grid and their relations. In this section we introduce those concepts with some illustrations and two case studies from Portland and Stockholm.

2.1   Smart Grid

The term smart grid is coined by Amin in 2005 [3]. Smart grid is a type of electrical grid which attempts to predict and intelligently respond to the behavior and actions of all electric power users connected to it - suppliers, consumers and those that do both - in order to efficiently deliver reliable, economical and sustainable electricity services [16]. Smart grid have three economic goals: to enhance the reliability, to reduce peak demand and to reduce total energy consumption. To achieve these goals, various technologies have been and are being developed and integrated in the electrical network. It is not intended to replace the current system of power grid but only to improve it.

Smart grid integrates advanced sensing technologies, control methods and integrated communications into current electricity grid both in transmission and distribution levels [17,2]. Smart grid should have following key characteristics: (i) self-healing; (ii) consumers motivation and participation; (iii) attack resistance; (iv) higher quality power; (v) different generation and storage options; (vi) flourished markets; (vii) efficiency and (viii) higher penetration of intermittent power generation sources. Those key characteristics guarantee a economical, efficient and reliable energy service, and by doing so, they also contribute to the energy and ecological sustainability. In the next subsection, we will discuss the relations between ecodistrict and smart grid.

2.2   Ecodistrict and Smart Grid

Ecodistrict is an urban planning aiming to integrate objectives of *sustainable development* and reduce the *ecological footprint* of the project. Sustainable development is a pattern of economic growth in which resource use aims to meet

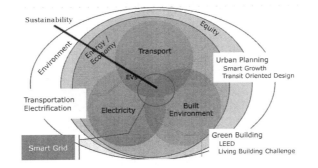

Fig. 1. Sustainability and Smart Grid

human needs while preserving the environment so that these needs can be met not only in the present, but also for generations to come [20]. The *ecological footprint* is a measure of human demand on the Earth's ecosystems. To attain those two goals, ecodistricts depend not only on technologies such as smart grid, waste treatment and green infrastructures, but also on active community participation.

Smart grid plays a significant role in ecodistrict, for it provides efficient, reliable and sustainable energy to the residents of the district. On the other hand, the active participation of the community facilitates the realization of smart grid goals such as energy usage efficiency and reliability. The relationship between sustainability and smart grid is shown in figure 1[1]. We can see that the intersection among transportation, built environment and electricity composes smart grid and sustainability lies at the heart of the intersection. In the next subsection, two ecodistrict cases will further illustrate the importance of smart grid in ecodistrict.

2.3   Ecodistrict Cases

Portland's Ecodistricts. To test the ecodistrict approach, Portland sustainability institute partnered with the city of Portland in 2009 to launch five Portland pilot districts as sustainability innovation zones where the latest in community organizing, business practices, technology and supportive public policy comes together to drive ambitious sustainability outcomes. Over a three-year period, district stakeholders have agreed to work with their neighbors to set rigorous goals, develop a road map and implement projects. Their approach emphasizes on the whole system integration, faster investment cycles and community engagements [10]. Even though the five ecodistricts are long term projects and far from being finished, they have already begun to provide valuable experiences and inspirations to other cities, attracted numerous investments, and achieved various economical and social successes. Besides, Portland sustainability institute organizes the annual *ecodistricts summit* since 2010, which

---

[1] From John Thornton's presentation: Solutions for Sustainability.
http://cleanfuture.us/2009/04/solutions-for-sustainability-2

is an international platform for experience exchange and discussion on ecodistricts.

Stockholm's Ecodistricts. The ecodistrict of *Hammarby Sjöstad*, situated to the south of Stockholm's city center, was one of the first such districts to implement a holistic environmental vision incorporating aspects relating to waste, energy and water as part of one sustainable system. This vision is encompassed within the "Hammarby Model" an eco-cycle which promotes the integration of various technical supply systems, so that the waste from one system becomes a resource for another. The district has since become a model for the application of integrated urban planning throughout the world [4].

The Norra Djursgardsstaden development, also known as Stockholm Royal Seaport, is a brownfield rehabilitation project whose planned completion date is 2025. The area will include 10,000 dwellings and 30,000 office spaces, following the same mixed use concept as Hammarby Sjöstad. In terms of planned infrastructure, it places the same emphasis as Hammarby Sjöstad on public transport links as well as cycling and walking. It includes in its design many cutting-edge technologies, such as smart grid, electric vehicle charging, and renewable energy integration. It aims to become fossil fuel free by 2030, before the rest of the city. Norra Djursgardsstaden's energy consumption target is higher than that of Hammarby Sjöstad, at 55 kWh/m2/year, and it reached a decision on the environmental profile of the area at the very beginning of the project, showing Stockholm derived some useful lessons from the implementation of the Hammarby Sjöstad project. In figure 2, a smart grid network consisting of elements in Nurra Djursgardsstaden is shown [26].

3   Theoretical Approaches

After familiarizing with smart grid and ecodistrict concepts, in this section we discuss the ecodistrict and smart grid modeling approaches and theoretical tools. Before going into the details, let us recall the complex system definition. New England complex system institute[2] gives a simple definition of complex system : "complex systems is a new field of science studying how parts of a system give rise to the collective behaviors of the system, and how the system interacts with its environment". Complex system study embraces not only traditional disciplines of science, but also engineering, management, and medicine as well.

Smart grid itself is regarded as a complex system due to its heterogeneous actors, dynamic, complex interactions among them, and self healing, self organizing characters [14,22], let alone adding ecodistrict specific elements like community behavior, waste treatment and green infrastructures. Thus we can consult complex system approach in modeling smart grid and ecodistrict. The rest of this section justifies our agent based approach in modeling smart grid and ecodistrict, then introduces how complex networks and game theory provide answers to different problems in smart grid modeling.

---

[2] http://www.necsi.edu: visited May 2012.

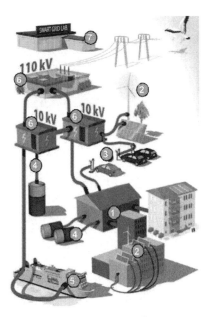

Fig. 2. Smart grid and Norra Djursgardsstaden ecodistricts: 1-smart home; 2-distributed generation; 3-electric vehicles; 4-energy storage; 5-smart harbor; 6-smart substation; 7-smart grid lab

3.1   Agent-Based Modeling and Simulation

Modeling an ecodistrict or smart grid will allow us through computer simulation to help decision making in designing and development of ecodistricts. Figure 3 shows the process of modeling and simulation [24,8]. Starting from a real world problem, first a conceptual model, which is the simplified representation of real world problem, is conceived. Then a computer model is created from this conceptual model, to simulate the problem. By studying the simulation results, we can have better understanding of the real world problem. We should also note that this process is not a linear one.

Agent based modeling and simulation (ABMS) is a relatively new modeling paradigm. It has strong roots in multi agent systems, and robotics in artificial intelligence, but its main roots are in modeling human social behavior and individual decision making [18]. In complex system science, the underlying notion that "systems are built from bottom-up" and the study of "how complex behaviors arise from among autonomous agents", make ABMS as a favorable candidate for modeling and simulation. Many different developments are achieved using ABMS in various disciplines like artificial intelligence, complexity science, game theory and social science, etc. Agents are the building blocks of ABMS, they have properties like: reactiveness, autonomous, spatial awareness, ability to learn, social ability, etc. Hence, agent based models are decentralized models. In ABMS, global system behaviors are defined at individual level and emerge

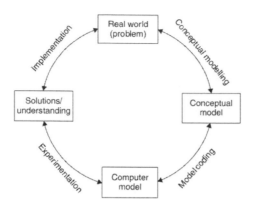

Fig. 3. The cycle of modeling and simulation

as a result of individual agent interactions. That is why ABMS is also called bottom-up approach [19,9]. This means that a system can be modeled starting from actors inside the system, and local interactions among them. This also explains why agent based modeling is widely used in modeling complex systems. To effectively and correctly model agent interactions and connections, we need to integrate other mathematical modeling tools into ABMS, which will be introduced in the following subsections.

3.2   Complex Networks Theory

Networks are representations of relation or interaction structure among entities. To study real world networks, different models have been proposed and studied thoroughly. Erdös and Rényi (1959) suggested the modeling of networks as random graphs [12]. In a random graph two pair of nodes are connected with probability p. This leads to a Poisson distribution when considering the numbers of connections of the nodes. However, the study of real world networks has shown that the degree distribution actually follows a power law [1]. Those networks are called scale free networks. For example, airline networks, metabolic networks, citation networks and even power grid networks are scale free networks.

In [28], the authors demonstrated that the real world power grid networks follow a exponential degree distribution, see figure 4. This indicates that complex networks can be used to model the topology of the power grids and information flow in smart grids. In other words, in an agent based model of smart grid, the agent networks can be effectively modeled by complex networks. The "Hammarby" model of Stockholm (section 2.3) shows several possible networks in an ecodistrict other than electricity network. Those should be taken into account in modeling according to the requirements of the system. Even if there is no relative studies about those networks, depending on their complexity a simple graph or complex networks approach should be enough to model them.

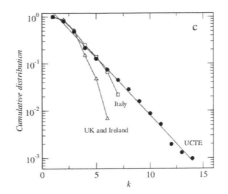

Fig. 4. Exponential degree distribution of Italian, UK and Ireland, European (UCTE) power grid networks [28]

In [30,13], the authors discussed the other applications of complex networks in smart grid systems, such as vulnerability analysis, data streaming mining, fault detection and early warning.

3.3   Game Theory

Game theory is the study of problems of conflict and cooperation among independent decision-makers. Its related areas are the differential games, the optimal control theory and the mathematical economics. It has played a substantial role in economics [5,21] and has been applied to application areas as defense strategy and public policy analysis among other areas. We can view the runs of a multi-agent models as a game where the agents are considered as the players of the game and their interactions as resulting of their own strategies. Agents can be in conflict of interest, which is modeled by game theory to find equilibrium and to compute expected outcome of any of the players: in this case, players do not cooperate and this kind of game is called non-cooperative game. If players accept to cooperate, we are lead from situations of conflict to situations of negotiation. In this case, the game is called a cooperative game, the players can form coalitions to achieve their outcome [23].

Saad *et al.* in [25] provided a comprehensive account of game theory application in smart grid systems, tailored to the interdisciplinary characteristics of these systems that integrate components from power systems, networking, communications and control. Authors overviewed the potential of applying game theory for addressing relevant and timely open problems in three emerging areas that pertain to smart grid: micro-grid systems, demand-side management, and communications. In [22,14], authors also emphasized the potential of game theory in smart grid modeling, notably the network of games. Another advantage of the game theory is that we can seamlessly integrate the game theoretic solutions into the agent based models at agent and network level to describe dynamic local rules as well as the *intelligence* of the agents.

4   Agent Based Model and ASPECS

In the previous section, we described theoretical approaches for modeling smart grid and ecodistrict. In this section we discuss a conceptual model for ecodistricts and smart grids, following ASPECS[3] guideline. ASPECS stands for "Agent-oriented Software Process for Engineering Complex Systems", so it is proposed to address the problems of modeling complex systems using agent based approach. ASPECS is based on a holonic organizational metamodel and provides a step-by-step guide from requirements to code allowing the modeling of a system at different levels of details using a set of refinement methods [11]. The life cycle of ASPECS consists in three phases: system requirement analysis, agent society design and implementation and deployment. Each phase is again divided into several steps, and work products are presented using UML, sometimes with extensions or modifications. In the rest of this section, we will discuss those three phases with an example. The model can not be discussed and presented in length, but it will provide a base for agent based ecodistrict and smart grid model.

4.1   System Requirement Analysis

The System Requirements phase aims at identifying a hierarchy of organizations, whose global behaviors may fulfill the system requirements under the chosen perspective [11]. This consist in describing domain requirements, problem ontology, organization, identifying capacities, roles, scenarios and interactions. The products of those steps are presented with UML diagrams.

Let's consider an example of modeling an ecodistrict with smart grid elements. After consulting experts and literature study, the requirements of the model has become clear, and it is described in the figure 5 From the figure, we can see that this model should be able to simulate mainly electricity usage and distribution, and it should take into account the environmental factors. The electricity usage and distribution is then divided into small parts like district electricity usage, railway station usage, distributed generation, and electricity storage, etc. It is also noted that the system should be able to save and visualize the results.

In figure 6, part of the possible ontology diagram is shown. The concepts and actions are represented with UML class, and their relationships are depicted with links. In the figure, concepts surrounding electric vehicle (EV) is shown. For example, EV belongs to "transport concept", and associated with "street" and "EV park". While an EV moves on a street, it consumes electricity. On the other hand, EV demands electricity from EV park, and EV park provides EV with electricity.

Continuing those diagrams, we may identify organizations, capacities, roles; describe scenarios and interactions between roles. Those steps help us clarify our own vision of the problem at hand, and pave the way to designing agents and their interactions in the next phase.

[3] http://www.aspecs.org

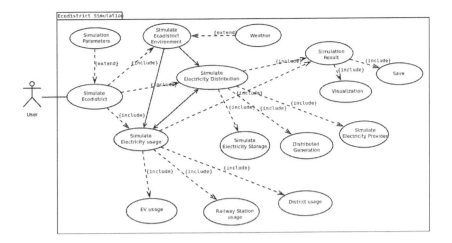

Fig. 5. Domain requirement description

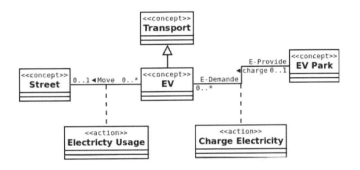

Fig. 6. Problem ontology description

4.2   Agent Society Design

The second phase is the Agent Society Design phase which aims at designing a society of agents whose global behaviors is able to provide an effective
solution to the problem described in the previous phase and to satisfy associated requirements. This phase is further divided into several steps as identifying
agents and their contents, describing role behaviors, communication protocols
and agent plans, etc. Again, the results of those products are presented with UML
diagrams.

In figure 7, agent identification diagram is shown. There are eight agents are
shown with their goals. For example, railway station agent can do three things:
providing electricity usage flexibility, demand electricity, and electricity usage.
According to the diagram, we can further define their inner architectures, their
interactions, and algorithms. At the end of this phase, we will have a complete
design of the model and will be ready to implement and deploy it.

316   M. Ahat, S. Ben Amor, and A. Bui

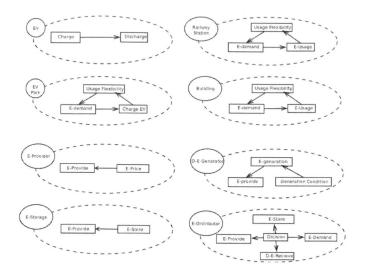

Fig. 7. Agent Identification

4.3   Implementation and Deployment

The third and last phase, namely Implementation and Deployment firstly aims
at implementing the agent-oriented solution designed in the previous phase by
deploying it to the chosen implementation platform. Secondly, it aims at detailing
how to deploy the application over various computational nodes. This phase also
depends on the platform of choice in which we will implement the model and on
which we will deploy it. Nevertheless, the various steps in this phase, provides
the designers and developers with thorough guideline to develop, test and deploy.
There are different multiagent platforms and designing tools available, some are
open source, some are propriety tools. Here we list two platforms, one is ASPECS
compatible, and another one is beginner friendly.

The first choice of development platform of ASPECS is *Janus*[4]. It is a rather
new multiagent platform, it supports the concepts discussed in ASPECS such
as roles, capacities and organizations. In addition, it natively manages the holon
concept to deal with hierarchical complex systems. It is open source and implemented in java 1.6. Only constraint of this platform is the lack of user interface
and visualization design toolkits. Another interesting platform is *Repast*[5], which
is also java based and open source platform. It is academically widely used,
and easy to design data visualization and graphical user interfaces. Thanks to
its modularity, it is easy to integrate other visualization tools, data analysis
libraries.

---

[4] http://www.janus-project.org
[5] http://repast.sourceforge.net

5   Conclusion and Perspectives

In this paper, we discussed the fundamentals for modeling smart grid enabled ecodistrict. After introducing the basic concepts of smart grid and ecodistrict, we discussed the theoretical approaches for modeling and simulation of ecodistricts and smart grids. Due to the complex system nature of ecodistrict and smart grid, we proposed agent based modeling approach. Besides, we discussed complex networks and game theory and their applicability in dealing with certain problems smart grid modeling. At the end, we discussed an example model of smart grid in an ecodistrict following ASPECS methodology. Even if some details are omitted from the example, it demonstrates how to design a model by thinking in agents, and it can be starting point for a smart grid or an ecodistrict model.

This work paves way to developing a real model of a smart grid or an ecodistrict. Here we list three possible continuation:

– Smart grid at a district level. This should permit to study possible smart grid implementation at a district level and its possible outcomes.
– Electirc vehicle to grid model [15]. This allows to study how to integrate electric vehicles into a smart grid and what are the benefits of integration.
– An ecodistrict model. The example can be extended by adding heating network and waste treatment network to model an ecodistrict. This will help decision making in planning and designing an ecodistrict.

References

1. Albert, R., Barabási, A.L.: Statistical mechanics complex networks. Reviews of Modern Physics 74, 47–97 (2002)
2. Amin, M., Stringer, J.: The electric power grid: Today and tomorrow. MRS Bulletin 33(4), 399–407 (2008)
3. Amin, M., Wollenberg, B.F.: Toward a smart grid: power delivery for the 21st century. IEEE Power and Energy Magazine 3(5), 34–41 (2005)
4. Ankersjö, P., Söderholm, G.: The swedish experience. In: Second Eco District Summit, Portland, Oregon (2011)
5. Aumann, R.J., Hart, S.: Handbook of Game Theory, vol. 1. Elsevier Science Publishers, North-Holland (1992)
6. Beaudoin, F., Henry, B., Henry, S., Lancaster, R., Miller, A.: Opportunities for the smart grid to integrate with and support ecodistricts and district energy system. In: Second Annual Interactive Conference on Planning the Smart Grid for Sustainable Communities, Portland, Oregon, US (2010)
7. Beceiro, J.: Austin's pecan street project. In: Second Annual Interactive Conference on Planning the Smart Grid for Sustainable Communities, Portland, Oregon, US (2010)
8. Boccara, N.: Modeling Complex Systems. Springer, New York (2004)
9. Borshchev, A., Filippov, A.: From system dynamics and discrete event to practical agent based modeling: Reasons, techniques, tools. In: The 22nd International Conference of System Dynamics, Oxford, England (2004)

10. Cole, N., Collins, P., Cowden, K., Gifford, C., Heinicke, S., Littlejohn, J.: Lessons from portland's ecodistrict pilots. In: Second Eco District Summit, Portland, Oregon, US (2011)
11. Cossentino, M., Gaud, N., Hilaire, V., Galland, S., Koukam, A.: Aspecs: an agent-oriented software process for engineering complex systems. Autonomous Agents and Multi-Agent Systems 20(2), 260–304 (2012)
12. Erdös, P., Rényi, A.: On random graphs. Publicationes Mathematicae (Debrecen) 6, 290–297 (1959)
13. Pagani, G.A., Aiello, M.: The power grid as a complex network: a survey. arXiv:1105.3338 (2011)
14. Guérard, G., Ben Amor, S., Bui, A.: Approche système complexe pour la modélisation des smart grids. In: 13th Roadef Conference, Angers, France, pp. 263–264 (2012)
15. Hermans, Y., Le Cun, B., Bui, A.: Individual decisions & schedule planner in a vehicle-to-grid context. In: 2012 IEEE International Electric Vehicle Conference, Greenville, US, pp. 1–6 (2012)
16. Keyhani, A.: Chapter 1: Smart Power Grids, pp. 1–25, Smart Power Grids 2011. Springer-Verlag, Berlin Heidelberg (2012)
17. National Energy Technology Laboratory: A vision for the modern grid v2.0. Technical report for the U.S. Department of Energy Office of Electricity Delivery and Energy Reliability (2009)
18. Macal, C.M., North, M.J.: Agent-based modeling and simulation: desktop abms. In: 39th Conference of Winter Simulation, WSC, Washington D.C., US, pp. 95–106 (2007)
19. Macal, C.M., North, M.J.: Tutorial on agent-based modeling and simulation. Journal of Simulation 4, 151–162 (2010)
20. United Nations: Report of the world commission on environment and development (1987)
21. Osborne, M.J., Rubinstein, A.: A Course in Game Theory. The MIT Press, Cambridge (1994)
22. Petermann, C., Ben Amor, S., Bui, A.: Approches théoriques pour la modélisation efficace de smart grid. In: 13th Roadef Conference, Angers, France, pp. 460–461 (2012)
23. Pham, T.L., Bui, M., Lamure, M.: Overview of game theory and using to model the knowledge of multi-agent system. In: RIVF 2003, Hanoi, Vietnam, pp. 99–103 (2003)
24. Robinson, S.: Simulation: The Practice of Model Development and Use. John Wiley & Sons, Chichester (2003)
25. Saad, W., Han, Z., Poor, H.V., Basar, T.: Game theoretic methods for the smart grid. CoRR, abs/1202.0452 (2012)
26. Sannino, A.: The role of power electronics in smart grids and renewable integration. Keynote Speech at Conference PCIM 2011, Nuremberg, Germany (2011)
27. A. Schurr, A.: What is the smart grid and how can it support a more sustainable society? Second Annual Interactive Conference on Planning the Smart Grid for Sustainable Communities, Portland, Oregon, US (2010)
28. Solé, R.V., Rosas-Casals, M., Corominas-Murtra, B., Valverde, S.: Robustness of the european power grids under intentional attack. Phys.Rev.E, 77(2) (2008)
29. Portland sustainability institute: The ecodistricts framework, version 1.2 (2012), http://www.pdxinstitute.org/
30. Yu, X.: Smart grids: A complex network view. Keynote Speech at Conference IECON 2011, Melbourne, Austrailia (2011)

Cloud-Based Data Warehousing Application Framework for Modeling Global and Regional Data Management Systems

Thanh Binh Nguyen

International Institute for Applied Systems Analysis (IIASA), Schlossplatz 1, A-2361
Laxenburg, Austria
nguyenb@iiasa.ac.at
http://www.iiasa.ac.at

Abstract. In this paper, a Cloud-based Data Warehousing Application
(CDWA) Framework is proposed to handle multi levels in structures
of data management systems, i.e. global data warehouse and its data
marts on the clouds. First, a Cloud-based Multidimensional Model, i.e.
dimensions and theirs related concepts, variables or facts, data cubes,
is specified in a very formal manner. Afterwards, to fulfill global and
regional specific requirements of data management systems, the Cloud-
based Data Warehousing Application Framework and its classes of services are modeled by using UML (Unified Modeling Language) to design
a centralized global data warehouse and its local regional data marts. In
this context, an implementation example is presented in order to proof
of our conceptual framework.

Keywords: Data Warehousing (DHW), OLAP, CDWA Framework, Multidimensional Data Model, UML, ETL.

1 Introduction

Multidimensional data analysis, as supported by OLAP (Online Analytical Processing), requires the computation of many aggregate functions over a large
volume of historically collected data [15]. In this context, as a collection of data
from multiple sources, integrated into a common repository and extended by
summary information, data warehousing (DWH) workloads usually consist of
a class of queries typically interleaved with group-by and aggregation OLAP
operators [10].

There arise a few data cubing architecture variations which make Web reporting and data analysis deployment scenarios an integral part of workgroup
collaboration [15], i.e. cubes can be dynamically generated, using parameters
specific to the user when the request is submitted. However, user requirements
and constraints frequently change over time and ever-larger volumes of data and
data modification, which may create a new level of complexity [16], maintain-
ing multiple copies of the same data across multiple cubes for different kinds
of business requirements. In order to achieve a truly open data place, we need

N.T. Nguyen, T. Van Do, and H.A. Le Thi (Eds.): *ICCSAMA 2013*, SCI 479, pp. 319–327.
DOI: 10.1007/978-3-319-00293-4_24   © Springer International Publishing Switzerland 2013

standardized and robust descriptions of data cubing services [16], making easier the processes of configuration, implementation and administration of data cubes in heterogeneous environments with various deployment support.

To solve the above challenge, cloud computing, which is a new computing paradigm to provide reliable, customized and QoS guaranteed dynamic computing environments for end-users [3,6], is considered as an option. Cloud computing delivers infrastructure, platform, and software (application) as services, which are made available as subscription-based services in a pay-as-you-go model to consumers. These services in industry are respectively referred to as Infrastructure as a Service (IaaS), Platform as a Service (PaaS), and Software as a Service (SaaS) [12]. According to [3], cloud computing, the long-held dream of computing as a utility, has the potential to transform a large part of the IT industry, making software even more attractive as a service.

In this paper, to handle multi levels in structures of data management systems, i.e. global data warehouse and its data marts on the clouds, a cloud-based data warehousing application framework, namely CDWA, is proposed. First a multi-dimensional data model is formulated in a very formal manner in the context of cloud-based data warehousing systems. Afterwards, to fulfill global and regional specific requirements of a class of data management systems, the Cloud-based Data Warehousing Application Framework and its classes of services are modeled by using UML (Unified Modeling Language) to design a centralized global data warehouse and its local regional data marts. In this context, an implementation example is presented in order to proof of our conceptual framework.

The rest of this paper is organized as follows: section 2 introduces some approaches and projects related to our work; in section 3, an introduction of cloud-based multidimensional data models and CDWA framework concepts will be presented. Section 4 will show our implementation results in term of typical case studies. At last, section 5 gives a summary of what have been achieved and future works.

2   Related Work

Our approach is rooted in several areas of cloud intelligence research, including the trends and concepts, the combined use of cloud computing and data warehousing technologies in supporting cloud intelligent systems. With the amount of data generated on the clouds increasing continuously, delivering the right and sufficient amount of information at the right time to the right business users has become more complicated and critical [13,14].

Cloud computing has developed quickly, with companies such as Amazon, Google and Salesforce.com getting ahead of the information technology infrastructure stalwarts such as HP, IBM, Microsoft, Dell, EMC, Sun and Oracle [3,12]. The latter are certainly participating, but doing so more behind the scenes notwithstanding some high profile press releases. More and more enterprise solutions and platforms for Business Intelligence have been developed such as IBM DB2 with Business Intelligence Tools, Microsoft SQL Server, Teradata Warehouse, SAS, iData Analyzer, Oracle, Cognos, Business Objects, etc.[3,4,11,13].

These applications and systems are aimed to empower businesses by providing direct access to information used to make decisions, create more effective plans and respond more quickly to problems and opportunities [14]. Thus, this approach effectively and efficiently leverages the data resources to satisfy their requirements for analysis, reporting and decision making process.

In the context of the GAINS model [2,1,6], the model covers a wider range of pollutants, and it also includes structural changes in the energy systems such as energy efficiency improvements and fuel substitution as means for emission reductions. It models the impacts of emission control measures on multiple pollutants (co-control) for a wide range of mitigation options in the energy, industrial, waste and agricultural sectors.

This paper focuses on integrating cloud computing and data warehousing technologies to facilitate the larger reusability of the CDWA framework. Specifically, we also introduce the modeling of conceptual architectural artifacts and their relationships, which formalization enables the design reusability and consistent development of GAINS systems.

3   Cloud-Based Multi Dimensional Data Model

On the basis of mathematical modeling, Cloud-based Multi Dimensional Data Model is formally defined, with the objectives of setting design alternatives in the context of Multi Dimensional Data Model previously described in [7]. The aim of this conceptual model is to provide an extension of the standard data model used in the literature by cloud-based modeling aspects, and to connect the defined model with GAINS-specific mathematical models in the implementation result section. The related concepts of: dimensions, i.e. dimension hierarchical domains, levels, schema; facts; and data cubes could be found in [7].

3.1   Conceptual Model

First we formulate a Cloud-based Multi Dimensional Data Model as follows:

$$G = \langle D, L, V, C \rangle \tag{1}$$

where:

- $D = \{D_1, .., D_n\}$ is a set of dimensions.
- $L = \{L_1, .., L_m\}$ is a set of dimension levels.
- $V = \{V_1, .., V_l\}$ is a set of variables or facts.
- $C = \{C_1, .., C_k\}$ is a set of data cubes.

Furthermore, a dimension D_i is a tuple $< DS_i, DM_i >$, where:

- $DS_i = \left\{ L_1^i \prec L_2^i.. \right\} \subset L$ is a dimension schema
- DM_i is a hierarchical domain of dimension D_i [7].

In this context, each data cube C^j could be seen as follows:

- C_j is a tuple $< CS_j, CM_j >$
- $CS_j = \{DS_1, .., DS_{\acute{n}}, V_1, ..V_{\acute{f}}\}$ is a data cube schema.
- CM_j is a hierarchical domain of data cube C_j [7].

The above specification allows us to define a set of cloud-based data marts, denoted by M, as follows:

$$\forall M^x \in M, M^x = \langle D^x, L^x, V^x, C^x \rangle \qquad (2)$$

3.2   Cloud-Based Data Mart Services

In this section, we define two main services, namely MSchema and MData. The two services will be applied at the global data warehouse to map from global schema to a specific data mart schema as well as to generate data cubes of the data marts based on the global data cube set. We formulate the two services as follows:

- $MSchema(M^x) = < D^x, L^x, V^x >$ where:
  - $D^x \subset D, L^x \subset L, V^x \subset V$.
  - Furthermore, $\forall D_i^x \in D^x, \exists D_i \in D$, and $DS_i^x = DS_i \cap L^x$.
- $MData(M^x) = C^x \subset C$.

3.3   Cloud-Based Data Warehousing Application Framework

In this section, a Cloud-based Data Warehousing Application Framework is introduced to specifying based components, including Global Multidimensional Schema and Database, Data Mart and Linked services. The framework is modeled by using UML as shown in figure 2. In this paper, we would like to focus in the global data warehouse and its specified data mart services. Specifications about data sources and ETL (extract-transform-load) services could be found in [5].

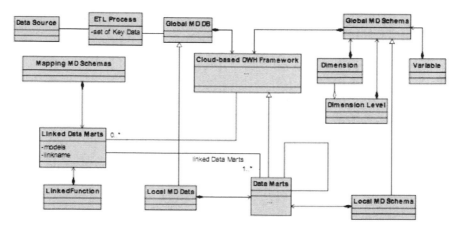

Fig. 1. The Cloud-based Data warehousing Application Framework

3.4 Global Data Warehousing Services

These services are used to model the global data warehouse and its components.

Global Schema Services. These services are used to specify global business multidimensional model, i.e. modeling dimensions, variables or facts. The specification of dimensions and variables has been introduced in [7] and summarized in section 3. In this context, a Global Schema object contains a set of Dimensions, each of which consists a set of Dimension Levels. Furthermore, the Global Schema object still has a set of Variable ones. The detail designs of Dimension and Variable classes could be found in [7].

Global Multidimensional Database Services. This class is used to generate fact tables and their related data cubes at a global data warehouse. Data from Data Sources could be integrated by using ETL Process. Full description about data sources as well as ETL process have been introduced in [8].

Data Mart Services. A data mart is designed as an instance of the Data Mart class. This means a data mart has it own Local Schema, which inherits from the Global Schema class. This allows that the structure of a regional data mart could be specified in this class by using the following steps:

- Local Schema definition. This step is used to specify business model of a local data mart. A subset of dimensions D and a subset of variables V will be assigned to the local data mart. Based on this information, data granularity (according to Dimensional Levels) will be selected.
- Dimension definition. Each dimension of the local data mart will be specified by selecting a subset of levels and by generating dimensional domain of its global dimension.
- Variable definition. A variable domain will be defined as a subset of it global one according to its related dimensions designed in the above step.
- Fact table generation. This step will generate fact table based on its dimensions, variables and its global fact table.
- Data Cube pre-calculated. Data cubes of the data mart will be calculated based on the fact table and regional specific requirements.

Linked Data Marts. A Linked Data Mart object holds information about a related set of data marts. In this context, a child data mart could be defined based on its parent one instead of based on the global data warehouse. This function is very useful for a group of regional data marts, e.g. from national to city levels.

4 GAINS Data Warehousing System: An Implementation Example of CDWF

In this section, the requirements of GAINS [2,9] will be introduced as a specific case study of CDWA. The GAINS model explores cost-effective emission control strategies to reduce greenhouse gases and/or improve local and regional air quality [2,9]. By selecting a smart mix of measures, countries can reduce air pollution control costs as well as cut greenhouse gas emissions. The above figure shows the GAINS cloud data warehouse portal which provides access to a GAINS global data warehouse and a number of GAINS regional data marts including:

- A global data warehouse, namely GAINS DWH is developed by collecting data from available international emission inventories and on national information supplied by individual countries. It is the global data warehouse in our cloud data warehouse application framework and used to configure and generate regional data warehouse(s) as required.
- For European countries where GAINS is used extensively by Member States of the EU and the European Commission to develop cost-effective strategies to reduce the environmental impact of air pollutions.
- For China and Asia, where GAINS DWH is being used to explore sustainable development pathways for the future. It is a typical example of developing Linked Data Mart concept, which is used to generate Chinese data mart from Asia one. Figure 3 shows the emission calculation by country in the Asia data mart, while figure 4 illustrates the Difference in emissions between the baseline and the Air quality scenarios in 2020 of Beijing calculated by Chinese data mart.

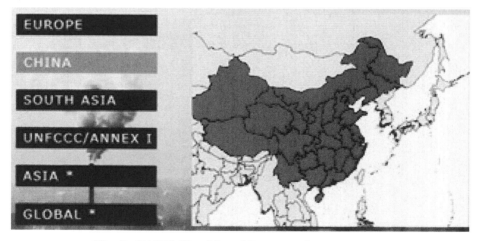

Fig. 2. GAINS Cloud-based Data warehousing systems

Total Emissions of SO2 by Country

Pollutant: SO2
Scenario: 450_WEO_2011 (ID: 450_WEO_2011)
Unit: [kt SO2/year]
User: ntbinh

Display table in an export format

Country	Abbr.	1990	1995	2000	2005	2010	2015	2020	2025	2030	2035	2040	2045	2050
CHINA	CHIN	15850.95	20225.35	23009.29	31383.81	35217.53	34214.92	29812.19	23903.18	19531.79	17995.25	4324.70	4324.70	4324.70
JAPAN	JAPA	943.60	787.92	679.14	754.77	578.03	548.91	518.70	495.95	477.91	463.12	86.59	86.59	86.59
Sum	Sum	16794.55	21013.27	23688.42	32138.58	35795.56	34763.83	30330.89	24399.13	20009.70	18458.37	4411.29	4411.29	4411.29

Fig. 3. Total Emissions of SO2 by Japan and China

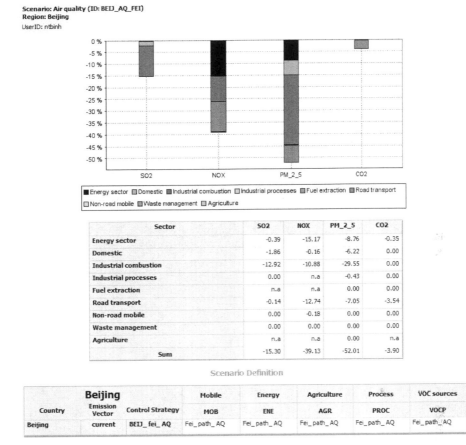

Co-benefits: Difference in emissions between the baseline and the reference scenarios in 2020

Scenario: Air quality (ID: BEIJ_AQ_FEI)
Region: Beijing
UserID: ntbinh

Sector	SO2	NOX	PM_2_5	CO2
Energy sector	-0.39	-15.17	-8.76	-0.35
Domestic	-1.86	-0.16	-6.22	0.00
Industrial combustion	-12.92	-10.88	-29.55	0.00
Industrial processes	0.00	n.a	-0.43	0.00
Fuel extraction	n.a	n.a	0.00	0.00
Road transport	-0.14	-12.74	-7.05	-3.54
Non-road mobile	0.00	-0.18	0.00	0.00
Waste management	0.00	0.00	0.00	0.00
Agriculture	n.a	n.a	0.00	n.a
Sum	-15.30	-39.13	-52.01	-3.90

Scenario Definition

	Beijing		Mobile	Energy	Agriculture	Process	VOC sources
Country	Emission Vector	Control Strategy	MOB	ENE	AGR	PROC	VOCP
Beijing	current	BEIJ_fei_AQ	Fei_path_AQ	Fei_path_AQ	Fei_path_AQ	Fei_path_AQ	Fei_path_AQ

Fig. 4. Difference in emissions between the baseline and the Air quality scenarios in 2020 of Beijing calculated by Chinese data mart

5   Conclusions and Future Works

In this paper, the CDWA framework with its concepts, i.e. cloud-based multi-dimensional data models, data marts, multi linked data marts with a variety of different domain-specific methodologies, has been introduced to support the experts for better business decisions. Moreover, the case studies of GAINS model, which has been introduced as an instance of just-specified framework, are presented as a suite to make available and to compare the implications of the outputs of different system models working at various spatial and temporal scales.

In the near future, the pursuit of semantic technologies will be used to enhance the efficiency and agility of CDWA solution, i.e. representation of data combination and constrains. Moreover, mathematical optimization and data mining algorithms will be adapted for multidimensional analysis of integrated data from heterogeneous sources in university environment.

References

1. Amann, M., Bertok, I., Borken-Kleefeld, J., Cofala, J., Heyes, C., Hoeglund Isaksson, L., Klimont, Z., Nguyen, B., Posch, M., Rafaj, P., et al.: Cost-eective control of air quality and greenhouse gases in europe: modeling and policy applications. Environmental Modelling & Software (2011)
2. Amann, M., Bertok, I., Borken, J., Cofala, J., Heyes, C., Hoglund, L., Klimont, Z., Purohit, P., Rafaj, P., Schoepp, W., Toth, G., Wagner, F., Winiwarter, W.: GAINS - potentials and costs for greenhouse gas mitigation in annex i countries. Technical report, International Institute for Applied Systems Analysis (IIASA), Laxenburg, Austria (2008)
3. Armbrust, M., Fox, A., Grffith, R., Joseph, A., Katz, R., Konwinski, A., Lee, G., Patterson, D., Rabkin, A., Stoica, I., et al.: Above the clouds: A berkeley view of cloud computing. Technical Report UCB/EECS-2009-28, EECS Department, University of California, Berkeley (2009)
4. Gangadharan, G., Swami, S.: Business intelligence systems: design and implementation strategies. In: 26th International Conference on Information Technology Interfaces, pp. 139–144 (2004)
5. Hoang, T.A.D., Nguyen, T.B.: State of the art and emerging rule-driven perspectives towards service-based business process interoperability, pp. 1–4. IEEE (2009)
6. Nguyen, T.B., Wagner, F., Schoepp, W.: Cloud intelligent services for calculating emissions and costs of air pollutants and greenhouse gases. In: Nguyen, N.T., Kim, C.-G., Janiak, A. (eds.) ACIIDS 2011, Part I. LNCS, vol. 6591, pp. 159–168. Springer, Heidelberg (2011)
7. Nguyen, T.B., Tjoa, A.M., Wagner, R.: Conceptual multidimensional data model based on MetaCube. In: Yakhno, T. (ed.) ADVIS 2000. LNCS, vol. 1909, pp. 24–33. Springer, Heidelberg (2000)
8. Nguyen, T.B., Ibrahim, M.T.: Metadata integration framework for managing forest heterogeneous information resources. In: DEXA Workshops, pp. 586–591 (2004)
9. Nguyen, T.B., Wagner, F., Schoepp, W.: GAINS – an interactive tool for assessing international GHG mitigation regimes. In: Kranzlmüller, D., Toja, A.M. (eds.) ICT-GLOW 2011. LNCS, vol. 6868, pp. 124–135. Springer, Heidelberg (2011)
10. Ponniah, P.: Data warehousing fundamentals: a comprehensive guide for IT professionals, vol. 1. Wiley-Interscience (2001)
11. Tvrdikova, M.: Support of decision making by business intelligence tools, pp. 364–368. IEEE (June 2007)
12. Wang, L., von Laszewski, G., Kunze, M., Tao, J.: Cloud computing: A perspective study. In: Proceedings of the Grid Computing Environments (GCE) Workshop. Held at the Austin Civic Center, Austin, November 16 (2008), https://ritdml.rit.edu/handle/1850/7821
13. Watson, H.J., Wixom, B.H.: The current state of business intelligence. Computer 40, 96–99 (2007)

14. Wei, X., Xiaofei, X., Lei, S., Quanlong, L., Hao, L.: Business intelligence based group decision support system. In: Proceedings of 2001 International Conferences on Info-tech and Info-net, ICII 2001, Beijing, vol. 5, pp. 295–300 (2001)
15. Insightreport.com: Data Publishing Architecture for the Extended Enterprise. Technical Report (March 2003)
16. IBM Software: Build high-speed, scalable analytics into the data warehouse. Technical Report (May 2010)

Part V

Logic Based Methods for Decision Making and Data Mining

A Tableau Method with Optimal Complexity for Deciding the Description Logic SHIQ

Linh Anh Nguyen[1,2]

[1] Institute of Informatics, University of Warsaw
Banacha 2, 02-097 Warsaw, Poland
nguyen@mimuw.edu.pl
[2] Faculty of Information Technology, College of Technology,
Vietnam National University, 144 Xuan Thuy, Hanoi, Vietnam

Abstract. We present the first tableau method with an EXPTIME (optimal) complexity for checking satisfiability of a knowledge base in the description logic $\mathcal{SHIQ}$. The complexity is measured using binary representation for numbers. Our method is based on global state caching and integer linear feasibility checking.

1 Introduction

Automated reasoning in description logics (DLs) has been an active research area for more than two decades. It is useful, for example, for the Semantic Web in engineering and querying ontologies. One of basic reasoning problems in DLs is to check satisfiability of a knowledge base in a considered DL. Most of other reasoning problems in DLs are reducible to this one. In this chapter we study the problem of checking satisfiability of a knowledge base in the DL $\mathcal{SHIQ}$, which extends the basic DL $\mathcal{ALC}$ with transitive roles, hierarchies of roles, inverse roles and quantified number restrictions.

Tobies in his PhD thesis [7] proved that the problem is EXPTIME-complete (even when numbers are coded in binary). On page 127 of [7] he wrote: *"The previous EXPTIME-completeness results for $\mathcal{SHIQ}$ rely on the highly inefficient automata construction of Definition 4.34 used to prove Theorem 4.38 and, in the case of knowledge base reasoning, also on the wasteful pre-completion technique used to prove Theorem 4.42. Thus, we cannot expect to obtain an implementation from these algorithms that exhibits acceptable runtimes even on relatively "easy" instances. This, of course, is a prerequisite for using SHIQ in real-world applications."*

Together with Horrocks and Sattler, Tobies therefore developed a tableau decision procedure for $\mathcal{SHIQ}$ [3,7]. Their decision procedure has a non-optimal complexity (NEXPTIME) when unary representation is used for numbers, and has a higher complexity (N2EXPTIME) when binary representation is used.

As $\mathcal{SHIQ}$ is a well-known and useful DL, developing a complexity-optimal tableau decision procedure for checking satisfiability of a knowledge base in $\mathcal{SHIQ}$ is very desirable. In this chapter we present the first tableau method with

N.T. Nguyen, T. Van Do, and H.A. Le Thi (Eds.): *ICCSAMA 2013*, SCI 479, pp. 331–342.
DOI: 10.1007/978-3-319-00293-4_25   © Springer International Publishing Switzerland 2013

an ExpTime (optimal) complexity for checking satisfiability of a knowledge base in the DL $\mathcal{SHIQ}$. The complexity is measured using binary representation for numbers. Our method is based on global state caching [2,5] and integer linear feasibility checking.

We are aware of only Farsiniamarj's master thesis [1] (written under supervision of Haarslev) as a work that directly combines tableaux with integer programming. Some related works are discussed in that thesis and we refer the reader there for details. In [1] Farsiniamarj presented a hybrid tableau calculus for the DL $\mathcal{SHQ}$ (a restricted version of $\mathcal{SHIQ}$ without inverse roles), which is based on the so-called atomic decomposition technique and combines arithmetic and logical reasoning. He stated that *"The most prominent feature of this hybrid calculus is that it reduces reasoning about qualified number restrictions to integer linear programming. [...] In comparison to other standard description logic reasoners, our approach demonstrates an overall runtime improvement of several orders of magnitude."* [1]. On the complexity matter, Farsiniamarj wrote *"the complexity of the algorithm seems to be characterized by a double-exponential function"* [1, page 79]. That is, his algorithm for $\mathcal{SHQ}$ is not complexity-optimal.

Our method of exploiting integer linear programming is different from Farsiniamarj's one [1]: in order to avoid nondeterminism, we only check feasibility but do not find and use solutions of the considered set of constraints.

Due to the lack of space, in this chapter we restrict ourselves to introducing the problem of checking satisfiability of a knowledge base in the description logic $\mathcal{SHIQ}$ and presenting some examples to illustrate our tableau method. For a full description of a tableau decision procedure with an ExpTime complexity for the problem together with proofs we refer the reader to our manuscript [6].

2   Notation and Semantics of $\mathcal{SHIQ}$

Our language uses a finite set $\mathbf{C}$ of *concept names*, a finite set $\mathbf{R}$ of *role names*, and a finite set $\mathbf{I}$ of *individual names*. We use letters like A and B for concept names, r and s for role names, and a and b for individual names. We refer to A and B also as *atomic concepts*, and to a and b as *individuals*.

For $r \in \mathbf{R}$, let r^- be a new symbol, called the *inverse* of r. Let $\mathbf{R}^- = \{r^- \mid r \in \mathbf{R}\}$ be the set of *inverse roles*. For $r \in \mathbf{R}$, define $(r^-)^- = r$. A *role* is any member of $\mathbf{R} \cup \mathbf{R}^-$. We use letters like R and S to denote roles.

An ($\mathcal{SHIQ}$) *RBox* $\mathcal{R}$ is a finite set of role axioms of the form $R \sqsubseteq S$ or $R \circ R \sqsubseteq R$. We say that R is a *subrole* of S (w.r.t. $\mathcal{R}$) if there are roles $R_0 = R$, $R_1, \ldots, R_{k-1}$, $R_k = S$, where $k \geq 0$, such that, for every $1 \leq i \leq k$, $R_{i-1} \sqsubseteq R_i$ or $R_{i-1}^- \sqsubseteq R_i^-$ is an axiom of $\mathcal{R}$. We say that R is a *transitive role* (w.r.t. $\mathcal{R}$) if $R \circ R \sqsubseteq R$ or $R^- \circ R^- \sqsubseteq R^-$ is an axiom of $\mathcal{R}$. A role is *simple* (w.r.t. $\mathcal{R}$) if it is neither transitive nor has any transitive subroles (w.r.t. $\mathcal{R}$).

Concepts in $\mathcal{SHIQ}$ are formed using the following BNF grammar, where n is a nonnegative integer and S is a simple role:

$$C, D ::= \top \mid \bot \mid A \mid \neg C \mid C \sqcap D \mid C \sqcup D \mid \exists R.C \mid \forall R.C \mid \geq n\,S.C \mid \leq n\,S.C$$

We use letters like C and D to denote arbitrary concepts.

A *TBox* is a finite set of axioms of the form $C \sqsubseteq D$ or $C \doteq D$.

An *ABox* is a finite set of *assertions* of the form $a\!:\!C$, $R(a,b)$ or $a \not\approx b$.

A *formula* is defined to be either a concept or an ABox assertion.

A *knowledge base* in $\mathcal{SHIQ}$ is a tuple $\langle \mathcal{R}, \mathcal{T}, \mathcal{A} \rangle$, where $\mathcal{R}$ is an RBox, $\mathcal{T}$ is a TBox and $\mathcal{A}$ is an ABox.

We say that a role S is *numeric* w.r.t. a knowledge base $KB = \langle \mathcal{R}, \mathcal{T}, \mathcal{A} \rangle$ if: it is simple w.r.t. $\mathcal{R}$ and occurs in a concept $\geq n\,S.C$ or $\leq n\,S.C$ in KB; or $S \sqsubseteq_\mathcal{R} R$ and R is numeric w.r.t. KB; or S^- is numeric w.r.t. KB. We will simply call such an S a numeric role when KB is clear from the context.

An *interpretation* $\mathcal{I} = \langle \Delta^\mathcal{I}, \cdot^\mathcal{I} \rangle$ consists of a non-empty set $\Delta^\mathcal{I}$, called the *domain* of $\mathcal{I}$, and a function $\cdot^\mathcal{I}$, called the *interpretation function* of $\mathcal{I}$, that maps each concept name A to a subset $A^\mathcal{I}$ of $\Delta^\mathcal{I}$, each role name r to a binary relation $r^\mathcal{I}$ on $\Delta^\mathcal{I}$, and each individual name a to an element $a^\mathcal{I} \in \Delta^\mathcal{I}$. The interpretation function is extended to inverse roles and complex concepts as follows, where $\sharp Z$ denotes the cardinality of a set Z:

$$(r^-)^\mathcal{I} = \{\langle x, y \rangle \mid \langle y, x \rangle \in r^\mathcal{I}\} \qquad \top^\mathcal{I} = \Delta^\mathcal{I} \qquad \bot^\mathcal{I} = \emptyset$$
$$(\neg C)^\mathcal{I} = \Delta^\mathcal{I} - C^\mathcal{I} \qquad (C \sqcap D)^\mathcal{I} = C^\mathcal{I} \cap D^\mathcal{I} \qquad (C \sqcup D)^\mathcal{I} = C^\mathcal{I} \cup D^\mathcal{I}$$
$$(\exists R.C)^\mathcal{I} = \{x \in \Delta^\mathcal{I} \mid \exists y [\langle x, y \rangle \in R^\mathcal{I} \text{ and } y \in C^\mathcal{I}]\}$$
$$(\forall R.C)^\mathcal{I} = \{x \in \Delta^\mathcal{I} \mid \forall y [\langle x, y \rangle \in R^\mathcal{I} \text{ implies } y \in C^\mathcal{I}]\}$$
$$(\geq n\,R.C)^\mathcal{I} = \{x \in \Delta^\mathcal{I} \mid \sharp\{y \mid \langle x, y \rangle \in R^\mathcal{I} \text{ and } y \in C^\mathcal{I}\} \geq n\}$$
$$(\leq n\,R.C)^\mathcal{I} = \{x \in \Delta^\mathcal{I} \mid \sharp\{y \mid \langle x, y \rangle \in R^\mathcal{I} \text{ and } y \in C^\mathcal{I}\} \leq n\}.$$

Note that $(r^-)^\mathcal{I} = (r^\mathcal{I})^{-1}$ and this is compatible with $(r^-)^- = r$.

The relational composition of binary relations R_1, R_2 is denoted by $R_1 \circ R_2$.

An interpretation $\mathcal{I}$ is a *model of an RBox* $\mathcal{R}$ if for every axiom $R \sqsubseteq S$ (resp. $R \circ R \sqsubseteq R$) of $\mathcal{R}$, we have that $R^\mathcal{I} \subseteq S^\mathcal{I}$ (resp. $R^\mathcal{I} \circ R^\mathcal{I} \subseteq R^\mathcal{I}$).

An interpretation $\mathcal{I}$ is a *model of a TBox* $\mathcal{T}$ if for every axiom $C \sqsubseteq D$ (resp. $C \doteq D$) of $\mathcal{T}$, we have that $C^\mathcal{I} \subseteq D^\mathcal{I}$ (resp. $C^\mathcal{I} = D^\mathcal{I}$).

An interpretation $\mathcal{I}$ is a *model of an ABox* $\mathcal{A}$ if for every assertion $a\!:\!C$ (resp. $R(a,b)$ or $a \not\approx b$) of $\mathcal{A}$, we have that $a^\mathcal{I} \in C^\mathcal{I}$ (resp. $\langle a^\mathcal{I}, b^\mathcal{I} \rangle \in R^\mathcal{I}$ or $a^\mathcal{I} \neq b^\mathcal{I}$).

An interpretation $\mathcal{I}$ is a *model of a knowledge base* $\langle \mathcal{R}, \mathcal{T}, \mathcal{A} \rangle$ if $\mathcal{I}$ is a model of all $\mathcal{R}$, $\mathcal{T}$ and $\mathcal{A}$. A knowledge base $\langle \mathcal{R}, \mathcal{T}, \mathcal{A} \rangle$ is *satisfiable* if it has a model.

3   A Tableau Method for $\mathcal{SHIQ}$

We define tableaux as rooted graphs. Such a graph is a tuple $G = \langle V, E, \nu \rangle$, where V is a set of nodes, $E \subseteq V \times V$ is a set of edges, $\nu \in V$ is the root, and each node $v \in V$ has a number of attributes. If there is an edge $\langle v, w \rangle \in E$ then we call v a *predecessor* of w, and call w a *successor* of v. Attributes of tableau nodes are:

- $Type(v) \in \{\text{state}, \text{non-state}\}$.
- $SType(v) \in \{\text{complex}, \text{simple}\}$ is called the subtype of v.
- $Status(v) \in \{\text{unexpanded}, \text{p-expanded}, \text{f-expanded}, \text{incomplete}, \text{closed}, \text{open}\}$, where p-expanded (resp. f-expanded) means partially (resp. fully) expanded.
- $Label(v)$ is a finite set of formulas. The label of a complex node consists of ABox assertions, while the label of a simple node consists of concepts.
- $RFmls(v)$ is a finite set of formulas, called the set of reduced formulas of v.
- $StatePred(v) \in V \cup \{\text{null}\}$ is called the state-predecessor of v. It is available only when $Type(v) = \text{non-state}$. If v is a non-state and G has no paths connecting a state to v then $StatePred(v) = \text{null}$. Otherwise, G has exactly one state u that is connected to v via a path not containing any other states, and we have that $StatePred(v) = u$.
- $ATPred(v) \in V$ is called the after-transition-predecessor of v. It is available only when $Type(v) = \text{non-state}$ and $SType(v) = \text{simple}$. In that case, if $v_0 = StatePred(v) \, (\neq \text{null})$ then there is exactly one successor v_1 of v_0 such that every path connecting v_0 to v must go through v_1, and we have that $ATPred(v) = v_1$. We define $\texttt{AfterTrans}(v) = (ATPred(v) = v)$. If $\texttt{AfterTrans}(v)$ holds then v has exactly one predecessor u and u is a state.
- $CELabel(v)$ is a tuple $\langle CELabelT(v), CELabelR(v), CELabelI(v) \rangle$ called the coming edge label of v. It is available only when $\texttt{AfterTrans}(v)$ holds.
  - $CELabelT(v) \in \{\text{testingClosedness}, \text{checkingFeasibility}\}$,
  - $CELabelR(v) \subseteq \mathbf{R} \cup \mathbf{R}^-$ (is a set of roles),
  - $CELabelI(v) \in \mathbf{I} \cup \{\text{null}\}$.
- $FmlsRC(v)$ is a set of formulas, called the set of formulas required by converse (inverse) for v. It is available only when $Type(v) = \text{state}$ and $Status(v) = \text{incomplete}$.
- $FmlFB(v)$ is a formula, called the formula for branching on at v. It is available only when $Type(v) = \text{state}$ and $Status(v) = \text{incomplete}$.
- $ILConstraints(v)$ is a set of integer linear constraints. It is available only when $Type(v) = \text{state}$. The constraints use variables x_w indexed by successors w of v with $CELabelT(w) = \text{checkingFeasibility}$. Such a variable x_w specifies how many copies of w will be used as successors of v.

We will use also new concept constructors $\preceq n\,R.C$ and $\succeq n\,R.C$. The difference between $\preceq n\,R.C$ and $\leq n\,R.C$ is that, for checking $\preceq n\,R.C$, we do not have to look to predecessors of the node. The aim for $\succeq n\,R.C$ is similar.

By the *local graph* of a state v we mean the subgraph of G consisting of all the paths starting from v and not containing any other states. Similarly, by the local graph of a non-state v we mean the subgraph of G consisting of all the paths starting from v and not containing any states.

We give here an intuition behind the structure of G:

- If u is a state of G and $v_1, \ldots, v_k$ are all the successors of u then:
  - the local graph of each v_i is a directed acyclic graph,
  - the local graphs of v_i and v_j are disjoint if $i \neq j$,
  - the local graph of u is a graph rooted at u and consisting of the edges from u to $v_1, \ldots, v_k$ and the local graphs of $v_1, \ldots, v_k$,

- if w is a node in the local graph of some v_i then $StatePred(w) = u$ and $ATPred(w) = v_i$.
- If u is a state of G then:
  - each edge from outside to inside of the local graph of u must end at u,
  - each edge outgoing from the local graph of u must start from a non-state and is the only outgoing edge of that non-state.
- The root ν of G is a complex non-state. G consists of: the local graph of the root ν, the local graphs of states, and edges coming to states.
- Each complex node of G is like an ABox (its label is an ABox), which can be treated as a graph whose vertices are named individuals. On the other hand, a simple node of G stands for an unnamed individual. If there is an edge from a simple non-state v to (a simple node) w then v and w stand for the same unnamed individual. An edge from a complex node v to a simple node w with $CELabelI(w) = a$ can be treated as an edge from the named individual a (an inner node of the graph representing v) to the unnamed individual corresponding to w, and that edge is via the roles from $CELabelR(w)$.
- G consists of two layers: the layer of complex nodes and the layer of simple nodes. The former layer consists of the local graph of the root ν together with a number of complex states and edges coming to them. The edges from the layer of complex nodes to the layer of simple nodes are exactly the edges outgoing from those complex states. There are no edges from the layer of simple nodes to the layer of complex nodes.

We apply global state caching: if v_1 and v_2 are different states then $Label(v_1) \neq Label(v_2)$. If v is a non-state such that `AfterTrans(v)` holds then we also apply global caching for the local graph of v: if w_1 and w_2 are different nodes of the local graph of v then $Label(w_1) \neq Label(w_2)$.

Assume that concepts are represented in negation normal form (NNF), where $\neg$ occurs only directly before atomic concepts. For simplicity, we treat axioms of the TBox $\mathcal{T}$ as concepts representing global assumptions. That is, we assume that $\mathcal{T}$ consists of concepts in NNF. Thus, an interpretation $\mathcal{I}$ is a model of $\mathcal{T}$ iff $\mathcal{I}$ validates every concept $C \in \mathcal{T}$.

Let $\langle \mathcal{R}, \mathcal{T}, \mathcal{A} \rangle$ be a knowledge base in NNF of the logic $\mathcal{SHIQ}$, with $\mathcal{A} \neq \emptyset$. We now present our method for checking satisfiability of $\langle \mathcal{R}, \mathcal{T}, \mathcal{A} \rangle$. A $C_{\mathcal{SHIQ}}$-tableau for $\langle \mathcal{R}, \mathcal{T}, \mathcal{A} \rangle$ is a rooted graph $G = \langle V, E, \nu \rangle$ constructed as follows:

Initialization: At the beginning, $V = \{\nu\}$, $E = \emptyset$, and ν is the root with $Type(\nu) = $ non-state, $SType(\nu) = $ complex, $Status(\nu) = $ unexpanded, $Label(\nu) = \mathcal{A} \cup \{(a{:}C) \mid C \in \mathcal{T}$ and a is an individual occurring in $\mathcal{A}\}$, $StatePred(\nu) = $ null, $RFmls(\nu) = \emptyset$.

Rules' Priorities and Expansion Strategies: The graph is then expanded by the following rules, which are specified in detail in [6]:

- (UPS): rules for updating statuses of nodes,
- (US): a unary static expansion rule,
- (KCC): rules for keeping converse compatibility,
- (NUS): a non-unary static expansion rule,

- (FS): forming-state rules,
- (TP): a transitional partial-expansion rule,
- (TF): a transitional full-expansion rule.

The transitional partial-expansion rule deals with "transitions" via non-numeric roles, while the transitional full-expansion rule deals with "transitions" via numeric roles.

Each of the rules is parametrized by a node v. We say that a rule is *applicable* to v if it can be applied to v to make changes to the graph. The rules (UPS), (US), (KCC) (in the first three items of the above list) have highest priorities, and are ordered decreasingly by priority. If none of them is applicable to any node, then choose a node v with status unexpanded or p-expanded, choose the first rule applicable to v among the rules in the last four items of the above list, and apply it to v. Any strategy can be used for choosing v, but it is worth to choose v for expansion only when v could affect the status of the root ν of the graph, i.e., only when there may exist a path from ν to v without any node of status incomplete, closed or open.

Termination: The construction of the graph ends when the root ν receives status closed or open or when every node that may affect the status of ν (i.e., reachable from ν via a path without nodes with status incomplete, closed or open) has been fully expanded (i.e., has status f-expanded).

To check satisfiability of $\langle \mathcal{R}, \mathcal{T}, \mathcal{A} \rangle$ one can construct a $C_{\mathcal{SHIQ}}$-tableau for it, then return "no" when the root of the tableau has status closed, or "yes" in the other case. In [6] we have proved that:

- A $C_{\mathcal{SHIQ}}$-tableau for $\langle \mathcal{R}, \mathcal{T}, \mathcal{A} \rangle$ can be constructed in ExpTime.
- If $G = \langle V, E, \nu \rangle$ is an arbitrary $C_{\mathcal{SHIQ}}$-tableau for $\langle \mathcal{R}, \mathcal{T}, \mathcal{A} \rangle$ then $\langle \mathcal{R}, \mathcal{T}, \mathcal{A} \rangle$ is satisfiable iff $Status(\nu) \neq$ closed.

4   Illustrative Examples

Example 4.1. This example is based on an example in the long version of our paper [5] on $\mathcal{SHI}$. Let

$$\mathcal{R} = \{r \sqsubseteq s, \quad r^- \sqsubseteq s, \quad s \circ s \sqsubseteq s\}$$
$$\mathcal{T} = \{\exists r.(A \sqcap \forall s. \neg A)\}$$
$$\mathcal{A} = \{a : \top\}.$$

In Figures 1 and 2 we illustrate the construction of a $C_{\mathcal{SHIQ}}$-tableau for the knowledge base $\langle \mathcal{R}, \mathcal{T}, \mathcal{A} \rangle$. At the end the root ν receives status closed, hence $\langle \mathcal{R}, \mathcal{T}, \mathcal{A} \rangle$ is unsatisfiable. As a consequence, $\langle \mathcal{R}, \mathcal{T}, \emptyset \rangle$ is also unsatisfiable.   □

Example 4.2. Let us construct a $C_{\mathcal{SHIQ}}$-tableau for $\langle \mathcal{R}, \mathcal{T}, \mathcal{A} \rangle$, where

$$\mathcal{A} = \{a : \forall r.A_1, a : \leq 3\, r.A_1, a : \exists r.A_2, a : \geq 2\, r.A_3, r(a, b), b : (\neg A_1 \sqcup \neg A_2), b : \neg A_3\},$$

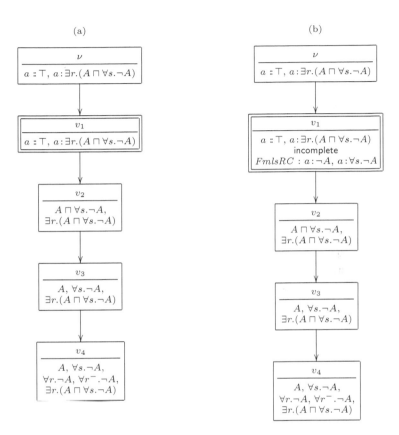

Fig. 1. An illustration for Example 4.1: part I. The graph (a) is the graph constructed until checking converse compatibility for v_1. In each node, we display the formulas of the node's label. The root ν is expanded by the forming-state rule (FS$_2$). The complex state v_1 is expanded by the transitional partial-expansion rule, with $CELabelT(v_2) = \text{testingClosedness}$, $CELabelR(v_2) = \{r, s, s^-\}$ (s and s^- are included because $r \sqsubseteq_{\mathcal{R}} s$ and $r \sqsubseteq_{\mathcal{R}} s^-$) and $CELabelI(v_2) = a$. The simple non-states v_2 and v_3 are expanded by the unary static expansion rule (the concepts $\forall r.\neg A$ and $\forall r^-.\neg A$ are added into $Label(v_4)$ because $\forall s.\neg A \in Label(v_3)$, $r \sqsubseteq_{\mathcal{R}} s$ and $r^- \sqsubseteq_{\mathcal{R}} s$). The node v_1 is the only state. We have, for example, $StatePred(v_4) = v_1$ and $ATPred(v_4) = v_2$. Checking converse compatibility for v_1 using v_4 (i.e., using the facts that $\{\forall s.\neg A, \forall r^-.\neg A\} \subset Label(v_4)$, $StatePred(v_4) = v_1$, $ATPred(v_4) = v_2$, $CELabelT(v_2) = \text{testingClosedness}$, $CELabelR(v_2) = \{r, s, s^-\}$, $CELabelI(v_2) = a$, $r^- \sqsubseteq_{\mathcal{R}} s$ and $trans_{\mathcal{R}}(R)$ holds), $Status(v_1)$ is set to incomplete and $FmlsRC(v_1)$ is set to $\{a : \neg A, a : \forall s.\neg A\}$. This results in the graph (b). The construction is then continued by re-expanding the node ν. See Figure 2 for the continuation.

338   L.A. Nguyen

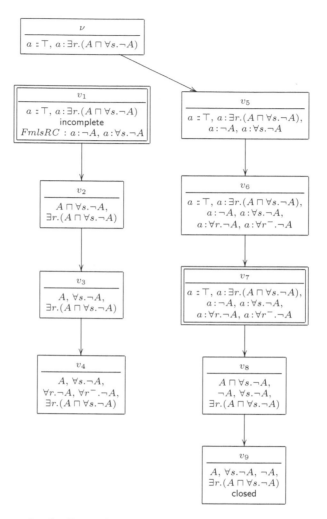

Fig. 2. An illustration for Example 4.1: part II. This is a $C_{\mathcal{SHIQ}}$-tableau for $\langle \mathcal{R}, \mathcal{T}, \mathcal{A} \rangle$. As in the part I, in each node we display the formulas of the node's label. The root ν is re-expanded by deleting the edge $\langle \nu, v_1 \rangle$ and connecting ν to a new complex non-state v_5. The node v_5 is expanded using the unary static expansion rule (the assertions $a : \forall r.\neg A$ and $a : \forall r^-.\neg A$ are added into $Label(v_6)$ because $a : \forall s.\neg A \in Label(v_5)$, $r \sqsubseteq_{\mathcal{R}} s$ and $r^- \sqsubseteq_{\mathcal{R}} s$). The complex non-state v_6 is expanded using the forming-state rule (FS$_2$). The complex state v_7 is expanded using the transitional partial-expansion rule, with $CELabelT(v_8) =$ testingClosedness, $CELabelR(v_8) = \{r, s, s^-\}$ (s and s^- are included because $r \sqsubseteq_{\mathcal{R}} s$ and $r \sqsubseteq_{\mathcal{R}} s^-$) and $CELabelI(v_8) = a$. The simple non-state v_8 is expanded using the unary static expansion rule. Since $\{A, \neg A\} \subset Label(v_9)$, the simple non-state v_9 receives status closed. After that the nodes $v_8, \ldots, v_5$ and ν receive status closed in subsequent steps. The nodes v_1 and v_7 are the only states.

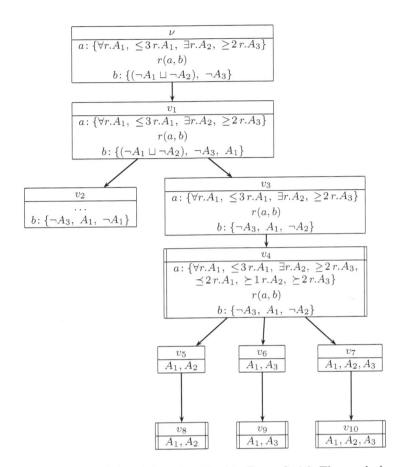

Fig. 3. An illustration of the tableau described in Example 4.2. The marked nodes v_4, v_8, v_9, v_{10} are states.

$\mathcal{R} = \emptyset$ and $\mathcal{T} = \emptyset$. An illustration is presented in Figure 3.

At the beginning, the graph has only the root ν which is a complex non-state with $Label(\nu) = \mathcal{A}$. Since $\{a : \forall r.A_1, r(a,b)\} \subset Label(\nu)$, applying the unary static expansion rule to ν, we connect it to a new complex non-state v_1 with $Label(v_1) = Label(\nu) \cup \{b : A_1\}$.

Since $b : (\neg A_1 \sqcup \neg A_2) \in Label(v_1)$, applying the non-unary static expansion rule to v_1, we connect it to two new complex non-states v_2 and v_3 with

$$Label(v_2) = Label(v_1) - \{b : (\neg A_1 \sqcup \neg A_2)\} \cup \{b : \neg A_1\}$$
$$Label(v_3) = Label(v_1) - \{b : (\neg A_1 \sqcup \neg A_2)\} \cup \{b : \neg A_2\}.$$

Since both $b : A_1$ and $b : \neg A_1$ belong to $Label(v_2)$, the node v_2 receives status closed. Applying the forming-state rule (FS$_2$) to v_3, we connect it to a new complex state v_4 with

$$Label(v_4) = Label(v_3) \cup \{a : \preceq 2\,r.A_1,\ a : \succ 1\,r.A_2,\ a : \succeq 2\,r.A_3\}.$$

The assertion $a : \preceq 2\, r.A_1 \in Label(v_4)$ is due to $a : \leq 3\, r.A_1 \in Label(v_3)$ and the fact that $\{r(a, b),\ b : A_1\} \subset Label(v_3)$. The assertion $a : \succeq 1\, r.A_2 \in Label(v_4)$ is due to $a : \exists r.A_2 \in Label(v_3)$ and the fact that $b : \neg A_2 \in Label(v_3)$. Similarly, the assertion $a : \succeq 2\, r.A_3 \in Label(v_4)$ is due to $a : \geq 2\, r.A_3 \in Label(v_3)$ and the fact that $b : \neg A_3 \in Label(v_3)$. When realizing the requirements for a at v_4, we will not have to pay attention to the relationship between a and b.

As r is a numeric role, applying the transitional partial-expansion rule to v_4, we just change its status to p-expanded. After that, applying the transitional full-expansion rule to v_4, we connect it to new simple non-states v_5, v_6 and v_7 with $CELabelT$ equal to checkingFeasibility, $CELabelR$ equal to $\{r\}$, $CELabelI$ equal to a, $Label(v_5) = \{A_1, A_2\}$, $Label(v_6) = \{A_1, A_3\}$ and $Label(v_7) = \{A_1, A_2, A_3\}$. The creation of v_5 is caused by $a : \succeq 1\, r.A_2 \in Label(v_4)$, while the creation of v_6 is caused by $a : \succeq 2\, r.A_3 \in Label(v_4)$. The node v_7 results from merging v_5 and v_6. Furthermore, $ILConstraints(v_4)$ consists of $x_{v_i} \geq 0$, for $5 \leq i \leq 7$, and

$$x_{v_5} + x_{v_6} + x_{v_7} \leq 2$$
$$x_{v_5} + x_{v_7} \geq 1$$
$$x_{v_6} + x_{v_7} \geq 2.$$

The set $ILConstraints(v_4)$ is feasible, e.g., with $x_{v_5} = 0$ and $x_{v_6} = x_{v_7} = 1$.

Applying the forming-state rule (FS_1) to v_5 (resp. v_6, v_7), we connect it to a new simple state v_8 (resp. v_9, v_{10}) with the same label.

Expanding the nodes v_8, v_9 and v_{10}, their statuses change to p-expanded and then to f-expanded. The graph cannot be modified anymore and becomes a $C_{\mathcal{SHIQ}}$-tableau for $\langle \mathcal{R}, \mathcal{T}, \mathcal{A} \rangle$, with $Status(\nu) \neq$ closed. Thus, the considered knowledge base $\langle \mathcal{R}, \mathcal{T}, \mathcal{A} \rangle$ is satisfiable. Using the solution $x_{v_5} = 0$, $x_{v_6} = x_{v_7} = 1$ for $ILConstraints(v_4)$, we can extract from the graph a model $\mathcal{I}$ for $\langle \mathcal{R}, \mathcal{T}, \mathcal{A} \rangle$ with $\Delta^{\mathcal{I}} = \{a, b, v_9, v_{10}\}$, $a^{\mathcal{I}} = a$, $b^{\mathcal{I}} = b$, $A_1^{\mathcal{I}} = \{b, v_9, v_{10}\}$, $A_2^{\mathcal{I}} = \{v_{10}\}$, $A_3^{\mathcal{I}} = \{v_9, v_{10}\}$ and $r^{\mathcal{I}} = \{\langle a, b \rangle, \langle a, v_9 \rangle, \langle a, v_{10} \rangle\}$. □

Example 4.3. Let us construct a $C_{\mathcal{SHIQ}}$-tableau for $\langle \mathcal{R}, \mathcal{T}, \mathcal{A} \rangle$, where

$$\mathcal{A} = \{a : \exists r.A,\ a : \forall r.(\neg A \sqcup \neg B),\ a : \geq 1000\, r.B,\ a : \leq 1000\, r.(A \sqcup B)\},$$

$\mathcal{R} = \emptyset$ and $\mathcal{T} = \emptyset$. An illustration of the tableau is given in Figure 4.

At the beginning, the graph has only the root ν which is a complex non-state with $Label(\nu) = \mathcal{A}$. Applying the forming-state rule (FS_2) to ν, we connect it to a new complex state v_1 with

$$Label(v_1) = Label(\nu) \cup \{a : \succeq 1\, r.A,\ a : \succeq 1000\, r.B,\ a : \preceq 1000\, r.(A \sqcup B)\}.$$

Applying the transitional partial-expansion rule to v_1, we change its status to p-expanded. After that, applying the transitional full-expansion rule to v_1, we connect it to new simple non-states $v_2, \ldots, v_7$ with $CELabelT$ equal to checkingFeasibility, $CELabelR$ equal to $\{r\}$, $CELabelI$ equal to a, and

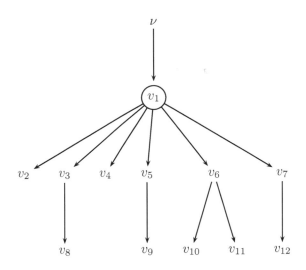

Fig. 4. An illustration of the tableau described in Example 4.3

$$Label(v_2) = \{A,\ \neg A \sqcup \neg B,\ A \sqcup B\}$$
$$Label(v_3) = \{A,\ \neg A \sqcup \neg B,\ \neg A \sqcap \neg B\}$$
$$Label(v_4) = \{B,\ \neg A \sqcup \neg B,\ A \sqcup B\}$$
$$Label(v_5) = \{B,\ \neg A \sqcup \neg B,\ \neg A \sqcap \neg B\}$$
$$Label(v_6) = \{A,\ B,\ \neg A \sqcup \neg B,\ A \sqcup B\}$$
$$Label(v_7) = \{A,\ B,\ \neg A \sqcup \neg B,\ \neg A \sqcap \neg B\}.$$

Note that $\neg A \sqcap \neg B$ is the NNF of $A \sqcup B$. The nodes v_2 and v_3 are created due to $a : \succeq 1\, r.A \in Label(v_1)$. The nodes v_4 and v_5 are created due to $a : \succeq 1000\, r.B \in Label(v_1)$. The node v_6 results from merging v_2 and v_4. The node v_7 results from merging v_3 and v_5. Furthermore, $ILConstraints(v_1)$ consists of $x_{v_i} \geq 0$, for $2 \leq i \leq 7$, and

$$x_{v_2} + x_{v_3} + x_{v_6} + x_{v_7} \geq 1$$
$$x_{v_4} + x_{v_5} + x_{v_6} + x_{v_7} \geq 1000$$
$$x_{v_2} + x_{v_4} + x_{v_6} \leq 1000.$$

Applying the unary static expansion rule to v_3, we connect it to a new simple non-state v_8 with $Label(v_8) = \{A, \neg A, \ldots\}$. The status of v_8 is then updated to closed and propagated back to make the status of v_3 also become closed, which causes addition of the constraint $x_{v_3} = 0$ into the set $ILConstraints(v_1)$.

Similarly, applying the unary static expansion rule to v_5, we connect it to a new simple non-state v_9 with $Label(v_9) = \{B, \neg B, \ldots\}$. The status of v_9 is then updated to closed and propagated back to make the status of v_5 also become closed, which causes addition of the constraint $x_{v_5} = 0$ into the set $ILConstraints(v_1)$.

342   L.A. Nguyen

Applying the non-unary static expansion rule to v_6, we connect it to new simple non-states v_{10} and v_{11} with $Label(v_{10}) = \{A, \neg A, \ldots\}$ and $Label(v_{11}) = \{B, \neg B, \ldots\}$. The statuses of v_{10} and v_{11} are then updated to closed and propagated back to make the status of v_6 also become closed, which causes addition of the constraint $x_{v_6} = 0$ into the set $ILConstraints(v_1)$.

Applying the unary static expansion rule to v_7, we connect it to a new simple non-state v_{12} with $Label(v_{12}) = \{A, B, \neg A, \neg B, \ldots\}$. The status of v_{12} is then updated to closed and propagated back to make the status of v_7 also become closed, which causes addition of the constraint $x_{v_7} = 0$ into the set $ILConstraints(v_1)$.

With $x_{v_3} = x_{v_5} = x_{v_6} = x_{v_7} = 0$, $ILConstraints(v_1)$ becomes infeasible, and $Status(v_1)$ becomes closed. Consequently, ν receives status closed. Thus, the considered knowledge base $\langle \mathcal{R}, \mathcal{T}, \mathcal{A} \rangle$ is unsatisfiable. □

5 Conclusions

We have presented the first tableau method with an EXPTIME (optimal) complexity for checking satisfiability of a knowledge base in the DL $\mathcal{SHIQ}$. The complexity is measured using binary representation for numbers. Our detailed tableau decision procedure for $\mathcal{SHIQ}$ given in the long version [6] of the current chapter has been designed to increase efficiency of reasoning. It is a framework, which can be implemented with various optimization techniques [4].

Acknowledgments. This work was supported by Polish National Science Centre grants 2011/01/B/ST6/02759 and 2011/01/B/ST6/02769.

References

1. Farsiniamarj, N.: Combining integer programming and tableau-based reasoning: a hybrid calculus for the description logic SHQ. Master's thesis, Concordia University (2008)
2. Goré, R., Widmann, F.: Sound global state caching for ALC with inverse roles. In: Giese, M., Waaler, A. (eds.) TABLEAUX 2009. LNCS, vol. 5607, pp. 205–219. Springer, Heidelberg (2009)
3. Horrocks, I., Sattler, U., Tobies, S.: Reasoning with individuals for the description logic SHIQ. In: McAllester, D. (ed.) CADE-17. LNCS, vol. 1831, pp. 482–496. Springer, Heidelberg (2000)
4. Nguyen, L.A.: An efficient tableau prover using global caching for the description logic ALC. Fundamenta Informaticae 93(1-3), 273–288 (2009)
5. Nguyen, L.A.: A cut-free ExpTime tableau decision procedure for the description logic SHI. In: Jędrzejowicz, P., Nguyen, N.T., Hoang, K. (eds.) ICCCI 2011, Part I. LNCS, vol. 6922, pp. 572–581. Springer, Heidelberg (2011), http://arxiv.org/abs/1106.2305
6. Nguyen, L.A.: Exptime tableaux for the description logic SHIQ based on global state caching and integer linear feasibility checking. arXiv:1205.5838 (2012)
7. Tobies, S.: Complexity results and practical algorithms for logics in knowledge representation. PhD thesis, RWTH-Aachen (2001)

An Approach to Semantic Indexing
Based on Tolerance Rough Set Model

Sinh Hoa Nguyen[1] and Hung Son Nguyen[2]

[1] Polish-Japanese Institute of Information Technology,
Koszykowa 86, 02-008, Warsaw, Poland
[2] Institute of Mathematics, Warsaw University
Banacha 2, 02-097, Warsaw, Poland
{hoa,son}@mimuw.edu.pl

Abstract. In this article we propose a general framework incorporating semantic indexing and search of texts within scientific document repositories, where document representation may include, excepts the content, some additional document meta-data, citations and semantic information. Our idea is based on application of Tolerance Rough Set Model, semantic information extracted from source text and domain ontology to approximate concepts associated with documents and to enrich the vector representation. We present the experiment performed over the freely accessed biomedical research articles from Pubmed Cetral (PMC) portal. The experimental results are showing the advantages of the proposed solution.

Keywords: Semantic indexing, tolerance rough set model, MeSH, PubMed.

1 Introduction

The main issues in semantic search engines are related to representation of domain knowledge in both offline document indexing process as well as the application of semantic indices in the online dialog with users. In practice, the searching process may show to be inaccurate because the user's intention behind the search is not clearly expressed by too general, short queries. In some searching scenario, the user provides the search engine with a phrase which is intended to denote an object about which the user is trying to research information. There is no particular document which the user knows about that she/he is trying to get to. Rather, the user is trying to locate a number of documents which together will give him/her the information she/he is trying to find. One of the forms of semantic search is closely related with exploratory search, in which the search results are organizing into clusters to facilitate a quick navigation through search results and helps users to specify their search intentions.

Document description enrichment is a pivotal problem in meta search engine and especially in Semantic Search Engine. In this paper, we discuss a framework of document description extension. We are concerned with applying domain knowledge and semantic similarity to document description enrichment. Several works have been devoted to a problem of document description enrichment. In

N.T. Nguyen, T. Van Do, and H.A. Le Thi (Eds.): *ICCSAMA 2013*, SCI 479, pp. 343–354.
DOI: 10.1007/978-3-319-00293-4_26 © Springer International Publishing Switzerland 2013

[1] a method for snippet extension was investigated. The main idea was based on application of collocation similarity measure to enrich the vector representation. In [2] the author presented a method of associating terms in document representation with similar concepts drawn from domain ontology.

This paper is a part of the ongoing project, called SONCA (Search based on ONtologies and Compound Analytics), realized at the Faculty of Mathematics, Informatics and Mechanics of the University of Warsaw. It is part of the project 'Interdisciplinary System for Interactive Scientific and Scientific-Technical Information' (www.synat.pl). SONCA is an application based on a hybrid database framework, wherein scientific articles are stored and processed in various forms. SONCA is expected to provide interfaces for intelligent algorithms identifying relations between various types of objects. It extends typical functionality of scientific search engines by more accurate identification of relevant documents and more advanced synthesis of information. To achieve this, concurrent processing of documents needs to be coupled with the ability to produce collections of new objects using queries specific for analytic database technologies.

Ultimately, SONCA should be capable of answering the user query by listing and presenting the resources (documents, Web pages, etc.) that correspond to it *semantically*. In other words, the system should have some *understanding* of the intention of the query and of the contents of documents stored in the repository as well as the ability to retrieve relevant information with high efficacy. The system should be able to use various knowledge sources related to the investigated areas of science. It should also allow for independent sources of information about the analyzed objects, such as, e.g., information about scientists who may be identified as the stored articles' authors.

In [3], we presented a generalized schema for the problem of semantic extension for snippets. We investigated two levels of extension. The first level was related to extending a concept space for data representation and the second one was related to associating terms in the concept space with semantically related concepts. The first extension level is performed by adopting semantic information extracted from a document content or from other meta-data like citations, authors, conferences. Term association is achieved by application of Tolerance Rough Set Model [4][5] and domain ontology, in order to approximate concepts existing in documents description and to enrich the vector representation.

This paper is an extension of [3]. We perform a deep analysis of the proposed approach on the database of articles on biomedical topics PubMed. We consider several data extension models including citations, shortened description and semantic information build using MeSH dictionary. We compare those models in document clustering problem based on documents available in PubMed biomedical articles database. By using extended representation better grouping results could be archived, we support this statement with experimental results.

The paper is organized as follows. In Section 2 we recall the general approach to enriching of snippet representation using *Generalized* TRSM. Section 3 presents the setting for experiments and validation method based on expert's tags. Section 4 is devoted to experiments, followed by conclusions in section 5.

2   Tolerance Rough Set Model

Tolerance Rough Set Model was developed in [6,7] as a basis to model documents and terms in Information Retrieval, Text Mining, etc. With its ability to deal with vagueness and fuzziness, Tolerance Rough Set Model seems to be a promising tool to model relations between terms and documents. In many Information Retrieval problems, especially in document clustering, defining the relation (i.e. similarity or distance) between document-document, term-term or term-document is essential. In Vector Space Model, is has been noticed [7] that a single document is usually represented by relatively few terms[1]. This results in zero-valued similarities which decreases quality of clustering. The application of TRSM in document clustering was proposed as a way to enrich document and cluster representation with the hope of increasing clustering performance.

In fact Tolerance Rough Set Model is a special case of a generalized approximation space, which has been investigated by Skowron and Stepaniuk [4]. as a generalization of standard rough set theory. Generalized approximation space utilizes every tolerance relation overs objects to determine the main concepts of rough set theory, i.e., lower and upper approximation.

The main idea of TRSM is to capture conceptually related index terms into classes. For this purpose, the tolerance relation R is determined as the co-occurrence of index terms in all documents from D. The choice of co-occurrence of index terms to define tolerance relation is motivated by its meaningful interpretation of the semantic relation in context of IR and its relatively simple and efficient computation.

2.1   Standard TRSM

Let $D = \{d_1, \ldots, d_N\}$ be a corpus of documents Assume that after the initial processing documents, there have been identified N unique terms (e.g. words, stems, N-grams) $T = \{t_1, \ldots, t_M\}$.

Tolerance Rough Set Model, or briefly TRSM, is an approximation space $\mathcal{R} = (T, I_\theta, \nu, P)$ determined over the set of terms T where:

- The parameterized **uncertainty function** $I_\theta : T \to \mathcal{P}(T)$ is defined by

$$I_\theta(t_i) = \{t_j \mid f_D(t_i, t_j) \geq \theta\} \cup \{t_i\}$$

where $f_D(t_i, t_j)$ denotes the number of documents in D that contain both terms t_i and t_j and θ is a parameter set by an expert. The set $I_\theta(t_i)$ is called the *tolerance class* of term t_i.
- **Vague inclusion function** $\nu(X, Y)$ measures the degree of inclusion of one set in another. The vague inclusion function is defined as $\nu(X, Y) = \frac{|X \cap Y|}{|X|}$. It is clear that this function is monotone with respect to the second argument.

---

[1] In other words, the number of non-zero values in document's vector is much smaller than vector's dimension – the number of all index terms.

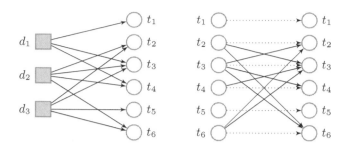

Fig. 1. Bag-of-words (left) determines the term co-location graph with $\theta = 2$ (right)

- **Structural function:** All tolerance classes of terms are considered as structural subsets: $P(I_\theta(t_i)) = 1$ for all $t_i \in T$.

In TRSM model $\mathcal{R} = (T, I, \nu, P)$, the membership function μ is defined by

$$\mu(t_i, X) = \nu(I_\theta(t_i), X) = \frac{|I_\theta(t_i) \cap X|}{|I_\theta(t_i)|}$$

where $t_i \in T$ and $X \subseteq T$. The lower and upper approximations of any subset $X \subseteq T$ can be determined by the same maneuver as in approximation space [4]:

$$\mathbf{L}_\mathcal{R}(X) = \{t_i \in T \mid \nu(I_\theta(t_i), X) = 1\} \qquad \mathbf{U}_\mathcal{R}(X) = \{t_i \in T \mid \nu(I_\theta(t_i), X) > 0\}$$

The standard TRSM was applied for document clustering and snippet clustering tasks (see [6], [7], [5], [3], [8]). In those applications, each document is represented by the upper approximation of its *set of words/terms*, i.e. the document $d_i \in D$ is represented by $\mathbf{U}_\mathcal{R}(d_i)$. For the example in Figure 1, the enriched representation of d_1 is $\mathbf{U}_\mathcal{R}(d_1) = \{t_1, t_3, t_4, t_2, t_6\}$.

2.2   Extended TRSM Using Semantic Concepts

Let $D = \{d_1, \ldots, d_N\}$ be a set of documents and $T = \{t_1, \ldots, t_M\}$ the set of *index terms* for D. Let C be the set of concepts from a given domain knowledge (e.g. the concepts from DBpedia or from a specific ontology).

The extended TRSM is an approximation space $\mathcal{R}_C = (T \cup C, I_{\theta,\alpha}, \nu, P)$, where C is the mentioned above set of concepts. The uncertainty function $I_{\theta,\alpha} : T \cup C \to \mathbb{P}(T \cup C)$ has two parameters θ and α is defined as follows:

- for each term $c_i \in C$ the set $I_{\theta,\alpha}(c_i)$ contains α top terms from the bag of terms of c_i calculated from the textual descriptions of concepts.
- for each term $t_i \in T$ the set $I_{\theta,\alpha}(t_i) = I_\theta(t_i) \cup C_\alpha(t_i)$ consists of the tolerance class of t_i from the standard TRSM and the set of concepts, whose description contains the term t_i as the one of the top α terms.

In extended TRSM, the document $d_i \in D$ is represented by

$$\mathbf{U}_{\mathcal{R}_C}(d_i) = \mathbf{U}_\mathcal{R}(d_i) \cup \{c_j \in C \mid \nu(I_{\theta,\alpha}(c_j), d_i) > 0\} = \bigcup_{t_j \in d_i} I_{\theta,\alpha}(t_i)$$

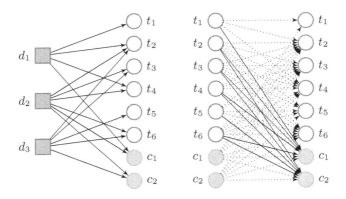

Fig. 2. Extended TRSM with $\theta = 1$, bag-of-words document representation (left) determines the structure of ESA model (right) when filtered to term $\rightarrow$ concept edges

2.3 Weighting Schema

Any text d_i in the corpus D can be represented by a vector $[w_{i1}, \ldots, w_{iM}]$, where each coordinate $w_{i,j}$ expresses the significance of j-th term in this document. The most common measure, called *tf-idf* index (term frequency-inverse document frequency) [9], is defined by:

$$w_{i,j} = tf_{i,j} \times idf_j = \frac{n_{i,j}}{\sum_{k=1}^{M} n_{i,k}} \times \log\left(\frac{N}{|\{i : n_{i,j} \neq 0\}|}\right), \tag{1}$$

where $n_{i,j}$ is the number of occurrences of the term t_j in the document d_i.

Both standard TRSM and extended TRSM are the conceptual models for the Information Retrieval. Depending on the current application, different extended weighting schema can be proposed to achieve as highest performance as possible. Let us recall some existing weighting scheme for TRSM:

1. The extended weighting scheme is inherited from the standard TF-IDF by:

$$w_{ij}^* = \begin{cases} (1 + \log f_{d_i}(t_j)) \log \frac{N}{f_D(t_j)} & \text{if } t_j \in d_i \\ 0 & \text{if } t_j \notin \mathbf{U}_{\mathcal{R}}(d_i) \\ \min_{t_k \in d_i} w_{ik} \frac{\log \frac{N}{f_D(t_j)}}{1 + \log \frac{N}{f_D(t_j)}} & \text{otherwise} \end{cases} \tag{2}$$

This extension ensures that each term occurring in the upper approximation of d_i but not in d_i itself has a weight smaller than the weight of any terms in d_i. Normalization by vector's length is then applied to all document vectors: $w_{ij}^{new} = w_{ij}^* / \sqrt{\sum_{t_k \in d_i} (w_{ij}^*)^2}$ (see [6], [7]). The example of standard TRSM is presented in Table 1.

2. Explicit Semantic Analysis (ESA) proposed in [10] is a method for automatic tagging of textual data with predefined concepts. It utilizes natural language

Table 1. Example snippet and its two vector representations in standard TRSM

Original vector		Enriched vector	
Term	Weight	Term	Weight
auditor	0.567	auditor	0.564
bankruptcy	0.4218	bankruptcy	0.4196
signalling	0.2835	signalling	0.282
EconPapers	0.2835	EconPapers	0.282
rates	0.2835	rates	0.282
versus	0.223	versus	0.2218
issue	0.223	issue	0.2218
Journal	0.223	Journal	0.2218
MODEL	0.223	MODEL	0.2218
prediction	0.1772	prediction	0.1762
Vol	0.1709	Vol	0.1699
		applications	0.0809
		Computing	0.0643

Title: EconPapers: Rough sets bankruptcy prediction models versus auditor

Description: Rough sets bankruptcy prediction models versus auditor signalling rates. Journal of Forecasting, 2003, vol. 22, issue 8, pages 569-586. Thomas E. McKee. ...

definitions of concepts from an external knowledge base, such as an encyclopedia or an ontology, which are matched against documents to find the best associations. Such definitions are regarded as a regular collection of texts, with each description treated as a separate document. The original purpose of Explicit Semantic Analysis was to provide means for computing semantic relatedness between texts. However, an intermediate result – weighted assignments of concepts to documents (induced by the term-concept weight matrix) may be interpret as a weighting scheme of the concepts that are assigned to documents in the extended TRSM.

Let $W_i = [w_{i,j}]_{j=1}^N$ be a bag-of-words representation of an input text d_i, where $w_{i,j}$ is a numerical weight of term t_j expressing its association to the text d_i (e.g., its tf-idf). Let $s_{j,k}$ be the strength of association of the term t_j with a knowledge base concept c_k, $k \in \{1,\ldots,K\}$ an inverted index entry for t_j. The new vector representation, called a *bag-of-concepts* representation of d_i, is denoted by $[u_{i,1},\ldots u_{i,K}]$, where:

$$u_{i,k} = \sum_{j=1}^N w_{i,j} s_{j,k}. \tag{3}$$

For practical reasons it is better to represent documents by the most relevant concepts only. In such a case, the association weights can be used to create a ranking of concept relatedness. With this ranking it is possible to select only top concepts from the list or to apply some more sophisticated methods that involve utilization of internal relations in the knowledge base. An example of top 20 concepts for an article from PubMed is presented in Figure 3.

The described above weighting scheme naturally utilized in Document Retrieval as a semantic index [11,12]. A user may query a document retrieval engine for documents matching a given concept. If the concepts are already assigned to

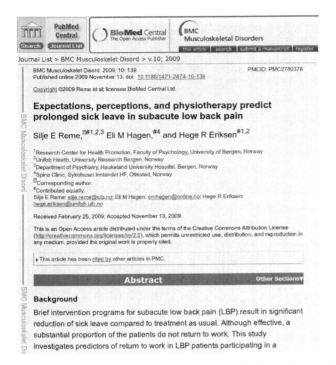

Fig. 3. An example of a document and the list of top 20 concepts assigned by the semantic tagging algorithm

documents, this problem is conceptually trivial. However such a situation is relatively rare, since employment of experts who could manually labelled documents from a huge repository is expensive. On the other hand, utilization of an automatic tagging method, such as ESA, allows to infer labeling of previously untagged documents. More sophisticated weighting schema have been proposed in, e.g. [13], [2].

2.4 Further Extensions of TRSM

The standard TRSM can be extended in many ways. In [3], an extended TRSM using citing information has been proposed to enrich the representation of documents. The extension method is similar to the extended TRSM using concepts.

The extended tolerance rough set model based on both citations and semantic concepts is a tuple:

$$\mathcal{R}_{Final} = (\mathcal{R}_T, \mathcal{R}_B, \mathcal{R}_C, \alpha_n)$$

where $\mathcal{R}_T$, $\mathcal{R}_B$ and $\mathcal{R}_C$ are tolerance spaces determined over the set of terms T, the set of bibliography items B and the set of concepts in the knowledge domain C, respectively. The function $\alpha_n : \mathcal{P}(T) \longrightarrow \mathcal{P}(C)$ is called the *semantic association* for terms. For any $T_i \subset T$, $\alpha_n(T_i)$ is the set of n concepts most associated with T_i, see [2].

In this model each document $d_i \in D$ is represented by a triple:

$$d_i \dashrightarrow (\mathbf{U}_{\mathcal{R}_T}(d_i), \mathbf{U}_{\mathcal{R}_B}(d_i), \alpha_n(T_i))\}$$

where T_i is the set of terms occurring in d_i and B_i is the set of bibliography items cited by d_i

3 Verification of TRSM

The standard TRSM as well as its extensions can be used to solve many tasks in Information Retrieval. We present some experiment results related to the application of the proposed method in search result clustering algorithms.

The aim of our experiments is to explore clusterings induced by different document representations (lexical, semantic and structural). An experiment path (from querying to search result clustering) consists of three stages:

- Search and filter documents matching to a query. Search result is a list of *snippets* and document identifiers.
- Extend representations of snippets and documents by *citations* and/or *semantically similar concepts* from MeSH ontology (terms used for clustering were automatically assigned by an algorithm[2], terms used for validation are were assigned by human experts).
- Cluster document search results.

In order to perform validation (and choose parameters of clustering algorithms) one needs a set of search queries that reflect actual user usage patterns. We extract 100 most frequent one-term queries from the daily log previously investigated by Herskovic et al. in [14].

Twenty five of these queries were used for initial fine-tuning of parameters. For each algorithm, we performed a set of experiments with various settings and picked parameters that lead (on the average) to best clusterings (with regard to RandIndex measure). In our remaining experiments (discussed in a further section of the article) we compared the performance of these algorithms for different extensions.

Medical Subject Headings (MeSH) is a controlled vocabulary created (and continuously being updated) by the United States National Library of Medicine. It is used by MEDLINE and PubMed databases.

The majority of documents in Pubmed Central are tagged by human experts using headings and (optionally) accompanying subheadings (qualifiers). A single document is typically tagged by 10 to 18 heading-subheading pairs.

There are approximately 25000 subject headings and 83 subheadings. There is a rich structural information accompanying headings: each heading is a node of (at least one) tree and is accompanied by further information (e.g. allowable qualifiers, annotation, etc.). Currently we do not use this information, but in some experiments we use a hierarchy of qualifiers[2] by exchanging a given qualifier by its (at most two) topmost ancestors or roots.

---

[2] http://www.nlm.nih.gov/mesh/subhierarchy.html

4   Results of Experiments

The aim of this section is to evaluate the semantic enrichment models and to select a proper clustering algorithm for each model. Our experiments can be divided into three groups. The first group of experiments is to examine whether the sematic extensions improve the clustering quality and to verify if it is robust against varying evaluation criterions. The second group of experiments is to select an optimal clustering algorithm for every type of sematic extensions. The last experiment investigates the time performance of clustering algorithms implemented in our system.

Our experiments are carried out on three clustering algorithms: (1) KMeans clustering, (2) Hierarchical clustering and (3) Singular Value Decomposition (SVD) based clustering (Lingo algorithm). In our experiments, we adopt Rand Index measure to clustering evaluation.

4.1   Quality of Enrichment Methods

The goal of this set of experiments is to examine if the semantic extension will improve the clustering quality and if it is robust against different evaluation criterions. For a fixed clustering algorithm, the clustering quality varies depending on a way of document representation. In experiments, we focus on four types of document representation: (1) **S**nippet based (**S**); (2) **S**nippet enriched by **i**nbound and **o**utbound citations (**SIO**); (3) **S**nippet enriched by semantic concepts adopted from MeSH - the domain ontology (**SM**) and (4) **S**nippet enriched by **i**nbound and **o**utbound citations and semantic concepts (**SIOM**). The document representation is better if the target clusters are more concordant with expert's clusters.

Figure 5 presents the quality comparison of the basic representation (Snippet) and extended representations (SIO,SIOM and SM). In all cases the extended representations show to be significantly better (the higher Rand index). When K-Means algorithm is used, the representation enriched by citations (SIO) was slightly better then the full extension (SIOM), but when hierarchical clustering or Lingo clustering algorithm was used, the full extension showed to be better then representation enriched only by citations.

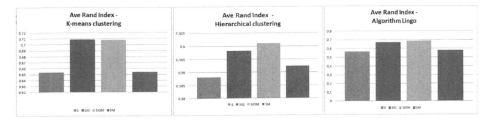

Fig. 4. Comparing the quality of standard presentation model and three enrichment models. An evaluation is based on 83 subheadings.

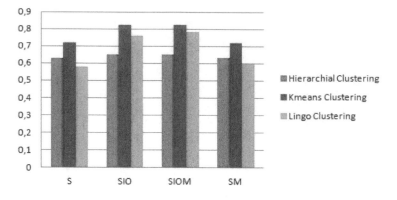

Fig. 5. The quality of clustering algorithms

To examine the robustness of enrichment models against varying evaluation criterions, we investigate the quality of these models over two another evaluation methods: based on *83 subheadings* and based on *headings* assigned by experts to the document.

Figure 4 presents the quality of different enrichment methods comparing to the standard (tf-idf) representation model. One can observe, in the both cases the semantic extensions (SIO and SIOM) show an advantage over the standard representation.

4.2   Quality of Clustering Algorithms

The aim of this section is to compare the quality of clustering algorithms impact with the extension types and to select the best algorithm for every kind of extension. We use the set of headings to defined a similarity of documents. In Figure 5, the quality of considered clustering algorithms are presented. In all cases the k-means algorithm show to be better. The Lingo algorithm is a little worse than k-means in the case of extended representation SIO and SIOM.

In particular, using SIO model with k-means algorithm one can improve the Rand Index above 25% (Rand Index raised from 0.65 to 0.82).

4.3   Time Performance of Clustering Algorithms

This set of experiments investigates the time required for each clustering algorithm. For every type of extension we calculate the average time for k-means, hierarchical i Lingo algorithm. The results are presented in Figure 6. In our experiments the time performance is expressed in milliseconds. One can observe that all three algorithms are effective, all cluster a set of 200 search results in less than 25 seconds. This time is feasible for an online clustering system. However the k-means and Lingo algorithm are about 10 time quicker than the hierarchical one. The average time of these algorithms is about 2 sec for one query. In Figure 6 we can see the time performance of presented clustering algorithms.

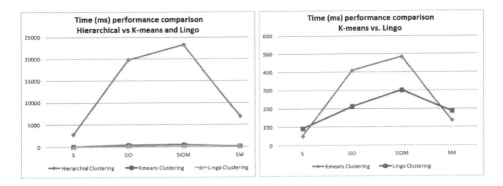

Fig. 6. The time performance of clustering algorithms

5 Conclusions and Further Work

We had propose a generalized Tolerance Rough Set Model for text data. The proposed method has been applied to extend the poor TF-IDF representation of snippets in the search result clustering problem. This functionality has been implemented within the SONCA system and tested on the biomedical documents, which are freely available from the highly specialized biomedical paper repository called PubMed Central (PMC).

Our preliminary experiments lead to several promising conclusions. The future plans are briefly outlined as follows:

- extend the experiments with different semantic indexing methods that are currently designed for SONCA system,
- analyse label quality of clusters resulting from different document representations,
- conduct experiments using other extensions (e.g. citations along with their context; information about authors, institutions, fields of knowledge or time),
- visualization of clustering results.

Acknowledgments. The authors are partially supported by the National Centre for Research and Development (NCBiR) under Grant No. SP/I/1/77065/10 by the strategic scientific research and experimental development program: "Interdisciplinary System for Interactive Scientific and Scientific-Technical Information", by the Polish National Science Centre under the grant 2012/05/B/ST6/03215 and by the Grant No. ST/SI/03/2012 from the Polish Japanese Institute of Information Technology (PJIIT).

References

1. Ngo, C.L., Nguyen, H.S.: A method of web search result clustering based on rough sets. In: Skowron, A., Agrawal, R., Luck, M., Yamaguchi, T., Mor-Mahoudeaux, P., Liu, J., Zhong, N. (eds.) Web Intelligence, pp. 673–679. IEEE Computer Society (2005)

2. Szczuka, M., Janusz, A., Herba, K.: Semantic clustering of scientific articles with use of DBpedia knowledge base. In: Bembenik, R., Skonieczny, L., Rybiński, H., Niezgodka, M. (eds.) Intelligent Tools for Building a Scient. Info. Plat. SCI, vol. 390, pp. 61–76. Springer, Heidelberg (2012)
3. Nguyen, S.H., Jaśkiewicz, G., Świeboda, W., Nguyen, H.S.: Enhancing search result clustering with semantic indexing. In: Proceedings of the Third Symposium on Information and Communication Technology, SoICT 2012, pp. 71–80. ACM, New York (2012)
4. Skowron, A., Stepaniuk, J.: Tolerance approximation spaces. Fundamenta Informaticae 27(2-3), 245–253 (1996)
5. Nguyen, H.S., Ho, T.B.: Rough document clustering and the internet. In: Pedrycz, W., Skowron, A., Kreinovich, V. (eds.) Handbook of Granular Computing, pp. 987–1004. Wiley & Sons (2008)
6. Kawasaki, S., Nguyen, N.B., Ho, T.B.: Hierarchical document clustering based on tolerance rough set model. In: Zighed, D.A., Komorowski, J., Żytkow, J.M. (eds.) PKDD 2000. LNCS (LNAI), vol. 1910, pp. 458–463. Springer, Heidelberg (2000)
7. Ho, T.B., Nguyen, N.B.: Nonhierarchical document clustering based on a tolerance rough set model. International Journal of Intelligent Systems 17(2), 199–212 (2002)
8. Virginia, G., Nguyen, H.S.: Investigating the effectiveness of thesaurus generated using tolerance rough set model. In: Kryszkiewicz, M., Rybinski, H., Skowron, A., Raś, Z.W. (eds.) ISMIS 2011. LNCS, vol. 6804, pp. 705–714. Springer, Heidelberg (2011)
9. Feldman, R., Sanger, J. (eds.): The Text Mining Handbook. Cambridge University Press (2007)
10. Gabrilovich, E., Markovitch, S.: Computing semantic relatedness using wikipedia-based explicit semantic analysis. In: Proceedings of the 20th International Joint Conference on Artifical Intelligence, IJCAI 2007, pp. 1606–1611. Morgan Kaufmann Publishers Inc., San Francisco (2007)
11. Hliaoutakis, A., Varelas, G., Voutsakis, E., Petrakis, E.G.M., Milios, E.: Information retrieval by semantic similarity. Int. Journal on Semantic Web and Information Systems (IJSWIS). Special Issue of Multimedia Semantics 3(3), 55–73 (2006)
12. Rinaldi, A.M.: An ontology-driven approach for semantic information retrieval on the web. ACM Trans. Internet Technol. 10:1–10:24 (2009)
13. Janusz, A., Świeboda, W., Krasuski, A., Nguyen, H.S.: Interactive document indexing method based on explicit semantic analysis. In: Yao, J., Yang, Y., Słowiński, R., Greco, S., Li, H., Mitra, S., Polkowski, L. (eds.) RSCTC 2012. LNCS, vol. 7413, pp. 156–165. Springer, Heidelberg (2012)
14. Herskovic, J.R., Tanaka, L.Y., Hersh, W., Bernstam, E.V.: A day in the life of pubmed: analysis of a typical day's query log. Journal of the American Medical Informatics Association, 212–220 (2007)

An Approach for Mining Concurrently Closed Itemsets and Generators

Anh Tran[1], Tin Truong[1], and Bac Le[2]

[1] Department of Mathematics and Computer Science, University of Dalat, Dalat, Vietnam
{Anhtn,tintc}@dlu.edu.vn
[2] University of Natural Science Ho Chi Minh, Ho Chi Minh, Vietnam
lhbac@fit.hcmus.edu.vn

Abstract. Closed itemsets and their generators play an important role in frequent itemset and association rule mining since they lead to a lossless representation of all frequent itemsets. The previous approaches discover either frequent closed itemsets or generators separately. Due to *their properties and relationship*, the paper proposes *GENCLOSE* that *mines them concurrently*. In a levelwise search, it enumerates the generators using *a necessary and sufficient condition for producing (i+1)-item generators from i-item ones*. The condition is designed based on object-sets which can be implemented efficiently using diffsets, is very convenience and *is reliably proved*. Along that process, *preclosed itemsets are gradually extended using three proposed expanded operators*. Also, we prove that they bring us *to expected closed itemsets*. Experiments on many benchmark datasets confirm the efficiency of *GENCLOSE*.

Keywords: Frequent closed itemsets, generators, concept lattice, data mining.

1 Introduction

Association rule mining [1] from transaction datasets was popularized particularly and is one of the essential and fundamental techniques in data mining. The task is to find out the association rules that satisfy the pre-defined minimum support and confidence from a given dataset. As usual, it includes two phases of (1) to extract all frequent itemsets of which the occurrences exceed the minimum support, and (2) to generate association rules for the given minimum confidence from those itemsets. If we know all frequent itemsets and their supports, the association rule generation is straightforward. Hence, most of the researchers concentrate on the studies of finding a solution for mining frequent itemsets. Many algorithms have been proposed in order to do that mining task such as *Apriori* [2], *D-Eclat* [14]. Those algorithms all show good performance with sparse datasets having short itemsets such as market data. For dense ones where contain many long frequent itemsets, the class of frequent itemsets can grow to be unwieldy [6]. Thus, discovering frequent itemsets usually gets much duplication, wastes much time and is then infeasible. Mining only maximal frequent itemsets is a solution to overcome that disadvantage. There are many proposed algorithms for mining them [6-7]. An itemset is maximal frequent if none of its immediate supersets is

N.T. Nguyen, T. Van Do, and H.A. Le Thi (Eds.): *ICCSAMA 2013*, SCI 479, pp. 355–366.
DOI: 10.1007/978-3-319-00293-4_27   © Springer International Publishing Switzerland 2013

frequent. The number of maximal itemsets is much smaller than the one of all frequent itemsets [16]. From them, we can generate all sub frequent itemsets. However, since frequent itemsets can come from different maximal ones, we spend much time for finding them again. Further, it is not easy to determine their supports exactly. Hence, maximal itemsets are not suitable for rule generation within the itemsets.

Mining frequent closed itemsets and generators has been received the attention of many researchers [8-10, 12, 13, 15-16] for their importance in the mining of frequent itemsets as well association rules. An itemset is closed if it is identical to its closure [17]. A generator of an itemset is its minimal proper subset having the same closure with it. Some studies [10, 11, 15] concerned only the generators of closed itemset as a mean to achieve the purpose of mining either closed itemsets or association rules. They are also called minimal generators [9, 15] or free-sets [8]. Dong et al. in [9] concentrated on the mining of non-redundant generators, but, it exceeds the scope of this paper. Which is the role of frequent closed itemsets and generators in the frequent itemset and association rule mining? First, one can see that a frequent itemset contains its generators and is contained in its closure. They all have the same closure, share the same set of transactions. Hence, *frequent closed itemsets and generators lead to a lossles representation of all frequent itemsets* in the sense that we can determine not only the support of any itemset [8, 15] but also a class of frequent itemsets having the same closure (so the same support) [4, 12]. Since their cardinality is typically orders of magnitude much fewer than all frequent itemsets [16], discovering them can be of a great help to purge many redundant itemsets. Therefore, they play *an important role in* many studies of *rule generation* such as [3, 5, 11, 15]. Pasquier et al. [11] used the minimal rules to generate a condensed representation for association rules. Zaki in [15] proposed a framework for reduction of association rule set [13] which is based on the non-redundant rules. Two those rule kinds are obtained from frequent closed itemsets and the generators. Also based on them, we showed some structures of association rule set. For mining frequent itemsets with constraint, the task of extracting frequent closed itemsets and generators restricted on contraint from the original ones is essential [4].

The problem of mining frequent closed itemsets and generators is stated as follows: *Given a transaction dataset and a minsup threshold, the task is to find all frequent closed itemsets together their generators.*

Related Work. *There have been several algorithms* proposed for *mining closed itemsets*. They can be divided into three approaches, namely *generate-and-test, divide-and-conquer* and *hybrid. Close* [10] is in the first which executes a level-wise progress. In each step, it does two phases. The first is to generate the candidates for generators by joining generators found in previous step. If the support of a candidate equals to the one of any its subset, it could not be a generator. The second computes the closures of generators, but unfortunately, it is very expensive since there are a big number of transaction intersection operations. The second approach uses divide-and-conquer strategy to search over tree structures, e.g., *Closet+*[13]. The hybrid one which combines two above approaches includes *Charm* [16] that executes a highly efficiently hybrid search that skips many levels of the *IT-Tree* to quickly identify the frequent closed itemsets, instead of having to enumerate many possible subsets. A hash-based approach is applied to speed up subsumption checking. To compute quickly frequency as well as to reduce the sizes of the intermediate tidsets, the diffset technique is used. Experiments in [16] showed that *Charm* outperforms the existing algorithms on many datasets. However, *very little studies* concentrate on *mining* the *generators* of frequent closed itemsets. *SSMG-Miner* was developed in [9] based on a depth-first search. While mining non-redundant generators in addition, it does not output the closed itemsets. Further, we need to access the dataset for generating local generators. Boulicaut et al. [8] presented *MineEX* for generating frequent free-sets. Unfortunately, the algorithm has to scan the dataset at each step for testing if a candidate is free. *Talky-G* [12] is presented for mining generators. Since it is stand-alone algorithm, we have to apply an algorithm, e.g. *Charm*, for mining closed itemsets and then group the generators of the same closed itemset. This combination is called *Touch* algorithm [12]. Further, we find that *Charm-L* [16] outputs explicitly the closed itemset lattice in a non-considerable additional amount of time compared to *Charm* (see [16]). *MinimalGenerator* [15] is an algorithm that discovers all generators of a frequent closed itemsets using only its immediate sub ones (in term of the set containment order). Hence, in a hybrid approach, it seems to be feasible to mine first the frequent closed itemset lattice by *Charm-L* and then to apply *MinimalGenerator* for discovering their generators. We call this hybrid algorithm *CharmLMG*.

Contribution. Almost of those algorithms discover either frequent closed itemsets or generators separately. The fact brings up a natural idea that it should mine concurrently both of them. *Close* is such an example, however, its execution is very expensive. In Section 2, we give some properties of closed sets, generators and their relationship. Those bring about our idea in the development of *GENCLOSE* presented in Section 3. It includes two following key features (reliably proven by the theorems of 1, 2):

1) In a level-wise search, it mines first the generators by breadth-first search using a necessary and sufficient condition to determine the class of $(i+1)$-item generators from the class of i-item ones ($i \geq 1$) based on the sets of transactions. On the other hand, *Close* uses a necessary condition based on closures.

2) Based on the way that *Charm* applies four properties of itemset-tidset pairs to extend itemsets to closed ones and the relationship of generators and closed itemsets, *we develop* and use *three new operators for expanding itemsets to their closures simultaneously with the mining of generators*.

The rest of the paper is organized as follows. We compare *GENCLOSE* against *CharmLMG* and *Touch* in Section 4. The conclusion is shown in Section 5.

2   Closed Itemset, Generator and Their Relationship

Given non-empty sets $\mathcal{O}$ containing *objects* (or a dataset of transactions) and A *items* (attributes) related to objects in $\mathcal{O}$. A set A of items in A, $A \subseteq A$, is called an *itemset*. A set O of objects in $\mathcal{O}$ is called an *object-set*. For $O \subseteq \mathcal{O}$, $\lambda(O)$ is the itemset that occurs in all transactions of O, defined as $\lambda(O) := \{a \in A \mid a \in o,\ \forall o \in O\}$. The set of objects, in which itemset A appears as subset, is named by $\rho(A)$, $\rho(A) := \{o \in \mathcal{O} \mid A \subseteq o\}$. Define h and h' as union mappings of λ and ρ: $h = \lambda_o \rho$, $h' = \rho_o \lambda$, we say $h(A)$ and $h'(O)$ the *closures* of A, O. An itemset A is called a *closed itemset* if and only if (iff for short) $A = h(A)$ [17]. Symmetrically, O is called a *closed object-set* iff $O = h'(O)$.

Let *minsup* be the user-given *minimum support*, *minsup* $\in [1; |\mathcal{O}|]$. The number of transactions containing A is called the *support* of A, $supp(A) = |\rho(A)|$. If $supp(A) \geq minsup$ then A is called a *frequent itemset* [1]. If a frequent itemset is closed, thus, we call it a *frequent closed itemset*. For two non-empty itemsets of G, A such that $G \subseteq A \subseteq A$, G is called a *generator* of A iff $h(G) = h(A)$ and $(\forall \emptyset \neq G' \subset G \Rightarrow h(G') \subset h(G))$. Let $Gen(A)$ be the class of all generators of A. Since it is finite, we can enumerate it $Gen(A) = \{G_i: i = 1, 2, ..., |Gen(A)|\}$.

Proposition 1. $\forall A, A_1, A_2 \in 2^A, \forall O, O_1, O_2 \in 2^O$, the following statements hold true:

(a) $A_1 \subseteq A_2 \Rightarrow \rho(A_1) \supseteq \rho(A_2); O_1 \subseteq O_2 \Rightarrow \lambda(O_1) \supseteq \lambda(O_2)$ (1)

(b) $A \subseteq h(A), O \subseteq h'(O)$ (2)

(c) $A_1 \subseteq A_2 \Rightarrow h(A_1) \subseteq h(A_2); O_1 \subseteq O_2 \Rightarrow h'(O_1) \subseteq h'(O_2)$ (3)

(d) $\rho(A_1)=\rho(A_2) \Leftrightarrow h(A_1)=h(A_2)$ and $\rho(A_1) \subset \rho(A_2) \Leftrightarrow h(A_1) \supset h(A_2)$ (4)

(e) $\rho(\cup_{i \in I} A_i) = \cap_{i \in I} \rho(A_i)$. Thus, $h(A) = \cap_{o: A \subseteq \lambda(o)} \lambda(\{o\}) = \cap_{o \in \rho(A)} \lambda(o)$ (5)

Proof: Due to space limit, we omit the proof.   □

Proposition 2. *(Features of generators).* Let $A \subseteq A$. At the same time:

(a) $Gen(A) \neq \emptyset$ (6)

(b) $G \in Gen(A) \Leftrightarrow [\rho(G)=\rho(A)$ and $\forall G' \subset G \Rightarrow \rho(G') \supset \rho(G)]$ (7)

(c) *If G is a generator then $\forall \emptyset \neq G' \subset G: G'$ is a generator of $h(G')$*   (8)

Proof:

(a) Assuming that: $|A| = m$. Let us consider finitely the subsets of A that each of them is created by deleting an item of A: $A_{i1} = A\backslash\{a_{i1}\}, a_{i1} \in A, \forall i_1=1..m$.

Case 1: If $\rho(A_{i1}) \supset \rho(A), \forall i_1=1..m$, then A is a generator of A.

Case 2: Otherwise, there exists $i_1=1..m$: $\rho(A_{i1})=\rho(A)$. The above steps are repeated for A_{ij} $(j=1..m-1)$ until:

- *case 1* happens, thus, we get the generator $A_{i,j-1}$ of A; or

- *case 2* comes when $|A_{i,m-1}|=1$, i.e. $\rho(A_{i1})=\rho(A_{i2})=..=\rho(A_{i,m-1})=\rho(A)$. Hence, $A_{i,m-1}$ is a generator of A. Since A is finite, we always get a generator of A.

(b) Based on property 4 of proposition 1, we have: $\lambda(\rho(G)) = h(G) = h(A) = \lambda(\rho(A)) \Leftrightarrow \rho(G) = \rho(h(G)) = \rho(h(A)) = \rho(A)$. Further, $\forall G' \subset G: \lambda(\rho(G')) = h(G') \subset h(G) = \lambda(\rho(G)) \Leftrightarrow \rho(G') = \rho(h(G')) \supset \rho(h(G)) = \rho(G)$.

(c) Supposing that the contrary happens. So, there exists a proper subset G'' of G', $G'' \neq \emptyset$ such that $\rho(G'') = \rho(G')$. For $G_0=G'\backslash G''$, $G_1=G\backslash G_0$, we have: $\emptyset \neq G_1 = (G\backslash G') +^1 (G'\backslash G_0) = (G\backslash G')+G'' \subset (G\backslash G')+G'=G$. Otherwise, by *(5)*, $\rho(G) = \rho(G\backslash G')\cap\rho(G') = \rho(G\backslash G')\cap\rho(G'') = \rho((G\backslash G')+G'') = \rho(G_1)$, a contradiction!   □

Property (c) of proposition 2 implies the *Apriori principle of generators*. Following from it, we find that any *(i+1)-item generator* $G=g_1g_2... g_{i-1}g_ig_{i+1}$ [2] is created by the combining two i-item generators of $G_1=g_1g_2...g_{i-1}g_i$, $G_2=g_1g_2...g_{i-1}g_{i+1}$. Theorem 1 proposed hereafter give us an efficient way to *mine the class of (i+1)-item generators from the class of i-item ones*.

---

[1] The notation "+" represents the union of two disjoint sets.

[2] We write the set $\{a_1, a_2, .., a_n\}$ simply $a_1a_2..a_n$.

Theorem 1. *(The necessary and sufficient condition to produce generators).* For $\varnothing \neq G \subseteq \mathcal{A}$, $G_g := G \backslash \{g\}$, $g \in G$. The following conditions are equivalent:

(a) *G is a generator of h(G)*

(b) $\rho(G) \notin \bigcup_{g \in G} \{\rho(G_g)\}$ \hfill (9)

(c) $|\rho(G)| < |\rho(G_g)|$ *(i.e. supp(G) < supp(G_g)), $\forall g \in G$* \hfill (10)

(d) $|h(G_g)| < |h(G)|$, $\forall g \in G$ \hfill (11)

(e) *not($G \subseteq h(G_g)$)*, $\forall g \in G$ \hfill (12)

Proof:

(a) ⇔ (b): "⇒": If G is a generator of $h(G)$ then every non-empty strict subset G' ($\varnothing \neq G' \subset G$, especially for $G'=G_g$) of G is also a generator of $h(G')$. Then, $\rho(G) \subset \rho(G_g)$. Hence, $\rho(G) \neq \rho(G_g)$, $\forall g \in G$. "⇐": On the contrary, suppose that G is not a generator of $h(G)$. Thus: $\exists\, G' \subset G$: $\rho(G') = \rho(G)$, $G' \neq \varnothing$. It follows from *(1)* that $|G \backslash G'| = 1$ and $G' = G_g$, with $g \in G$. Therefore, $\rho(G_g) = \rho(G)$. That contradicts to *(9)*!

(b) ⇔ (c) ⇔ (d): By *(1)*, we have: $\rho(G) \subseteq \rho(G_g) \Leftrightarrow h(G) \supseteq h(G_g)$, $|\rho(G)| \leq |\rho(G_g)|$ and $|h(G)| \geq |h(G_g)|$, for any $g \in G$. Since $\mathcal{O}$ is finite, $\rho(G) = \rho(G_g) \Leftrightarrow |\rho(G)| = |\rho(G_g)| \Leftrightarrow |h(G)| = |h(G_g)|$, i.e. *not(9) ⇔ not(10) ⇔ not(11)*. Hence, we have: *(9) ⇔ (10) ⇔ (11)*.

(b) ⇔ (e): *not(9) ⇔ $h(G) \in \cup_{g \in G}\, h(G_g)$ (since (4)) ⇔ $\exists g \in G$: $G \subseteq h(G_g) \Leftrightarrow$ not(12)* (*). We first have $g \in G \subseteq h(G)$ by property (b) of proposition 1. If $G \subseteq h(G_g)$ or $g \in h(G_g)$ then, based on *(3)*, *(4)* and *(5)*, $h(\{g\}) \subseteq h(G) \subseteq h(G_g) \subseteq h(G)$ and $h(G) = h(h(G_g) \cup h(\{g\})) = h(G_g))$. Thus, (*) is hold. Therefore, we have *(9) ⇔ (12)*. □

Remark 1. For $G_1 = g_1 g_2 \ldots g_{i-1} g_i$, $G_2 = g_1 g_2 \ldots g_{i-1} g_{i+1}$, $G = G_1 \cup G_2$, $G_k \in Gen(h(G_k))$, $k=1,2$, the sufficient condition "$\rho(G) \neq \rho(G_g)$, for $g \in G_1 \cap G_2$" in *(9)* can not be skipped! This follows that the union of two generators could be not a generator. In fact, let us consider the dataset containing four transactions of *abcd*, *abc*, *abd* and *bc*. It is easy to see that $G_1 = bd$ is a generator of *abd* and $G_2 = bc$ a generator of *bc*. But, their union $G = G_1 \cup G_2 = bcd$ is not a generator since $\rho(bcd) = \rho(cd)$ but $cd \subset bcd$.

Remark 2. Let us review *consequence 2* in [10]. We find that the conclusion $h(I) = h(s_a)$ is not true because of the assumption that I is an i-generator and $\varnothing \neq s_a \subset I$. In fact, I is an i-generator (i-item generator) iff ($\forall \varnothing \neq s_a \subset I \Rightarrow h(s_a) \subset h(I)$). The consequence will become a necessary condition to an itemset is an i-item generator, if it is corrected as follows: *Let I be an i-itemset and S={s_1, s_2, .., s_j} a set of (i-1)-subsets of I where* $\bigcup_{s \in S} s = I$. *If $\exists s_a \in S$ such as $I \subseteq h(s_a)$, i.e. $\underline{h(I) = h(s_a)}$, then $\underline{I\ is\ not\ a\ generator}$.* We showed that *(12)* is more general and is also a sufficient one. This condition uses the closed itemset $h(G)$. But, at the time of discovering the generators, we do not know their closures. Hence, it seems reasonable to use object-set $\rho(G)$. Both *(9)* and *(10)* are designed for discovering generators, but *(10)* is simpler than *(9)*, especially on the datasets which object-set sizes can grow considerably!

Remark 3. In *Talky-G*, we need to check to see if a potential generator, which passed two tests of frequency and tidset, includes *a proper subset* with the same support in the set of mined generators (see *getNextGenerator* function [12]). Though a special hash table is used for doing the task, it seems to be very time-consuming. Condition *(10)* shows that we only need to consider the candidate with its *immediate subsets* – the generators mined from previous step in a level-wise progress.

3   GENCLOSE Algorithm

This section presents *GENCLOSE* that executes a breadth-first search over an *ITG-tree* (itemset-tidset-generator tree), like *Charm* which discovers an *IT-tree*, to discover generators. Simultaneously the corresponding closed itemsets are gradually explored. In each step, *GENCLOSE* tests the necessary and sufficient condition *(10)* for producing new generators from the generators of the previous step. Unlike *Close* that needs to scan the database, we propose three expanded operators (described formally in *3.2*) and apply them for discovering the closures along the process of mining generators.

Table 1. Dataset $\mathcal{D}_1$

Objects	Items
o_1	a b c e g u
o_2	a c d f u
o_3	a d e f g u
o_4	b c e f g u
o_5	b c e
o_6	b c

Itemset-Tidset-Generator Tree. *Fig. 1* shows *ITG-tree* created from the execution of *GENCLOSE* on dataset $\mathcal{D}_1$, as shown in *Table 1*, with *minsup=1*. *This figure is used for the examples in the rest of the paper.* A tree node includes the following fields. The first, namely *generator set (GS)*, contains the generators. The second, called *pre-closed itemset*, or *PC* for short, is the itemset that shares the same objects to them. *PC* is gradually enlarged to its closure – *h(PC)*. The third field *O* is the set of those objects. Thus, we have: $PC \subseteq \lambda(O)=h(G)$ and $\rho(PC)=O=\rho(G)$, $G \in GS$ (in implementation, we save their differences). The last one, called *supp*, stores the cardinality of *O*, i.e. *supp(PC)*. Initializing by *Root (Level 0)*, its *PC* is assigned by $\varnothing$. By the convention, $\rho(\varnothing):=\mathcal{O}$, *Root.O=$\mathcal{O}$*. The *ITG-tree* is expanded level by level. Each one splits into the *folders*. Level *1*, called *L[1]*, includes only the folder containing the nodes as *Root*'s children. If the combination of two nodes *X* and Y_1 in level *i* (called *L[i]*) creates node *Z* at *L[i+1]*, we say *X* the *left-parent* of *Z*. If *X* and Y_2 also in *L[i]* form *T*, then *Z* and *T* have the *same left-parent X*, i.e. they are in the *same folder* according to *X*. However, *the nodes of the same folder* can have *different left-parents*!

3.1 GENCLOSE Algorithm

The task of *GENCLOSE* is to output $\mathcal{LCG}$– the *list of all frequent closed itemsets together the corresponding generators and supports.* Its pseudocode is shown in *Fig. 2*. We start by eliminating non-frequent items from $\mathcal{A}$ and sorting in ascending order them first by their supports and then by their weights [16]. Each item $a \in \mathcal{A}^F$ forms a node in *L[1]*. Starting with *i=1, the i-th step* is broken into three phases as follows.

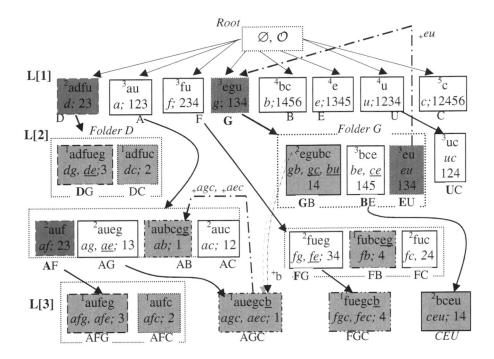

Fig. 1. *ITG-tree* created from $\mathcal{D}_1$ with *minsup=1*

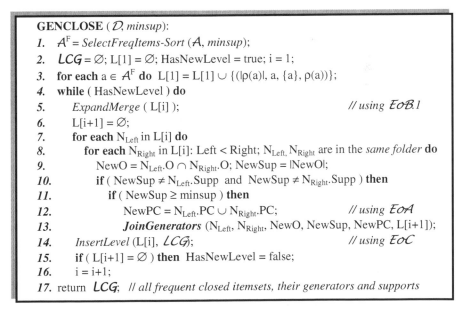

GENCLOSE ($\mathcal{D}$, minsup):

1. A^F = SelectFreqItems-Sort (A, minsup);
2. LCG = ∅; L[1] = ∅; HasNewLevel = true; i = 1;
3.   for each a ∈ A^F do L[1] = L[1] ∪ {(|ρ(a)|, a, {a}, ρ(a))};
4.   while (HasNewLevel) do
5.   ExpandMerge (L[i]);   // using $\mathcal{EOB}.1$
6.   L[i+1] = ∅;
7.   for each N_{Left} in L[i] do
8.   for each N_{Right} in L[i]: Left < Right; N_{Left}, N_{Right} are in the *same folder* do
9.   NewO = N_{Left}.O ∩ N_{Right}.O; NewSup = |NewO|;
10.   if (NewSup ≠ N_{Left}.Supp and NewSup ≠ N_{Right}.Supp) then
11.   if (NewSup ≥ minsup) then
12.   NewPC = N_{Left}.PC ∪ N_{Right}.PC;   // using $\mathcal{EOA}$
13.   JoinGenerators (N_{Left}, N_{Right}, NewO, NewSup, NewPC, L[i+1]);
14.   InsertLevel (L[i], LCG);   // using $\mathcal{EOC}$
15.   if (L[i+1] = ∅) then HasNewLevel = false;
16.   i = i+1;
17. return LCG; // all frequent closed itemsets, their generators and supports

Fig. 2. *GENCLOSE* algorithm

The first phase is called by *ExpandMerge* procedure for *extending pre-closed item-sets of the nodes at L[i]* as well as *merging them* using operator $\mathcal{EoB}.1$ (see *3.2*). Let N_{Left} and N_{Right} be two nodes at $L[i]$ that *Left* comes before *Right*. It tests which of three following cases is satisfied. If $N_{Left}.O \subset N_{Right}.O$, since property 4 of proposition 1, $h(N_{Left}.PC) \supset h(N_{Right}.PC)$. Toward to closures, $N_{Left}.PC$ should be extended by $N_{Right}.PC$. If $N_{Left}.O \supset N_{Right}.O$, $N_{Right}.PC := N_{Right}.PC \cup N_{Left}.PC$. Otherwise, we push all generators in $N_{Right}.GS$ into $N_{Left}.GS$ and add $N_{Right}.PC$ to $N_{Left}.PC$. Then, we *move the nodes having the same folder* with N_{Right} to the folder including N_{Left} and discard N_{Right}.

The second phase *produces the nodes at L[i+1]* by considering each pair *(N_{Left}, N_{Right})*, written N_{Left}-N_{Right}, at *Lines 7, 8*. Since $NewO = N_{Left}.O \cap N_{Right}.O$ is a new closed object-set, if its cardinality exceeds *minsup*, we need to determine new frequent closed itemset $\lambda(NewO)$. First, we create its core, called *NewPC*. We then jump into *JoinGenerators (Fig. 3)* for mining its generators from i-item generators containing in N_{Left}-N_{Right}. We check *(10)* only for $g \in G_0$ since it is always satisfied for $g=g_i$ and $g=g_{i+1}$. *Case i=1*, we have immediately 2-item generators. *Otherwise*, if we touch a value of $g \in G_0$ such that $G_g = G \backslash \{g\}$ is not an *i*-item generator, it is obviously that *G* is not an *(i+1)*-item generator of *NewPC*. But we should not be in the hurry. It is necessary to consider the latter for *expanding* pre-closed itemsets. Let $Node_g$ be the node including G_g as an *i*-item generator. If it does not exist, i.e. *G* is not a generator, then we move to the next value of *g*. Otherwise, we check if $Node_g.Supp = NewSup$. If yes, i.e. *G* is not a generator, $Node_g.PC$ at $L[i]$ is enlarged by *NewPC*. If no, we expand *NewPC* by $Node_g.PC$. Clearly, *not only new pre-closed itemsets but also the ones at L[i] are expanded*. If *(10)* is satisfied, *G* is a new generator produced from (G_l, G_r). We return to *Line 18* to process next generator pairs. If there exists at least a new *(i+1)*-item generator, we get new node $^{NewSup}NewPC$-$NewGS$-$NewO$ at $L[i+1]$ and then return to *Lines 7, 8* for considering the remaining node pairs.

```
JoinGenerators (N_Left, N_Right, NewO, NewSup, NewPC, L[i+1]):
18.  for each (G_l∈N_Left.GS, G_r∈N_Right.GS: |G_l|=i, |G_r|=i and |G_l∩G_r|=i-1) do
19.      G = G_l∪G_r; G_0 = G_l∩G_r; G_is_Generator = true;
20.      for each g∈G_0 do
21.          G_g = G\{g};
22.          Node_g = SearchNodeWithGenerator (G_g);
23.          if ( Node_g is null or NewSup = Node_g.Supp ) then
24.          {
25.              G_is_Generator = false;
26.              if ( Node_g is not null ) then       // NewSup = |Node_g.O|
27.                  Node_g.PC = Node_g.PC ∪ NewPC;        // using EoB.2
28.          }
29.          else       // G can be a generator!
30.              NewPC = NewPC ∪ Node_g.PC;        // using EoA
31.      if ( G_is_Generator ) then NewGS = NewGS ∪ {G};
32.  if ( |NewGS| ≥ 1 ) then
33.      L[i+1] = L[i+1] ∪ {(NewSup, NewPC, NewGS, NewO)};
```

Fig. 3. *JoinGenerators* procedure

At the **last phase**, we will *finish the extension* for the *PC*s of the nodes of *L[i]* by *InsertLevel* that inserts in turn the nodes at *L[i]* into $\mathcal{LCG}$. Before adding a node *X*, *GENCLOSE* makes a check *if X.PC is closed*. If *there exists P* in $\mathcal{LCG}$ such that $supp(P)=X.Supp$ and $P \supseteq X.PC$, *P* is the closure that *X.PC* wants to touch. We *push* those *new generators* into the generator list of *P*. In contrast, since *X.PC* is a *new closed itemset*, we insert *X.PC*, its generators and support to $\mathcal{LCG}$.

An Example. *Fig. 1* is used through the example. For short, a tree node is written in the form $^{Supp}PC\text{-}GS\text{-}O$. At the beginning, we have the following nodes $D:=^{2}d\text{-}\{d\}\text{-}23$ [3], $A:=^{3}a\text{-}\{a\}\text{-}123$, $F:=^{3}f\text{-}\{f\}\text{-}234$, $G:=^{3}g\text{-}\{g\}\text{-}134$, *B, E, U* and *C*. First, *D* is considered with *A, F, U* and it becomes $^{2}dafu\text{-}\{d\}\text{-}23$ since *D.PC* is contained in their *PC*s. By the similar computations, we get *L[1]*. Next, we obtain *L[2]* by combining the node pairs of *L[1]*. Then, $\mathcal{LCG}$ is $\{^{2}dafu_{d}, {}^{3}au_{a}, {}^{3}fu_{f}, {}^{3}geu_{g}, {}^{4}bc_{b}, {}^{4}e_{e}, {}^{4}u_{u}, {}^{5}c_{c}\}$ where the generators were written in the right at the bottom. At the next step, *AG.PC* and *FG.PC* are taken into the *PC*s of *AB* and *FB*. Then, we merge *DE* into *DG*, thus, *DG* becomes $^{1}adfueg\text{-}\{dg, \underline{de}\}\text{-}3$. We try to consider the next combinations and find that the nodes *AE, FE, GC, BU, EC* should be merged into *AG, FG, GB, BE*, accordingly. Right after, *EC* is moved to the folder including *BE* and then this folder is merged into folder *G*. We look at pair *AG-AC*, we find out two new generators *agc, aec*. Here, *NewPC*, which is initially *aueg+auc*, *is enlarged by* *GB.PC* and becomes *auegcb*. Then, *GENCLOSE* calls *InsertLevel* to insert *L[2]* into $\mathcal{LCG}$. Since *AF.PC=auf* is not closed, *af* is a new 2-item generator of *adfu*. Then, we take into $\mathcal{LCG}$ the *PC*s of the remaining nodes and their corresponding generators. The next computations of *GENCLOSE* are straightforward.

3.2 The Expanded Operators $\mathcal{E}o\mathcal{A}$, $\mathcal{E}o\mathcal{B}$ and $\mathcal{E}o\mathcal{C}$

We call h_F, O_F the pre-closed itemset and object-set according to generator F. Suppose that $\mathcal{LCG}$ contained all frequent closed itemsets that includes *(i-1)*-item generators.

Let G be an i-item generator created by joining two nodes at L[i-1]:
$\mathcal{E}o\mathcal{A}$: h_G *is formed by* ($\forall g \in G: G_g = G\backslash\{g\}$ is an *(i-1)*-item generator):

$$h_G \leftarrow h_G \cup \bigcup_{g \in G} h_{G_g} \tag{13}$$

$\mathcal{E}o\mathcal{B}.1$: h_G *was extended by:*
$$h_G \leftarrow h_G \cup \bigcup_{G\sim \in L[i]:O_G \subseteq O_{G\sim}} h_{G\sim} \tag{14}$$

After joining the nodes at L[i]:
$\mathcal{E}o\mathcal{B}.2$: Let *{a}∪G*, called *aG*, be a new candidate *(i+1)*-item generator. If $|O_{aG}|=|O_G|$, then *aG is not an (i+1)-item generator*. Thus: $h_G \leftarrow h_G \cup \bigcup_{a:|O_{aG}|=|O_G|} h_{aG} \tag{15}$

$\mathcal{E}o\mathcal{C}$: Consider *how to insert* (sup_G, h_G, *GS*) into $\mathcal{LCG}$. First, we check to see if *there exists a closed itemset P being in* $\mathcal{LCG}$ *such that:*

---

[3] We write the set of objects simply their identifiers.

$$supp(P) = supp(h_G) \text{ and } P \supseteq h_G \qquad (16)$$

1) If *yes*, h_G is not a new closed itemset. Thus, we *add i-item generators* to P. 2) If *no*, we *insert new closed itemset* h_G together its *i-item generators* and support into LCG.

Theorem 2. *(The completeness of the expanding operators). After we use those operators for every node at level i containing i-item generators, h_G is closed. Thus, LCG is added by frequent closed itemsets having i-item generators or only those generators.*

Proof:

Case i=1: After we consider the set containment relation for the object-sets according to *1*-item generators: $\forall b \in A: \forall a \in h(b) \Leftrightarrow h(a) \subseteq h(b) \Leftrightarrow O_a \equiv \rho(a) \supseteq O_b \equiv \rho(b))$, therefore: $h_b \leftarrow h_b \cup \{a\}$. Then, $h(b) \subseteq h_b$. Clearly, $h(b) \supseteq h_b$. Hence $h(b) = h_b$.

Case i>1: suppose that the conclusion is true for *1, 2, .., i-1, i≥2*. After EoA, EoB finishes: $h_G = \bigcup_{g \in G} h_{G_g} \cup \bigcup_{G \sim : O_G \subseteq O_{G^-}} h_{G^-} \cup \bigcup_{a:|O_{aG}|=|O_G|} h_{aG}$.

We make the assumption that h_G is *not closed*, $h_G \subset h(G)$. For $\forall a \in h(G) \backslash h_G$ (**), we will prove that there exists $P \in LCG$ such that any present generator of P has at most *i-1* items: $h(P)=h(G) \supset h_G \Leftrightarrow [h(P) \supset h_G \text{ and } \rho(P)=O_G] \Leftrightarrow [h(P) \supset h_G \text{ and } |\rho(P)|=|O_G|]$.

- First, we prove that aG_g is not an i-item generator of $h(aG_g)$. Conversely, joining G with aG_g generates aG. Since $h(G)=h(aG)$, so aG is not a generator. However, since EoB was used, a was added to h_G. That implies $a \in h_G$ which contradicts to (**)!

- Since aG_g is not an *i*-item generator of $h(aG_g)$, there exists a minimal generator $\emptyset \neq G' \subset aG_g: h(G') = h(aG_g)$. (A) If $a \notin G'$, then $G' \subseteq G_g$. Hence, $h(aG_g) = h(G') \subseteq h(G_g) \subseteq h(aG_g)$. It follows that $a \in h(aG_g)=h(G_g)$, i.e. a contradiction to the fact that $a \notin h_G$ (**)! (B) Therefore, $a \in G'$ and there exists $G'=aG_{gB}$, with $\emptyset \neq B \subseteq G_g$: $h(aG_{gB})=h(aG_g)$ $(G_{gB}=G_g \backslash B=G \backslash (gB))$. Since $0 \leq |G_{gB}| \leq i-2$, $1 \leq |G'| \leq i-1$. In other words, generator G' of $h(aG_g)$ has at most i-1 items.

- What is left is to show that $\exists g \in G: h(aG_{gB})=h(aG_g)=h(G) \supset h_G$. It means that we will find out the closure $h(G)$ of h_G from the ancestor nodes $h(G')$ ($\supset h_G$ and $|G'| \leq i-1$) in LCG. Assume that the conversion comes: $\forall g \in G: h(aG_g) \subset h(G)$ $(\Leftrightarrow \rho(G) \subset \rho(aG_g))$ (***) and $\exists g' \in B \subseteq G_g: G'=aG_{gB}$ is a generator of $h(aG_g) = h(aG_{gB})$. Clearly, $g \neq g' \notin G'$. Then $G' = aG_{gB} \subseteq aG_{g'}$. But, we have also: $\rho(aG_{g'}) \subseteq \rho(G_{g'})$, $\rho(aG_{gB}) = \rho(aG_g) \subseteq \rho(G_g)$. Taking the intersection of two sides, we have: $\rho(aG_{g'})=\rho(aG_{g'}) \cap \rho(aG_{gB}) \subseteq \rho(G_{g'}) \cap \rho(G_g)=\rho(G_{g'} \cup G_g)=\rho(G) \subset \rho(aG_{g'})$ (by (***)). The contradiction happens! □

4   Experimental Results

The experiments below were carried out on a i5-2400 CPU, 3.10 GHz @ 3.09 GHz, with 3.16 GB RAM, running under Linux, Cygwin. To test the performance and correctness of *GENCLOSE*, we compare it against *CharmLMG* (http://www.cs.rpi.edu/~zaki) and *D-Touch* (a fast implementation of *Touch*, http://coron.wikidot.com/) on six following benchmark datasets at http://fimi.cs.helsinki.fi/data/: *C20d10k, C73d10k, Pumsb, Pumsbstar, Mushroom* and

Connect. They are highly correlated and dense datasets in terms that they produce many long frequent itemsets as well as only a small fraction of them is closed. The dataset characteristics can be found in [10, 16]. We did not choose the sparse ones since in which almost frequent itemsets are closed and they are generators themselves. We decided to get seven small values of *minsup* thresholds for each dataset, computed on percentages, ranged from: 80% down to 30% for *Connect*, 95% down to 70% for *Pumsb*, 75% down to 45% for *C73d10k*, 40% down to 20% for *Pumsbstar*, 15% down to 0.1% for *C20d10k*, and 18% down to 0.5% for *Mushroom*. The reason is that, for the big ones, the mining processes are often short, so, there is no difference in the performances of *GENCLOSE, CharmLMG* and *D-Touch*.

The experiments show that *the output of GENCLOSE* is *identical to* the ones of *CharmLMG* and *D-Touch*. Let *Time-GENCLOSE, Time-CharmLMG* and *Time-DTouch* be their running times, computed in seconds. *Fig. 4* shows them for *Connect*. For each dataset, we get the average number of the ratios of the running times of *CharmLMG* and *D-Touch* compared to *GENCLOSE (Time-CharmLMG / Time-GENCLOSE* and *Time-DTouch / Time-GENCLOSE)* on all minsup values. *Fig. 5* shows those numbers for all datasets. We find that, in general, *GENCLOSE are over many times faster than CharmLMG* and *D-Touch*. The reductions in the execution time of *GENCLOSE*, compared to *D-Touch* are lowered for *C20d10k* and *Mushroom*. But, one note that, *D-Touch* can not execute: for *C73d100k* with minsups of *50%* and *40%*; for *Pumsb* with minsups *75%, 70%*; for *Pumsb** with minsup = *20%*.

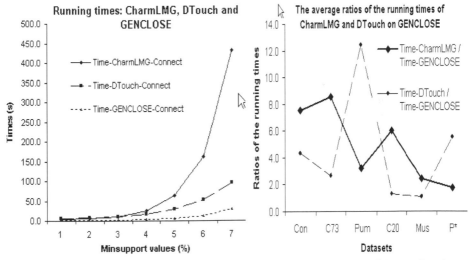

Fig. 4. The running times of *CharmLMG, D-Touch* and *GENCLOSE* on *Connect*

Fig. 5. Average ratios of the running times of *CharmLMG, D-Touch* on *GENCLOSE*

5   Conclusions

We gave some properties of closed sets and generators as well as their relations. Based on them, we developed *GENCLOSE*, an efficient algorithm for mining concurrently frequent closed itemsets and their generators. The background of the algorithm included the necessary and sufficient condition for producing generators and the operators for expanding itemsets were proven reliably. Many tests on benchmark datasets showed its efficiency compared to *CharmLMG* and *D-Touch*.

For mining either closed itemsets or generators separately, the depth-first algorithms usually outperform the level-wise ones. An interesting extension is to develop a depth-first miner based on the proposed approach.

References

1. Agrawal, R., Imielinski, T., Swami, N.: Mining association rules between sets of items in large databases. In: Proceedings of the ACM SIGMOID, pp. 207–216 (1993)
2. Agrawal, R., Srikant, R.: Fast algorithms for mining association rules. In: Proceedings of the 20th International Conference on Very Large Data Bases, pp. 478–499 (1994)
3. Tran, A., Truong, T., Le, B.: Structures of Association Rule Set. In: Pan, J.-S., Chen, S.-M., Nguyen, N.T. (eds.) ACIIDS 2012, Part II. LNCS (LNAI), vol. 7197, pp. 361–370. Springer, Heidelberg (2012)
4. Tran, A., Duong, H., Truong, T., Le, B.: Mining Frequent Itemsets with Dualistic Constraints. In: Anthony, P., Ishizuka, M., Lukose, D. (eds.) PRICAI 2012. LNCS (LNAI), vol. 7458, pp. 807–813. Springer, Heidelberg (2012)
5. Balcazar, J.L.: Redundancy, deduction schemes, and minimum-size base for association rules. Logical Methods in Computer Sciences 6(2:3), 1–33 (2010)
6. Bayardo, R.J.: Efficiently Mining Long Patterns from Databases. In: Proceedings of the SIGMOD Conference, pp. 85–93 (1998)
7. Burdick, D., Calimlim, M., Gehrke, J.: MAFIA: A maximal frequent itemset algorithm for transactional databases. In: Proceedings of ICDE 2001, pp. 443–452 (2001)
8. Boulicaut, J., Bykowski, A., Rigotti, C.: Free-Sets: A Condensed Representation of Boolean Data for the Approximation of Frequency Queries. Data Mining and Knowledge Discovery 7, 5–22 (2003)
9. Dong, G., Jiang, C., Pei, J., Li, J., Wong, L.: Mining Succinct Systems of Minimal Generators of Formal Concepts. In: Zhou, L., Ooi, B.-C., Meng, X. (eds.) DASFAA 2005. LNCS, vol. 3453, pp. 175–187. Springer, Heidelberg (2005)
10. Pasquier, N., Bastide, Y., Taouil, R., Lakhal, L.: Efficient mining of association rules using closed item set lattices. Information Systems 24(1), 25–46 (1999)
11. Pasquier, N., Taouil, R., Bastide, Y., Stumme, G., Lakhal, L.: Generating a condensed representation for association rules. J. of Intelligent Infor. Sys. 24(1), 29–60 (2005)
12. Szathmary, L., Valtchev, P., Napoli, A., Godin, R.: Efficient Vertical Mining of Frequent Closures and Generators. In: Adams, N.M., Robardet, C., Siebes, A., Boulicaut, J.-F. (eds.) IDA 2009. LNCS, vol. 5772, pp. 393–404. Springer, Heidelberg (2009)
13. Wang, J., Han, J., Pei, J.: Closet+: Searching for the best strategies for mining frequent closed itemsets. In: Proceedings of ACM SIGKDD 2003 (2003)
14. Zaki, M.J., Gouda, K.: Fast Vertical Mining Using Diffsets. In: Proc. 9th ACM SIGKDD Int'l Conf. Knowledge Discovery and Data Mining (2003)
15. Zaki, M.J.: Mining non-redundant association rules. Data Mining and Knowledge Discovery 9, 223–248 (2004)
16. Zaki, M.J., Hsiao, C.J.: Efficient algorithms for mining closed itemsets and their lattice structure. IEEE Trans. Knowledge and Data Engineering 17(4), 462–478 (2005)
17. Wille, R.: Concept lattices and conceptual knowledge systems. Computers and Math. with App. 23, 493–515 (1992)

An Efficient Algorithm for Mining Frequent Itemsets with Single Constraint

Hai Duong[1], Tin Truong[1], and Bac Le[2]

[1] Department of Mathematics and Computer Science, University of Dalat, Dalat, Vietnam
{Haidv,tintruong}@dlu.edu.vn
[2] University of Natural Science Ho Chi Minh, Ho Chi Minh, Vietnam
lhbac@fit.hcmus.edu.vn

Abstract. Towards the user, it is necessary to find the frequent itemsets which include a set C_0, especially when C_0 is changed regularly. Our recent studies showed that the frequent itemset mining with often changed constraints should be based on closed itemsets lattice and generators instead of database directly. In this paper, we propose a unique representation of frequent itemsets restricted on constraint C_0 using closed frequent itemsets and their generators. Then, we develop the efficient algorithm to quickly and non-repeatedly generate all frequent itemsets contain C_0. Extensive experiments on a broad range of many synthetic and real datasets showed the effectiveness of our approach.

Keywords: Frequent itemset, closed frequent itemset, generator, constraints mining, closed itemsets lattice.

1 Introduction

Mining frequent itemsets is one of the important tasks in data mining. Although the set of all frequent itemsets is quite huge, users only take care of a small number of them which satisfy some given constraints. As a remedy, the model of constraint-based mining has been developed [5, 17, 21]. Constraints help to focus on interesting knowledge and to reduce the number of patterns extracted to those of potential interest. In addition, they are used for decreasing the search space and enhancing the mining efficiency as well. The two important types of constraints which have been studied by many authors are the anti-monotone constraint [17] denoted as C_{am}, and the monotone constraint [19] denoted as C_m. The constraint C_{am} is simple and suitable with Apriori-like algorithms, so it is often integrated into them to prune candidates. On the contrary, the C_m is more complicated to exploit and less effective to prune the search space.

In this paper, we consider a type of C_m as follows. For a database T included in the set A of all items, let A^F be the set of all frequent ones and C_0 be a constraint subset ($C_0 \subseteq A$). The task is to find out all the frequent itemsets containing C_0 (constraint C_m). For example, users need to know frequent keyword sets including some given keywords. They can quickly lead users to the desired documents. A simple approach

N.T. Nguyen, T. Van Do, and H.A. Le Thi (Eds.): *ICCSAMA 2013*, SCI 479, pp. 367–378.
DOI: 10.1007/978-3-319-00293-4_28   © Springer International Publishing Switzerland 2013

is to mine first frequent itemsets by one of the algorithms, such as Apriori [1, 16], Eclat [24], FP-growth [20], and then the ones containing C_0 are filtered out in a post-processing step. This approach is inefficient because of the following disadvantages. First, it often has to test a huge number of itemsets in the last step. Second, we need to execute the algorithms again whenever the constraint is changed. Thus, our system is hard to immediately return frequent itemsets to users. A solution is to mine and save only once all frequent itemsets for the small values of minimum support. Then, whenever they are changed, the system extracts the satisfied ones. The cost for this second step seems to be very high because the number of frequent itemsets generated from the first step is usually enormous. Moreover, we have to use a lot of memory to store them.

Another promising solution is to combine the advantages of both constrained mining and condensed representation of frequent itemsets. Instead of mining all frequent itemsets, only a small number of the condensed ones are extracted. Condensed representation has three primary advantages. First, it is easier to store because the number of condensed ones is much smaller than the size of the class of all frequent ones, especially on dense databases. Second, we exploit it only once even when the constraints are changed. And last, the condensed representation can be used to generate all frequent ones and this generation can be performed without any access to database T. A type of condensed representation is maximal frequent itemsets [13, 15]. Since their cardinality is very small, so they can be discovered quickly. Further, we can generate all frequent itemsets from the maximal ones. However, that generation produces duplications. In addition, the generated frequent subsets can lose information about the support. Therefore, it needs to be recomputed when mining association rules. The other type of condensed representation is the closed frequent itemsets and their generators [9, 10, 11, 18] in which the generators are minimum and the closed ones are maximum. Each maximum one represents a class of all frequent itemsets having the same closure.

Based on the approach which uses the lattice of the closed itemsets and their generators, we recently applied successfully in mining frequent itemsets FI with both C_{am} and C_m separately. More details, in [2, 3] we used the lattice to mine all frequent itemsets contained in a subset C of set of all items on a given database (FI $\subseteq$ C). And in [4] we applied it to find all ones containing at least an item of a subset C, i.e., intersection of FI and C is not empty (FI $\cap$ C $\neq \varnothing$). This paper continues to use the lattice for mining frequent itemsets including above constraint subset C_0 ($C_0 \subseteq$ FI). The model proposed in the paper for this constraint can be described abstractly as follows. First, we mine only once the class LG_A containing the lattice of closed itemsets and their generators from T. Second, when C_0 is changed, we quickly determine from LG_A the class $FCS_{\supseteq C0}$ of closed frequent itemsets including C_0 and generators. Finally, from $FCS_{\supseteq C0}$, we completely, quickly and non-repeatedly generate all frequent itemsets with constraint using a structure and unique representation of frequent itemsets. The efficient algorithm is also proposed to execute the corresponding steps in that model.

The rest of the paper is organized as follows. Section 2 recalls some basic concepts in frequent itemset mining and some notations. In section 3, we present a unique representation of frequent itemsets with constraint and propose the efficient algorithm to exploit all frequent ones satisfying constraint C_0. Experimental results will be discussed in section 4. Finally, a short conclusion is presented in section 5.

2   Some Concepts and Notations

For a database T, let $\mathcal{O}$ be a non-empty set containing transactions, A be a set of items appearing in those transactions and $\mathcal{R}$ a binary relation on $\mathcal{O}$ x A. A set of items is called an itemset. Consider two operators: $\lambda: 2^{\mathcal{O}} \to 2^{A}$ and $\rho: 2^{A} \to 2^{\mathcal{O}}$ defined as follows ($\lambda(\varnothing) := A$ and $\rho(\varnothing) := \mathcal{O}$): $\forall O \subseteq \mathcal{O}, \forall A \subseteq A, \lambda(O) = \{a \in A \mid (o, a) \in \mathcal{R}, \forall o \in O\}$, $\rho(A) = \{o \in \mathcal{O} \mid (o, a) \in \mathcal{R}, \forall a \in A\}$. A closed operator h in 2^{A} [6, 26] is defined: $h = \lambda$ o ρ. Denote h(A) as the closure of subset A $\subseteq A$. A is called closed itemset if h(A)=A. The class of all closed itemsets is denoted as CS. The support of itemset A in T, denoted supp(A), is the number of the ratio of $|\rho(A)|$ to $|\mathcal{O}|$, i.e., supp(A) = $|\rho(A)| / |\mathcal{O}|$. The minimum support threshold is denoted s_0, with $s_0 \in [1/|\mathcal{O}|;$ 1]. An itemset A is called *frequent* if its support is no less than s_0, i.e., $s_0 \leq$ supp(A). Let FS and FCS be the classes of all frequent itemsets and all closed frequent itemsets with s_0. For any two sets G, A: $\varnothing \neq G \subseteq A \subseteq A$, G is called a generator [16] of A if h(G) = h(A) and ($\forall \varnothing \neq G' \subset G \Rightarrow h(G') \subset h(G)$). Let $G(A)$ be the class of all generators of A and LG_A the lattice of all closed itemsets and their generators.

Given that C_0 is constraint subset with $\varnothing \neq C_0 \subseteq A$. We denote $FS_{\supseteq C0} = \{L' \subseteq A:$ $L' \supseteq C_0$, supp(L') $\geq s_0\}$ the class of all frequent itemsets including C_0, $FCS_{\supseteq C0} = \{$ L $\in FCS \mid$ L $\supseteq C_0\}$ the class of all closed ones including C_0, $FS(L) = \{L' \subseteq L: h(L') = h(L)\}$ the class of all ones having the same closure L with L $\in FCS$, and $FS_{\supseteq C0}(L) = \{L' \subseteq L: L' \supseteq C_0, h(L') = h(L)\}$ the class of all frequent itemsets in $FS(L)$ including C_0, with L $\in FCS_{\supseteq C0}$.

Remark 1. $\forall L \in FCS, \forall L' \in [L]$, if $C_0 \subseteq L'$ then supp(C_0) $\geq$ supp(L') = supp(L) $\geq s_0$. Thus, we only consider constraint subsets C_0 which are frequent.

3   Mining Frequent Itemsets with Single Constraint

Definition 1. [22] (*Equivalence relation $\sim_h$ over 2^A*). $\forall A, B \in FS$:

$$A \sim_h B \Leftrightarrow h(A) = h(B).$$

Using this relation, we partition FS into the disjoint equivalence classes. Each class contains frequent itemsets having the same closure L $\in FCS$. We have the following theorem:

Theorem 1. [22] (*A partition of* FS).

$$FS = \sum_{L \in FCS} [L].$$

This partition allows us to independently exploit frequent itemsets with constraint in each equivalence class.

Example 1. Let us consider database T that is used in the rest of the paper in Fig. 1.a. For $s_0 = ¼$, using *Charm-L* [25] and *MinimalGenerators* [23] we mine the lattice of all closed frequent itemsets and their generators. The result is showed in Fig. 1.b. Then, we have FS = [adfg] + [deg] + [bcf] + [bce] + [dg] + [f] + [e] + [bc].[1] Thanks to this disjoint partition, we can significantly reduce the duplication in the mining process of frequent itemsets. However, theorem 2 in [2] showed that frequent itemsets generated in each class can be still duplicated.

Trans	Items
1	bce
2	adfg
3	bcf
4	deg

(a) *Database* T

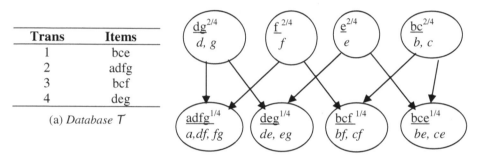

(b) The lattice of all closed itemsets (underline) and their generators (italic)

Fig. 1. Database and the corresponding lattice of closed itemsets

3.1   Partition the Class of Frequent Itemsets with Single Constraint

Based on the idea of the above partition, we have the proposition as follows:

Proposition 1. (*The disjoint partition the class of all frequent itemsets with constraint*). $\forall C_0 \subseteq A$ and $s_0 \le \text{supp}(C_0)$:

$$FS_{\supseteq C_0} = \sum_{L \in FCS_{\supseteq C_0}} FS_{\supseteq C_0}(L).$$

Proof:

- The sets in the right side are disjoint. In fact, $\forall L'_i \in FS_{\supseteq C_0}(L_i)$, where $L_i \in FCS_{\supseteq C_0}$, i = 1, 2 and $L_1 \ne L_2$. We have $h(L'_1) = L_1 \ne L_2 = h(L'_2)$. Thus, L'_1 and L'_2 are in the different equivalence classes $[L_1]$ and $[L_2]$. Hence, $L'_1 \ne L'_2$.

---

[1] The symbol + is denoted as the union of two disjoint sets.

- "⊆": $\forall L' \in FS_{\supseteq C0}$, assign that $L = h(L')$, we have: $supp(L') = supp(L) \geq s_0$. So $L \in FCS$. Moreover, we know that $C_0 \subseteq L' \subseteq L$. Therefore, $L \in FCS_{\supseteq C0}$. We conclude that $L' \in FS_{\supseteq C0}(L)$.
- "⊇": $\forall L \in FCS_{\supseteq C0}$, $L' \in FS_{\supseteq C0}(L)$, we have $C_0 \subseteq L'$ and $h(L') = h(L) = L$. Thus, $supp(L') = supp(L) \geq s_0$. Hence, $L' \in FS_{\supseteq C0}$.   □

Remark 2. The disjoint partition of the class of all frequent itemsets allows us to design parallel algorithms which can independently exploit each class. Then, the mining time can be reduced significantly.

Example 2. Fixed $s_0 = \frac{1}{4}$, we consider $L = adfg$, $G(L) = \{L_1 = a, L_2 = df, L_3 = fg\}$. From theorem 3 in [2], we obtain: $X_U = adfg$, $X_{U,1} = dfg$, $X_{U,2} = ag$, $X_{U,3} = ad$, $X_- = \varnothing$. Thus, $[adfg] = \{a, ad, af, ag, adf, adg, afg, adfg, df, dfg, fg\}$. For $C_0 = dg$, we test frequent itemsets L' of $[adfg]$ by the condition $C_0 \subseteq L'$ and get $FS_{\supseteq C0}(adfg) = \{adg, dfg, adfg\}$. Similarly, we have: $FS_{\supseteq C0}(deg) = \{deg\}$, $FS_{\supseteq C0}(dg) = \{dg\}$. Thus, $FS_{\supseteq C0} = FS_{\supseteq C0}(adfg) + FS_{\supseteq C0}(deg) + FS_{\supseteq C0}(dg) = \{adg, dfg, adfg, deg, dg\}$.

Following example 2, we find that the test of the condition $C_0 \subseteq L'$ is very expensive. In the next section, we propose the method to efficiently mine the frequent itemsets with constraint in each equivalence class.

3.2 Mining all Frequent Itemsets with Single Constraint

In this section, we point out the unique representation and structure of frequent itemset with constraint. For each $L \in FCS_{\supseteq C0}$, let $K_{min, C_0} := Minimal\{K_i = L_i \setminus C_0 \mid L_i \in G(L)\}$ be the class of all the minimum itemsets of $\{L_i \setminus C_0 \mid L_i \in G(L)\}$ in terms of the set containment order. Assign that $K_{U,C_0} := \bigcup_{K_i \in K_{min, C_0}} K_i$, $K_{U,C_0,i} := K_{U,C_0} \setminus K_i$,

$K_{-,C_0} := L \setminus (K_{U,C_0} + C_0)$, let us define:

$$FS^*_{\supseteq C0}(L) = \{L' = C_0 + K_i + K'_i + K^\sim \mid K_i \in K_{min, C_0}, K'_i \subseteq K_{U,C_0,i}, K^\sim$$

$$\subseteq K_{-,C_0} \text{ and } (K_j \not\subseteq K_i + K'_i, \forall K_j \in K_{min, C_0} : 1 \leq j < i)\}. \qquad (*)$$

Remark 3. If there exists $L_i \in G(L)$ such that $K_i = L_i \setminus C_0 = \varnothing$ then $FS^*_{\supseteq C0}(L) = \{L' = C_0 + L'', L'' \subseteq L \setminus C_0\}$. In this way, the frequent itemsets with constraint are generated more quickly. Indeed, if there exists $L_i \in G(L)$ such that $K_i = L_i \setminus C_0 = \varnothing$ which implies that $L_i \subseteq C_0$, then $K_{min, C_0} = \varnothing$, $K_{U,C_0} = \varnothing$, $K_{U,C_0,i} = \varnothing$. It deduce that $K_{-,C_0} = L \setminus C_0$. Thus, $L' = C_0 + L'' \in FS^*_{\supseteq C0}(L)$, where $L'' \subseteq L \setminus C_0$.

Theorem 2. (*Generating non-repeatedly the elements of $FS^*_{\supseteq C0}(L)$*).

For each $L \in \mathcal{FCS}_{\supseteq C0}$, we have:

a) $\mathcal{FS}_{\supseteq C0}(L) = \mathcal{FS}^*_{\supseteq C0}(L)$.

b) *The elements of* $\mathcal{FS}^*_{\supseteq C0}(L)$ *are generated non-repeatedly.*

Proof. a)

- "⊆": If $L' \in \mathcal{FS}_{\supseteq C0}(L)$, we always select the lowest index i such that $L_i \in \mathcal{G}(L)$ and $K_i = L_i \backslash C_0 \in K_{min, C_0}$ is the minimum set. According to theorem 3 in [2], we have

  $L' = L_i + L'_i + L^\sim$, where $L_i \in \mathcal{G}(L)$, $L_U = \bigcup_{L_i \in \mathcal{G}(L)} L_i$, $L'_i \subseteq L_{U,i} = L_U \backslash L_i$, $L^\sim \subseteq L = L \backslash L_U$. Thus, $L' = C_0 \cap (L_i + L'_i + L^\sim) + (L_i \backslash C_0) + (L'_i \backslash C_0) + (L^\sim \backslash C_0) = C_0 + K_i + (L'_i \backslash C_0) + (L^\sim \backslash C_0) = C_0 + K_i + K'_i + K^\sim$, where $K'_i = (L'_i \backslash C_0) \cap K_{U,C_0} \subseteq K_{U,C_0} \cap (L_U \backslash L_i) \subseteq K_{U,C_0} \backslash L_i \subseteq K_{U,C_0} \backslash K_i = K_{U,C_0,i}$ and $K^\sim = [(L_i' \backslash C_0) \backslash K_{U,C_0} + (L^\sim \backslash C_0)] \subseteq (L \backslash (C_0 + K_{U,C_0})) + (L \backslash K_U) \backslash C_0 \subseteq L \backslash (C_0 + K_{U,C_0}) = K_{\_,C_0}$. Hence, $L' \in \mathcal{FS}^*_{\supseteq C0}(L)$.

- "⊇": If $L' \in \mathcal{FS}^*_{\supseteq C0}(L)$, there exists $K_i = L_i \backslash C_0 \in K_{min, C_0}$, $L_i \in \mathcal{G}(L)$, $K'_i \subseteq K_{U,C_0,i}$ and $K^\sim \subseteq K_{\_,C_0}$: $C_0 \subseteq L' = C_0 + K_i + K'_i + K^\sim \subseteq L$. Thus, $h(L') \subseteq h(L)$. On the other hand, since $L_i = K_i + (L_i \cap C_0) \subseteq K_i + C_0 \subseteq L'$, so $h(L) = h(L_i) \subseteq h(L')$. Therefore, $h(L') = h(L)$. Hence, $L' \in \mathcal{FS}_{\supseteq C0}(L)$.

b) Assume that there exists i, k with $i > k \geq 1$ such that $L'_k \equiv L'_i$, where $L'_k := C_0 + K_k + K'_k + K^\sim_k$, $L'_i := C_0 + K_i + K'_i + K^\sim_i$, $K_k, K_i \in K_{min, C_0}$, $K_k \neq K_i$, $K^\sim_k, K^\sim_i \subseteq K_{\_,C_0}$, $K'_k \subseteq K_{U,C_0,k}$, $K'_i \subseteq K_{U,C_0,i}$. Since $K_k \cap C_0 = \emptyset$ and $K_k \cap K^\sim_i = \emptyset$, so $K_k \subset K_i + K'_i$ (the equality does not occur because K_i and K_k are two different minimum sets). It contradicts to the way that we select i. Therefore, all elements of $\mathcal{FS}^*_{\supseteq C0}(L)$ are generated non-repeatedly.

Example 3. For $s_0 = \frac{1}{4}$ and $C_0 = dg$. Consider $L = adfg$, $\mathcal{G}(L) = \{L_1 = a, L_2 = df, L_3 = fg\}$. We have: $K_1 = L_1 \backslash C_0 = a$, $K_2 = L_2 \backslash C_0 = f$, $K_3 = L_3 \backslash C_0 = f$, $K_{min, C_0} = $ Minimal$\{K_1, K_2, K_3\} = \{K_1, K_2\} = \{a, f\}$, $K_{U,C_0} = af$, $K_{U,C_0,1} = f$, $K_{U,C_0,2} = a$, $K_{\_,C_0} = \emptyset$. Consider K_1: dg+a, dg+af $\in \mathcal{FS}_{\supseteq C0}(adfg)$. Consider K_2: dg+f $\in \mathcal{FS}_{\supseteq C0}(adfg)$. Thus, $\mathcal{FS}_{\supseteq C0}(adfg) = \{adg, afdg, fdg\}$.

According to theorem 2, we abtain procedure *MFS-Contain-IC-OneClass* of which the pseudo code is presented in Fig. 2 for mining frequent itemsets with single constraint in a class. Using proposition 1 and this procedure, we propose algorithm *MFS-Contain-IC*, shown in Fig. 3, for mining all frequent ones with single constraint.

```
FS⊇c0(L)   MFS-Contain-IC-OneClass(L, G(L), C0)
 1.  ExistsLiInC0 = False;  FS⊇c0(L)= ∅;
 2.  for each (Li ∈ G(L)) do {
 3.         Ki = Li \ C0;
 4.         if (Ki= ∅) then {
 5.                ExistsLiInC0 = True; break;
 6.         }
 7.  }
 8.  if (ExistsLiInC0 = True) then {  //remark 2
 9.         For each (L'' ⊆ L\C0) do
10.                FS⊇c0(L) = FS⊇c0(L) + {C0 + L''};
11.  }
12.  else
13.  { Kmin,c0= Minimal{Ki} ;
```

$$14. \quad K_{U,C_0} = \bigcup_{K_i \in K_{min,C_0}} K_i \; ; \quad K_{-,C_0} = L \setminus (K_{U,C_0} + C_0);$$

```
15.   for (i=1; Ki ∈Kmin,c0; i++) do {
16.      KU,c0,i = KU,c0 \ Ki ;
17.       for each (K'i ⊆ KU,c0,i) do {
18.       IsDuplicate = false;   //Test condition(*)
19.       for (k=1; k<i; k++) do
20.          if (Kk ⊂ Ki+K'i) then
21.               { IsDuplicate = true; break; }
22.       if (not(IsDuplicate)) then
23.         for each (K˜ ⊆ K_, c0 ) do
24.             FS⊇c0(L) = FS⊇c0(L) + {C0 + Ki + K'i + K˜};
25.       }
26.   }
27.  }
28.  return  FS⊇c0(L);
```

Fig. 2. *MFS-Contain-IC-OneClass* procedure

```
MFS-Contain-IC(C0, s0)
1. FS⊇c0 = ∅;
2. if (supp(C0) < s0) then return FS⊇c0;  // Remark 1
3. For each (L∈ CS)  do
4.   if (supp(L)≥s0 and C0 ⊆ L) then {
5.   FS(L)⊇c0 = MFS-Contain-IC-OneClass(L,G(L),C0);
6.        FS⊇c0 = FS⊇c0 + FS(L)⊇c0;
7.   }
8. return  FS⊇c0;
```

Fig. 3. *MFS-Contain-IC* algorithm

4   Experimental Results

Experiments were performed on a i5-2400 CPU, 3.10GHz@ 3.09GHz PC with 3.16GB of memory, running Windows XP. Algorithm were coded in $C^{\#}$. To compare the performance, we used the source for *Charm-L, MinimalGenerators* and *dEclat* [28], converted to C#. *Charm-L* and *MinimalGenerators* are used in order to mine the lattice of the closed itemsets and their generators. *dEclat* is to exploit all frequent itemsets. Here, we modified it towards the mining frequent itemsets with constraint. More details, we filter the output thereof to determine frequent itemsets including C_0. This new version is called *SC- dEclat*. In a post-processing approach, we also create *SC-GenItemsets* from *Gen_Itemsets* [2].

For the performance test, we chose several benchmark databases in FIMDR [27]: Pumsb, Connect, Mushroom, T10I4D100K and T40I10D100K. Pumsb, Connect, Mushroom are real and dense ones, i.e. they produce many long frequent itemsets even for very high support values. The others are synthetic and sparse. Table 1 shows their characteristics.

Table 1. Databases' characteristics

Database	#Items	#Records	Avg. Length
Connect (C)	129	67557	43
Mushroom (M)	119	8124	23
Pumsb (P)	7117	49046	74
T10I4D100K (T10)	1000	100.000	10
T40I10D100K (T40)	1000	100.000	40

With dense databases, for each pair of database (DB) and minimum support (MS), the size of C_0 ranges from 4% to 14% of $|A^F|$ (step 2%). For sparse ones, its size ranges from 0.2% to 2% of $|A^F|$ (step 0.2%). For each C_0's size, we select 10 constraints for the dense ones and 6 constraints for the sparse ones (each of them contains items randomly selected from A^F). Let T_MCI, T_SCG and T_SCE be the average execution time of *MFS-Contain-IC*, *SC-GenItemsets* and *SC-dEclat* on 60 selected constraints.

Table 2 contains the experimental evaluation of *MFS-Contain-IC* against *SC-GenItemsets* and *SC-dEclat*, where: the average number of the percent ratios of T_MCI to T_SCG and T_SCE are shown in columns R_MG and R_ME (%). We call N the number of all frequent itemsets and NNC the number of frequent itemsets which do not contain C_0. Column RR is used to indicate the average number of the percent ratios of NNC to N. Comparing with *SC-GenItemsets*, for dense databases, we find that *MFS-Contain-IC* executes more quickly. The time can be reduced from 30.9% to 6.1%. In other words, we can save the amounts of the time ranging from 69.1% to 93.9%. In sparse ones, this reduction is lower. The reason is that the number of the frequent itemsets is small and their size is short, leading to a low cost to test the constraints. Comparing with *SC-dEclat*, in all databases, *MFS-Contain-IC* runs much more quickly. The time reductions are fluctuated from 13.8% to 2.3%. The reason is that there are a huge number of candidates (RR ranges from 98.4% to 99.9%) which fail the last test of *SC-dEclat*, leading to lower performance thereof.

Table 2. The time reductions of *MFS-Contain-IC*, compared to *SC-GenItemsets* and *SC-dEclat*

DB, MS	R_MG (%)	R_ME (%)	RR (%)	DB, MS	R_MG (%)	R_ME (%)	RR (%)
M,15	8.11	5.8	98.6	M,5	6.1	3.9	99.8
M,10	6.1	5.2	99.6	M,3	6.3	3.9	99.5
P,75	25.1	3.7	99.8	T10,0.09	57.3	3.9	99.6
P,70	20.6	5.5	99.7	T10,0.07	59.3	3.8	99.9
P,68	22.1	7.5	98.4	T10,0.04	63.1	3.5	99.7
C,75	15.5	2.3	99.9	T40,2	95.6	3.5	99.9
C,7	14.3	2.6	99.9	T40,0.9	94.3	11.9	99.6
C,65	30.9	6.1	99.4	T40,0.6	94.5	13.8	99.4

Fig. 4 contains the comparison of the average execution times over many support values. It is realized that there are the considerable enhancements in performance and scalability of *MFS-Contain-IC* in comparison to *SC-GenItemsets* and *SC-dEclat*.

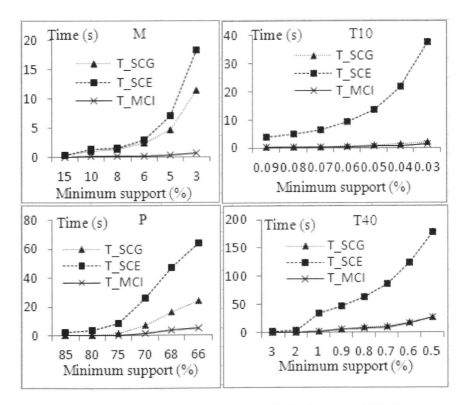

Fig. 4. *MFS-Contain-IC* in comparison with *SC-GenItemsets* and *SC-dEclat*

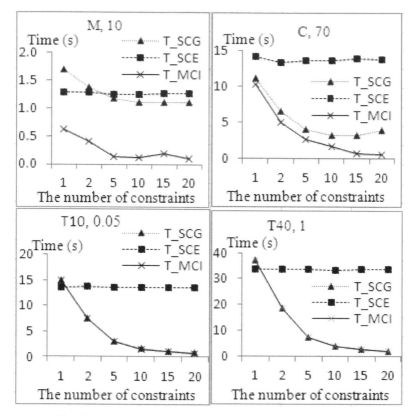

Fig. 5. *MFS-Contain-IC* and *SC-GenItemsets* against *SC-dEclat*

Further, Fig. 5 shows the performance of *MFS-Contain-IC* along the changes of the number of constraints. It is clear that the performance gap between *MFS-Contain-IC* and *SC-dEclat* widens for a higher number of constraints. The main reason is that, when the constraints change, *MFS-Contain-IC* executes without the exploration of closed itemset lattice and their generators again from database.

To conclude, one can see that *MFS-Contain-IC* outperforms both *SC-GenItemsets* and *SC-dEclat*, especially as minimum support is lowered and the number of constraints is increased.

5   Conclusion

We presented the unique representation and structure of frequent itemsets with constraint. The correctness of the theoretical results was reliably proven. The corresponding efficient algorithm was developed for exploiting all of them. The tests on the benchmark databases showed the efficiency of our approach. Moreover, the tests also showed the outstanding advantages of the algorithm when the minimum support values are very low, especially when constraint is changed regularly. In the future, we will study frequent itemset mining on more complicated types of constraints and exploit association rules based on them.

References

1. Agrawal, R., Srikant, R.: Fast algorithms for mining association rules. In: Proc. of the 20th International Conference on Very Large Data Bases, pp. 478–499 (1994)
2. Anh, T., Hai, D., Tin, T., Bac, L.: Efficient Algorithms for Mining Frequent Itemsets with Constraint. In: Proc. of the Third International Conference on Knowledge and Systems Engineering, pp. 19–25 (2011)
3. Tran, A., Duong, H., Truong, T., Le, B.: Mining Frequent Itemsets with Dualistic Constraints. In: Anthony, P., Ishizuka, M., Lukose, D. (eds.) PRICAI 2012. LNCS (LNAI), vol. 7458, pp. 807–813. Springer, Heidelberg (2012)
4. Anh, T., Tin, T., Bac, L., Hai, D.: Mining Association Rules Restricted on Constraint. In: Proc. IEEE-RIVF 2012, pp. 51–56 (2012)
5. Bayardo, R.J., Agrawal, R., Gunopulos, D.: Constraint-Based Rule Mining in Large, Dense Databases. Data Mining and Knowledge Discovery 4, 217–240 (2000)
6. Birkhoff, G.: Lattice theory. American Mathematical Society, New York (1948)
7. Bonchi, F., Giannotti, F., Mazzanti, A., Pedreschi, D.: Examiner: Optimized level-wise frequent pattern mining with monotone constraints. In: Proc. IEEE ICDM 2003, pp. 11–18 (2003)
8. Bonchi, F., Giannotti, F., Mazzanti, A., Pedreschi, D.: ExAnte: Anticipated data reduction in constrained pattern mining. In: Lavrač, N., Gamberger, D., Todorovski, L., Blockeel, H. (eds.) PKDD 2003. LNCS (LNAI), vol. 2838, pp. 59–70. Springer, Heidelberg (2003)
9. Bonchi, F., Lucchese, C.: On closed constrained frequent pattern mining. In: Proc. IEEE ICDM 2004 (2004)
10. Boulicaut, J.F., Bykowski, A.: Frequent closures as a concise representation for binary Data Mining. In: Terano, T., Chen, A.L.P. (eds.) PAKDD 2000. LNCS, vol. 1805, pp. 62–73. Springer, Heidelberg (2000)
11. Boulicaut, J.F., Bykowski, A., Rigotti, C.: Free-sets: a condensed representation of boolean data for the approximation of frequency queries. Data Mining and Knowledge Discovery 7, 5–22 (2003)
12. Bucila, C., Gehrke, J.E., Kifer, D., White, W.: Dualminer: A dual-pruning algorithm for itemsets with constraints. Data Mining and Knowledge Discovery 7, 241–272 (2003)
13. Burdick, D., Calimlim, M., Gehrke, J.: MAFIA: A maximal frequent itemset algorithm for transactional databases. In: Proc. IEEE ICDE 2001, pp. 443–452 (2001)
14. Jeudy, B., Boulicaut, J.F.: Optimization of association rule mining queries. Intelligent Data Analysis 6, 341–357 (2002)
15. Lin, D.I., Kedem, Z.M.: Pincer search: An efficient algorithm for discovering the maximum frequent sets. IEEE Trans. on Knowledge and Data Engineering 14, 553–566 (2002)
16. Mannila, H., Toivonen, H.: Levelwise search and borders of theories in knowledge discovery. Data Mining and Knowledge Discovery 1, 241–258 (1997)
17. Nguyen, R.T., Lakshmanan, V.S., Han, J., Pang, A.: Exploratory Mining and Pruning Optimizations of Constrained Association Rules. In: Proc. of the 1998 ACM-SIG-MOD Int'l Conf. on the Management of Data, pp. 13–24 (1998)
18. Pasquier, N., Taouil, R., Bastide, Y., Stumme, G., Lakhal, L.: Generating a condensed representation for association rules. J. Intelligent Information Systems 24, 29–60 (2005)
19. Pei, J., Han, J., Lakshmanan, L.V.S.: Mining frequent itemsets with convertible constraints. In: Proc. IEEE ICDE 2001, pp. 433–442 (2001)
20. Pei, J., Han, J.: Constrained Frequent Pattern Mining: A Pattern-Growth View. In: Proc. ACM SIGKDD Explorations (Special Issue on Constraints in Data Mining), vol. 4, pp. 31–39 (2002)

21. Srikant, R., Vu, Q., Agrawal, R.: Mining association rules with item constraints. In: Proc. KDD 1997, pp. 67–73 (1997)
22. Truong, T.C., Tran, A.N.: Structure of set of association rules based on concept lattice. In: Nguyen, N.T., Katarzyniak, R., Chen, S.-M. (eds.) Advances in Intelligent Information and Database Systems. SCI, vol. 283, pp. 217–227. Springer, Heidelberg (2010)
23. Zaki, M.J.: Mining non-redundant association rules. Data Mining and Knowledge Discovery, 223–248 (2004)
24. Zaki, M.J., Parthasarathy, S., Ogihara, M., Li, W.: New algorithms for fast discovery of association rules. In: Proc. 3rd Int. Conf. on Knowledge Discovery and Data Mining (KDD 1997), pp. 283–296 (1997)
25. Zaki, M.J., Hsiao, C.J.: Efficient algorithms for mining closed itemsets and their lattice structure. IEEE Trans. Knowledge and Data Engineering 17, 462–478 (2005)
26. Wille, R.: Concept lattices and conceptual knowledge systems. Computers and Math. with App. 23, 493–515 (1992)
27. Frequent Itemset Mining Dataset Repository (FIMDR),
    http://fimi.cs.helsinki.fi/data/ (accessed 2009)\
28. http://www.cs.rpi.edu/~zaki/wwwnew/pmwiki.php/Software/
    Software#patutils (acessed 2010)

Mining Frequent Weighted Closed Itemsets

Bay Vo[1], Nhu-Y Tran[2], and Duong-Ha Ngo[2]

[1] Information Technology College, Ho Chi Minh City, Viet Nam
vdbay@itc.edu.vn
[2] Department of Information Technology, Ho Chi Minh City of Food Industry, Viet Nam
{ytn,hand}@cntp.edu.vn

Abstract. Mining frequent itemsets plays an important role in mining association rules. One of methods for mining frequent itemsets is mining frequent weighted itemsets (FWIs). However, the number of FWIs is often very large when the database is large. Besides, FWIs will generate a lot of rules and some of them are redundant. In this paper, a method for mining frequent weighted closed itemsets (FWCIs) in weighted items transaction databases is proposed. Some theorems are derived first, and based on them, an algorithm for mining FWCIs is proposed. Experimental results show that the number of FWCIs is always smaller than that of FWIs and the mining time is also better.

Keywords: data mining, frequent weighted support, frequent weighted itemsets, frequent weighted closed itemsets.

1   Introduction

Classical association rule mining does not take into consideration the relative benefit of items. However, in some applications, we are interested in relative benefit (weighted value) associated with each item. For example, suppose bread incurs profit 20 cents and a bottle of milk 40 cents. It is thus desirable to identify new methods for applying Association Rule Mining (ARM) techniques to this kind data so that such relative benefits are taken into account.

In 1998, Ramkumar et al. [12] proposed a model for describing the concept of Weighted Association Rules (WARs) and presented an Apriori-based algorithm for mining Frequent Weighted Itemsets (FWIs). Some further ideas concerning with the mining of WARs are also discussed in [2]. Since then many Weighted Association Rule Mining (WARM) techniques have been proposed [14, 17, 20]. Khan et al. [4] extended the concept of WARM and proposed a method for mining fuzzy WARs.

The advantage offered by adopting FWCI mining is that there are fewer FWCI's than frequent itemsets, and hence computational efficiencies may be introduced. However, for mining efficient association rules, we need mine FWCIs [1, 9-10, 16, 21-22]. This paper proposes a technique for mining FWCIs. Some theorems are also proposed. Based on them and WIT-tree [5-6, 17], we develop an algorithm for fast mining FWCIs.

N.T. Nguyen, T. Van Do, and H.A. Le Thi (Eds.): *ICCSAMA 2013*, SCI 479, pp. 379–390.
DOI: 10.1007/978-3-319-00293-4_29   © Springer International Publishing Switzerland 2013

The rest of this paper is organized as follows. Section 2 presents some related work concerning the mining of FWIs and WARs. Section 3 presents a proposed modification of WIT-tree [5-6, 17] for compressing the database into a tree structure. An algorithm for mining FWCIs using WIT-trees is discussed in section 4. Experimental results are presented in section 5. Conclusions and future work will present in section 6.

2   Related Work

2.1   Weighted Items Transaction Databases

A weighted items transaction database (D) is defined as follows: D comprises a set of transactions $T = \{t_1, t_2, ..., t_m\}$, a set of items $I = \{i_1, i_2, ..., i_n\}$ and a set of positive weights $W = \{w_1, w_2, ..., w_n\}$ corresponding to each item in I. For example, consider the data presented in Table 1 and Table 2. Table 1 presents a data set comprising six transactions $T = \{t_1, ..., t_6\}$, and five items $I = \{A, B, C, D, E\}$. The weights of these items are presented in Table 2, $W = \{0.6, 0.1, 0.3, 0.9, 0.2\}$.

Table 1. The transaction database

Transactions	Bought items
1	A, B, D, E
2	B, C, E
3	A, B, D, E
4	A, B, C, E
5	A, B, C, D, E
6	B, C, D

Table 2. Item weights

Items	Weight
A	0.6
B	0.1
C	0.3
D	0.9
E	0.2

a) Galois connection

Let $\delta \subseteq I \times T$ be a binary relation, where I is a set of items and T is a set of transactions contained in the database D. Let $X \subseteq I$ and $Y \subseteq T$. Let $P(S)$ include all subsets of S. Two mappings between $P(I)$ and $P(T)$ are called Galois connections as follows [22]:

i) $t : P(I) \mapsto P(T), \ t(X) = \{y \in T \mid \forall x \in X, x \delta y\}$

ii) $i : P(T) \mapsto P(I), \ i(Y) = \{x \in I \mid \forall y \in Y, x \delta y\}$

The mapping $t(X)$ is the set of transactions in the database which contain X, and the mapping $i(Y)$ is an itemset that is contained in all the transactions Y.

Given X, X_1, $X_2 \in P(I)$ and Y, Y_1, $Y_2 \in P(T)$. The Galois connection satisfies the following properties [22]:

i) $X_1 \subset X_2 \Rightarrow t(X_1) \supseteq t(X_2)$

ii) $Y_1 \subset Y_2 \Rightarrow i(Y_1) \supseteq i(Y_2)$

iii) $X \subseteq i(t(X))$ and $Y \subseteq t(i(Y))$

b) Mining weighted association rules [17]

Definition 1. The transaction weight (tw) of a transaction t_k is defined as follow:

$$tw(t_k) = \frac{\sum\limits_{i_j \in t_k} w_j}{|t_k|}$$

Definition 2. The weighted support of an itemset is defined as follow:

$$ws(X) = \frac{\sum\limits_{t_k \in t(X)} tw(t_k)}{\sum\limits_{t_k \in T} tw(t_k)}$$

where T is the list of transactions in the database.

Example 1. Consider tables 1, 2, and definition 1, we can compute the $tw(t_1)$ value as follow:

$$tw(t_1) = \frac{0.6 + 0.1 + 0.9 + 0.2}{4} = 0.45$$

Table 3 shows all tw values of transactions in Table 1.

Table 3. Transaction weights for transactions in Table 1

Transactions	tw
1	0.45
2	0.2
3	0.45
4	0.3
5	0.42
6	0.43
Sum	2.25

From tables 1, 3, and definition 2, we can compute the $ws(BD)$ value as follow:

Because BD appears in transactions $\{1, 3, 5, 6\}$, $ws(BD)$ is computed:

$$ws(BD) = \frac{0.45 + 0.45 + 0.42 + 0.43}{2.25} \approx 0.78$$

2.2   Mining Frequent Closed Itemsets

An itemset X is called a frequent closed itemset if it is frequent, and it does not exist any frequent itemset Y such that $X \subset Y$ and $\sigma(X) = \sigma(Y)$. There are many methods proposed for mining frequent closed itemsets (FCIs) from data. They are divided into the following four categories [18]:

i) **Generate-and-test:** These methods are founded on the Apriori algorithm that uses a level-wise approach to discover FCIs. Some example algorithms include Close [10] and A-Close [9].

ii) **Divide-and-conquer:** These methods adopt a divide-and-conquer strategy and use compact data structures extended from a frequent-pattern (FP) tree to mine FCIs. Example algorithms include Closet [11], Closet+ [19] and FPClose [3].

iii) **Hybrid approaches:** These methods integrate both of the above two strategies to mine FCIs, which first transform the data into a vertical data format, and then develop properties and use a hash-table to prune non-closed itemsets. Example methods that use the hybrid approach include CHARM [23] and CloseMiner [13] also belong to them.

iv) **Hybrid approaches without duplication:** These methods differ from those using the hybrid approach in that they do not use the subsumption-checking technique, so identified FCIs need not be stored in the main memory. Methods within this category also do not use the hash-table technique as in the case of CHARM [23]. Example algorithms include DCI-Close 7], LCM [15] and PGMiner [8].

3   WIT-Tree Data Structure

In [5], authors proposed the WIT-tree (Weighted Itemset-Tidset tree) data structure, an expansion of the IT-tree proposed in [22], to mine high utility itemsets. To mine FWCIs, we modify the WIT-tree by changing *twu* to *ws* property. Using the WIT-tree, our proposed algorithm (see Section 4) only scans the data once because it is based on the intersection of Tidsets to compute the weighted support in next steps. Thus, it saves the time for the database scan and makes the algorithm to be done faster.

Each vertex in a WIT tree includes 3 fields:

    i. X: an itemset.
    ii. $t(X)$: the set of transaction contains X.
    iii. ws: the weighted support of X.

The vertex is denoted $\underset{ws(X)}{X \times t(X)}$.

The value of $ws(X)$ is computed by summing all *tw* values of transactions, $t(X)$, which their *tids* belong to and then dividing this by the sum of all *tw* values. Thus, computing of $ws(X)$ is achieved using the Tidset. Arcs connect vertices at k^{th} level (called X) with vertices at the $(k+1)^{th}$ level (called Y).

Definition 3. [23] – The equivalence class

Let I be a set of items and $X \subseteq I$, a function $p(X,k) = X[1:k]$ as the k length prefix of X and a prefix-based equivalence relation θ_K on itemsets as follows: $\forall X, Y \subseteq I, X \equiv_{\theta_k} Y \Leftrightarrow p(X,k) = p(Y,k)$.

The set of all itemsets which having the same prefix X is called an equivalence class, and denoted as the equivalence class with prefix X is [X].

Example 2: Consider tables 1 and 3 above, the associated WIT-tree for mining frequent weighted itemsets is as follows: The root node of the WIT-tree contains all 1-itemset nodes. All nodes at level 1 belong to the same equivalence class with prefix {} (or [∅]). Each node at level 1 will become a new equivalence class using its item as the prefix. With each node in the same prefix, it will join with all nodes following it to create a new equivalence class. The process will be done recursively to create new equivalence classes in the higher levels. For example, nodes {A}, {B}, {C}, {D}, {E} belong to the equivalence class [∅]. Consider node {A}, this node will join with all nodes following it ({B}, {C}, {D}, {E}) to create a new equivalence class [A] = {{AB}, {AC}, {AD}, {AE}}. [AB] will become a new equivalence class by also joining with all nodes following it ({AC}, {AD}, {AE}); and so on.

We can see [17] for more details about WIT-tree applying mining FWIs.

4   Mining Frequent Weighted Closed Itemsets

Definition 4: Let $X \subseteq I$ be a frequent weighted itemset, X is called a frequent weighted closed itemset if and only if it does not exist the frequent weighted itemset Y such that $X \subset Y$ and $ws(X) = ws(Y)$.

From the definition 4, there are a lot of FWIs that are not closed. For example, A, AB, AE are not closed because ABE has the same ws values with them. The purpose of the section is mining **FWCIs** from weighted items transaction databases fast.

Theorem 1. Given two itemsets X, Y where $X \subseteq Y$, $ws(X) = ws(Y) \Leftrightarrow t(X) = t(Y)$.

Proof:

✦   if $ws(X) = ws(Y)$ then $t(X) = t(Y)$

We have $ws(X) = \dfrac{\sum_{t_k \in t(X)} tw(t_k)}{\sum_{t_k \in T} tw(t_k)}$, $ws(Y) = \dfrac{\sum_{t_k \in t(Y)} tw(t_k)}{\sum_{t_k \in T} tw(t_k)}$ and $ws(X) = ws(Y)$

$$\Rightarrow \frac{\sum_{t_k \in t(X)} tw(t_k)}{\sum_{t_k \in T} tw(t_k)} = \frac{\sum_{t_k \in t(Y)} tw(t_k)}{\sum_{t_k \in T} tw(t_k)}$$

$$\Rightarrow \sum_{t_k \in t(X)} tw(t_k) = \sum_{t_k \in t(Y)} tw(t_k) \qquad (1)$$

According to property i) of Galois connection, we also have $X \subseteq Y \Rightarrow t(X) \supseteq t(Y)$
It impies that $\sum_{t_k \in t(X)} tw(t_k) = \sum_{t_k \in t(Y)} tw(t_k) + \sum_{t_k \in t(X) \backslash t(Y)} tw(t_k)$ (2)

From (1) and (2) we have $\sum_{t_k \in t(Y)} tw(t_k) + \sum_{t_k \in t(X) \backslash t(Y)} tw(t_k) = \sum_{t_k \in t(Y)} tw(t_k)$

$\Rightarrow \sum_{t_k \in t(X) - t(Y)} tw(t_k) = 0$

$\Rightarrow t(X) \backslash t(Y) = \emptyset$ (Because $tw(t_k) > 0$).

$\Rightarrow t(X) \subseteq t(Y)$ or $t(X) = t(Y)$.

$+$ if $t(X) = t(Y)$ then $ws(X) = ws(Y)$

According to Theorem 4.1 [17].

By theorem 1, we can use tidset to check the itemset is closed or not.

Theorem 2. Let $\underset{ws(X)}{X \times t(X)}$ and $\underset{ws(Y)}{Y \times t(Y)}$ are two nodes in the equivalence class [P], we have:

i)   If $t(X) = t(Y)$ then X, Y are not closed.

ii)   If $t(X) \subset t(Y)$ then X is not closed.

iii)   If $t(X) \supset t(Y)$ then Y is not closed.

Proof:

i)   We have $t(X \cup Y) = t(X) \cap t(Y) = t(X) = t(Y)$ (because $t(X) = t(Y)$) $\Rightarrow$ according to theorem 1, we have $ws(X) = ws(Y) = ws(X \cup Y) \Rightarrow X$ and Y are not closed.

ii)   We have $t(X \cup Y) = t(X) \cap t(Y) = t(X)$ (because $t(X) \subset t(Y)$) $\Rightarrow$ according to theorem 1, we have $ws(X) = ws(X \cup Y) \Rightarrow X$ is not closed.

iii)   We have $t(X \cup Y) = t(X) \cap t(Y) = t(Y)$ (because $t(X) \supset t(Y)$) $\Rightarrow$ according to theorem 1, we have $ws(Y) = ws(X \cup Y) \Rightarrow Y$ is not closed.

When we sort nodes in equivalence class P by increasing order according to cardinality of tidset, condition iii) of theorem 2 will not occur, so that we only consider conditions i) and ii).

In the process of mining **FWCIs**, considering nodes in the same equivalence class will consume a lot of time. Thus, we need to group nodes that satisfy condition i) at level 1 of WIT-tree together. In the process of creating a new equivalence class, we will group nodes that satisfy condition i) also. This reduces the cardinality of the equivalence class, so that the mining time decreases significantly. This approach differs from Zaki's approach [23] in that it decreases significantly the number of nodes, and it need not to remove the nodes that satisfy condition i) in an equivalence class.

4.1   The Algorithm

Input: A database D and *minws*
Output: **FWCIs** contains all frequent weighted closed itemsets that satisfy *minws* from D.
Method:
WIT-FWCIs()
1. $[\emptyset] = \{i \in I: ws(i) \geq minws\}$

2. **FWCIs** = ∅
3. **SORT**([∅])
4. **GROUP**([∅])
5. **FWCIs-EXTEND** ([∅])

FWCIs-EXTEND([P])
6. for all $l_i \in [P]$ do
7. $[P_i] = ∅$
8.   for all $l_j \in [P]$, with j > i do
9.   *if* $t(l_i) \subset t(l_j)$ then
10. $l_i = l_i \cup l_j$
11.   *else*
12. $X = l_i \cup l_j$
13. $Y = t(l_i) \cap t(l_j)$
14. $ws(X)$ = **COMPUTE-WS** (Y)
15.   if $ws(X) \geq minws$ then
16.   if $X \times Y$ is not subsumed then
17.   Add { $\underset{ws(X)}{X \times Y}$ } to $[P_i]$ //sort in increasing by |Y|
18.   Add $(l_i, ws(l_i))$ to **FWCIs** if it is closed
19.   **FWCIs-EXTEND** ([P_i])

Fig. 1. WIT-FWCIs algorithm for mining frequent weighted closed itemsets

The algorithm (in Fugure 1) commences with an empty equivalence class which contains simple items with their *ws* values satisfying *minws* (line 1). The algorithm then sorts nodes in equivalence class [∅] by increasing order according to cardinality of tidset (line 3). After that, it groups all nodes which have the same *tids* into the unique node (line 4), and calls procedure **FWCIs-EXTEND** with parameter [∅] (line 5). Procedure **FWCIs-EXTEND** uses equivalence class [P] as an input value, it considers each node in the equivalence class [P] with equivalence classes following it (lines 6 and 8). With each pair l_i and l_j, the algorithm considers condition ii) of theorem 2, if it satisfies (line 9) then the algorithm replaces equivalence class [l_i] by [$l_i \cup l_j$] (line 10), otherwise the algorithm creates a new node and adds it into equivalence class [P_i] (initially it is assigned by empty value, line 7). When tidset of X (i.e., Y) is identified, we need to check whether it is subsumed by any node or not (line 16), if not, then it is added into [P]. Adding a node $\underset{ws(X)}{X \times Y}$ into [P_i] is performed similarly as level 1 (i.e., consider it with nodes in [P_i], if exists the node that has the same tidset, then they are grouped together, line 17). After consider l_i with all nodes following it, the algorithm will add l_i and its *ws* into **FWCIs** (line 18). Finally, the algorithm is called recursively to generate equivalence classes after [l_i] (line 19).

Two nodes in the same equivalence class do not satisfy condition i) of theorem 2 because the algorithm groups these nodes into one node whenever they are added into [P]. Similarly, condition iii) does not occur because the nodes in the equivalence class [P] are sorted according to increasing order of cardinality of tidset.

4.2 Example

Using the example data presented in Tables 1 and 3, we illustrate the **WIT-FWCIs** algorithm with *minws* = 0.4 as follows. First of all, [∅] = {A, B, C, D, E}. After sorting and grouping, we have the result as $[\varnothing] = \{C, D, A, E, B\}$. Then, the algorithm calls the function **FWCIs-EXTEND** with input nodes $\{C, D, A, E, B\}$.

With the equivalence class $[C]$:

Consider C with D: we have a new itemset $CD\times56$ with $ws(CD) = 0.38 < minws$.

Consider C with A: we have a new itemset $CA\times45$ with $ws(CA) = 0.32 \quad < minws$.

Consider C with E: we have a new itemset $CE\times245$ with $ws(CE) = 0.41 \Rightarrow [C] = \{CE\}$.

Consider C with B: we have $t(C) \subset t(B)$ (satisfy the condition ii) of theorem 2) $\Rightarrow$ Replace $[C]$ by $[CB]$. It means that all equivalence classes following $[C]$ are replaced C into CB. Therefore, $[CE]$ is replaced into $[CBE]$.

After making the equivalence class $[C]$ (become $[CB]$ now), CB is added to **FWCIs** $\Rightarrow$ **FWCIs** = $\{CB\}$. The algorithm will be called recursively to create all equivalence classes following it.

Consider the equivalence class $[CBE] \in [CB]$: Add CBE to **FWCIs** $\Rightarrow$ **FWCIs** = $\{CB, CBE\}$.

With the equivalence class $[D]$:

Consider D with A: we have a new itemset $DA\times135$ with $ws(DA) = 0.59 \Rightarrow [D] = \{DA\}$.

Consider D with E: we have a new itemset $DE\times135 \Rightarrow$ Group DA with DE into $DAE \Rightarrow [D] = \{DAE\}$.

Consider D with B: we have $t(D) \subset t(B)$ (satisfy the condition ii) of theorem 2) $\Rightarrow$ Replace $[D]$ by $[DB]$. It means that all equivalence classes following $[D]$ are replaced D into DB. Therefore, $[DAE]$ is replaced into $[DBAE]$.

After making the equivalence class $[D]$ (become $[DB]$ now), DB is added to **FWCIs** $\Rightarrow$ **FWCIs** = $\{CB, CBE, DB\}$. The algorithm will be called recursively to create all equivalence classes following it.

Consider the equivalence class $[DBAE] \in [DB]$: Add $DBAE$ to **FWCIs** $\Rightarrow$ **FWCIs** = $\{CB, CBE, DB, DBAE\}$.

Similar to equivalence classes $[A]$, $[E]$, $[B]$. We have all **FWCIs** = $\{CB, CBE, DB, DBAE, AEB, EB, B\}$.

Results show that the number of FWCIs is smaller than FWIs (7 compare to 19), and the number of search levels in a tree by WIT-FWCIs is also less than that of WIT-FWIs (2 compared with 4). Thus, we can say that FWCIs mining is more efficient than FWIs.

5   Experimental Results

All experiments described below were performed on a Centrino core 2 duo (2×2.53 GHz), 4GBs RAM memory, Windows 7, using C# 2008. The experimental data sets used for the experimentation were downloaded from http://fimi.cs.helsinki.fi/data/. Some statistical information regarding these data sets is given in Table 4.

Table 4. Experimental databases

Databases (DB)	#Trans	#Items	Remark
Mushroom	8124	120	Modified
Connect	67557	130	Modified

Each database is modified by creating a table to store weighted values of items (value in the range of 1 to 10).

5.1   Number of Itemsets

Results from Table 5 show that the number of FWCIs is always smaller than FWIs. For example, consider Mushroom database with $minws = 20\%$, the number of FWIs is 53,513 and the number of FWCIs is 1199, the ratio is $1199/53513 \times 100\% \approx 2.24\%$.

Table 5. Compare the number of FWCIs with the number of FWIs

Databases	*minws(%)*	#FWIs	#FWCIs
Connect	97	512	297
	96	1147	513
	95	2395	861
	94	4483	1284
Mushroom	35	1159	252
	30	2713	423
	25	5643	696
	20	53513	1199

5.2   The Mining Time

Experimental results from Figures 2 and 3 show the efficiency of FWCIs mining. The time of FWCIs mining is faster than FWIs mining in these two databases. Especially, when *minws* is low, mining FWCIs is more efficient than mining FWIs. For example, consider Mushroom database, when $minws = 35\%$, the time for mining FWIs is 0.284(s) while the time for mining FWCIs is 0.15(s). When we decrease *minws* to 20%, the time for mining FWIs is 5.88(s) while the time for minng FWCIs is 0.541(s).

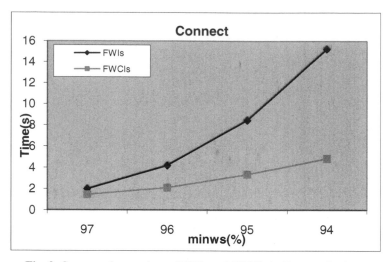

Fig. 2. Compare the run time of FWIs and FWCIs in Connect database

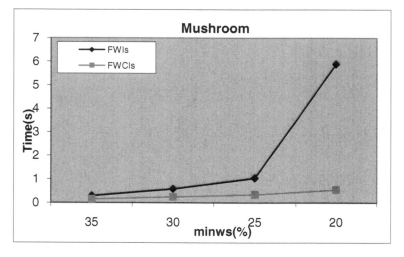

Fig. 3. Compare the run time of FWIs and FWCIs in Mushroom database

6   Conclusions and Future Work

This paper has proposed the method for mining frequent weighted itemsets and frequent weighted closed itemsets from weighted items transaction databases, and the efficient algorithm is also developed. As above mentioned, the number of FWCIs is always smaller than that of FWIs and the mining time is also better. By WIT-tree, the proposed algorithm only scans the database one and uses tidset to fast determine the weighted support of itemsets.

In the future, we will study how to mine efficient association rules from frequent weighted closed itemsets.

Acknowledgment. This work was supported by Vietnam's National Foundation for Science and Technology Development (NAFOSTED) under Grant Number 102.01-2012.47.

References

[1] Bastide, Y., Pasquier, N., Taouil, R., Stumme, G., Lakhal, L.: Mining minimal non-redundant association rules using frequent closed itemsets. In: Palamidessi, C., Moniz Pereira, L., Lloyd, J.W., Dahl, V., Furbach, U., Kerber, M., Lau, K.-K., Sagiv, Y., Stuckey, P.J. (eds.) CL 2000. LNCS (LNAI), vol. 1861, pp. 972–986. Springer, Heidelberg (2000)

[2] Cai, C.H., Fu, A.W., Cheng, C.H., Kwong, W.W.: Mining Association Rules with Weighted Items. In: Proceedings of International Database Engineering and Applications Symposium (IDEAS 1998), pp. 68–77 (1998)

[3] Grahne, G., Zhu, J.: Fast Algorithms for Frequent Itemset Mining Using FP-Trees. IEEE Transaction on Knowledge and Data Engineering 17(10), 1347–1362 (2005)

[4] Khan, M.S., Muyeba, M., Coenen, F.: A Weighted Utility Framework for Mining Association Rules. In: Proceedings of Second UKSIM European Symposium on Computer Modeling and Simulation Second UKSIM European Symposium on Computer Modeling and Simulation, pp. 87–92 (2008)

[5] Le, B., Nguyen, H., Cao, T.A., Vo, B.: A novel algorithm for mining high utility itemsets. In: The First Asian Conference on Intelligent Information and Database Systems, pp. 13–16 (2009) (published by IEEE)

[6] Le, B., Nguyen, H., Vo, B.: An Efficient Strategy for Mining High Utility Itemsets. International Journal of Intelligent Information and Database Systems 5(2), 164–176 (2011)

[7] Lucchese, B., Orlando, S., Perego, R.: Fast and Memory Efficient Mining of Frequent Closed Itemsets. IEEE Transaction on Knowledge and Data Engineering 18(1), 21–36 (2006)

[8] Moonestinghe, H.D.K., Fodeh, S., Tan, P.N.: Frequent Closed Itemsets Mining using Prefix Graphs with an Efficient Flow-based Pruning Strategy. In: Proceedings of 6th ICDM, Hong Kong, pp. 426–435 (2006)

[9] Pasquier, N., Bastide, Y., Taouil, R., Lakhal, L.: Discovering Frequent Closed Itemsets for Association Rules. In: Beeri, C., Bruneman, P. (eds.) ICDT 1999. LNCS, vol. 1540, pp. 398–416. Springer, Heidelberg (1999)

[10] Pasquier, N., Bastide, Y., Taouil, R., Lakhal, L.: Efficient Mining of Association Rules using Closed Itemset Lattices. Information Systems 24(1), 25–46 (1999)

[11] Pei, J., Han, J., Mao, R.: CLOSET: An Efficient Algorithm for Mining Frequent Closed Itemsets. In: Proc. of the 5th ACM-SIGMOD Workshop on Research Issues in Data Mining and Knowledge Discovery, Dallas, Texas, USA, pp. 11–20 (2000)

[12] Ramkumar, G.D., Ranka, S., Tsur, S.: Weighted Association Rules: Model and Algorithm. In: SIGKDD 1998, pp. 661–666 (1998)

[13] Singh, N.G., Singh, S.R., Mahanta, A.K.: CloseMiner: Discovering Frequent Closed Itemsets using Frequent Closed Tidsets. In: Proc. of the 5th ICDM, Washington DC, USA, pp. 633–636 (2005)

[14] Tao, F., Murtagh, F., Farid, M.: Weighted Association Rule Mining using Weighted Support and Significance Framework. In: SIGKDD 2003, pp. 661–666 (2003)

[15] Uno, T., Asai, T., Uchida, Y., Arimura, H.: An Efficient Algorithm for Enumerating Closed Patterns in Transaction Databases. In: Suzuki, E., Arikawa, S. (eds.) DS 2004. LNCS (LNAI), vol. 3245, pp. 16–31. Springer, Heidelberg (2004)

[16] Vo, B., Le, B.: Mining Minimal Non-redundant Association Rules using Frequent Itemsets Lattice. Journal of Intelligent Systems Technology and Applications 10(1), 92–106 (2011)

[17] Vo, B., Coenen, F., Le, B.: A new method for mining Frequent Weighted Itemsets based on WIT-trees. Expert Systems with Applications 40(4), 1256–1264 (2013)

[18] Vo, B., Hong, T.P., Le, B.: Mining most generalization association rules based on frequent closed itemsets. Int. J. of Innovative Computing Information and Control 8(10B), 7117–7132 (2012)

[19] Wang, J., Han, J., Pei, J.: CLOSET+: Searching for the Best Strategies for Mining Frequent Closed Itemsets. In: ACM SIGKDD International Conference on Knowledge Discovery and Data Mining, pp. 236–245 (2003)

[20] Wang, W., Yang, J., Yu, P.S.: Efficient Mining of Weighted Association Rules. In: SIGKDD 2003, pp. 270–274 (2003)

[21] Zaki, M.J.: Generating Non-Redundant Association Rules. In: Proc. of the 6th ACM SIGKDD International Conference on Knowledge Discovery and Data Mining, Boston, Massachusetts, USA, pp. 34–43 (2000)

[22] Zaki, M.J.: Mining Non-Redundant Association Rules. Data Mining and Knowledge Discovery 9(3), 223–248 (2004)

[23] Zaki, M.J., Hsiao, C.J.: Efficient Algorithms for Mining Closed Itemsets and Their Lattice Structure. IEEE Transactions on Knowledge and Data Engineering 17(4), 462–478 (2005)

Author Index